Urban Foraging in the Changing World

Shalini Dhyani • Mallika Sardeshpande
Editors

Urban Foraging in the Changing World

Springer

Editors
Shalini Dhyani
National Environmental Engineering
Research Institute, NM
Council Scientific & Industrial Research
Nagpur, Maharashtra, India

Mallika Sardeshpande
University of KwaZulu-Natal
Pietermaritzburg, South Africa

ISBN 978-981-97-0344-9 ISBN 978-981-97-0345-6 (eBook)
https://doi.org/10.1007/978-981-97-0345-6

This Springer imprint is published by the registered company Springer Nature Singapore Pte Ltd.
The registered company address is: 152 Beach Road, #21-01/04 Gateway East, Singapore 189721, Singapore

Foreword

One late summer afternoon in 2004, I was enjoying a stroll in Edinburgh's Braid Hills. It was a great year for brambles (*Rubus* sp.), and I had been plucking the berries from bushes, popping them into my mouth as I explored the trails. Coming down out of the hills and heading back toward my flat, I met two people who had been more intentional than I about foraging in this lovely park on the edge of the city. They were headed home with baskets filled with brambles destined to become a delicious finale for their evening meal. By this point in time, I had already been studying foraging in rural environments in the USA for a decade. In fact, at the invitation of UK Forest Research, I was in Scotland to study foraging throughout that country. Nevertheless, I continued to interview only rural foragers and it was another 4 years before it occurred to me that urban foraging might merit focused research effort.

I suppose I can forgive myself for this temporary oversight. After all, when I began my investigations of foraging in rural US regions, public land managers told me firmly and emphatically that nobody did that anymore. One could only imagine how much firmer the conviction would be about foraging in urban environments. And yet, I had seen it with my own eyes, even engaged in the practice myself. Surely there was something to the idea that warranted a closer look. In pursuit of an answer to that question, my colleagues Patrick Hurley, Rebecca McLain, Melissa Poe, and I began to review the literature in search of existing research on urban foraging. We found that there was plenty of adjacent research, but nothing in the English language literature we could find explicitly addressed urban foraging. And so, we began exploratory studies in Seattle and New York City, which gradually extended to studies conducted in the cities of Baltimore and Philadelphia. It quickly became clear that urban foraging was ripe for research.

Once we began to seek it out, evidence of urban foraging was abundant in both physical and virtual spaces. Colleagues in New York City sent us photographs of people harvesting dandelion (*Taraxacum* spp.) leaves on early spring mornings. The Himalayan blackberries (*Rubus armeniacus*) that are so abundant in Seattle and its surrounding exurban areas were being picked by city residents in quantities both large and small. A group of artists in Los Angeles had discovered laws in that city that allow the harvest of fruit, both wild and domesticated, in public rights of way and had begun offering tours doing just that. There were apps crowdsourcing information about the locations of forageable species in a variety of cities in North

America and western Europe, as well as dozens of groups of people using online forums to share information and organize foraging. Further, judging by media attention to the 1986 arrest of a man for picking a dandelion in New York's Central Park, urban foraging was far from a new phenomenon.

Meanwhile, continents away other scholars also had taken up the study of urban foraging. Notably Prof. Charles Shackleton (Rhodes University, South Africa) and his students were examining foraging in diverse environments in southern Africa, including cities and townships. Dr. Eefke Molle (University of Aberystwyth, United Kingdom) conducted a brilliant study of foraging in Kampala, Uganda, noting a bimodal pattern in the economic status of individuals most likely to engage in foraging. At roughly the same time, Dr. Harini Nagendra (Azim Premji University, India) and her students were looking at the effects of exclusionary development on foraging and other self-provisioning activities in Bengaluru. And Prof. Ingo Kowarik (Technische Universitat Berlin, Germany), together with his students, were conducting detailed ethnographic and ecological studies of foraging in parks and informal green spaces in Berlin. Here and there, individual graduate students all over the world were examining urban foraging through diverse disciplinary lenses, including political economy and sociology.

It is noteworthy that this research begins to emerge around 2008, the year in which the United Nations Population Fund estimated that for the first time more than half of humanity was living in cities. Independently, each individual scholar and team of researchers had realized that this ancient human practice was finding continued expression in the novel ecologies of cities. Not only was this intriguing, but it was also almost certainly, socially, and ecologically significant. But how? We applied our diverse theories and methods, with results that were at once somewhat disparate and yet revealed striking commonalities across cultures, economies, and urban structures. A picture of urban foraging as a ubiquitous and significant human practice began to emerge from points of data scattered across the globe.

Notwithstanding these advances in our understanding, much remains to be learned about foraging for wild plants and fungi in cityscapes from central business districts to parks both large and small, to peripheries and exurban spaces. These gaps are geographical and topical. Geographically, with some important exceptions, a disproportionate volume of the literature to date (at least that published in English) looks at urban foraging in the cities of North America and Western Europe. And much of that literature (again, with a growing number of important exceptions) examines the foraging practices of the dominant cultures of those nations. We are left, then, with some glaring holes in our knowledge. We know almost nothing about how international migrants, many of whom are from rural areas, may adapt knowledge and practices originating in distant places and cultures to their new urban homes. Likewise, there are several regions outside the global North where focused research on urban foraging appears to be entirely lacking. Topically, health and safety are crying out for additional research. A small number of studies have tested urban foraged fruits and greens for heavy metals and other contaminants, but many more are needed. The severe disruptions to economies and supply chains during the global SARS CoV-2 pandemic lend additional salience to improved understanding of the health and nutrition outcomes of consuming urban foraged foods. Explorations of governance structures that support or impede foraging in cities have only just begun, with special need to interrogate their outcomes in terms of equity and sustainability. And this list is far from an exhaustive tally of the areas of opportunity and need for further research on urban foraging.

Happily, the book you have before you makes an essential contribution to filling multiple gaps in our knowledge. From chapters documenting a foraging field course in the coastal United Kingdom to the practices and perceptions of foragers in Lagos, Nigeria, from the role of foraging in health and nutrition in the major cities of Uttar Pradesh, India, to the implications of conservation policy for urban foraging in Canadian cities, this book augments areas of existing knowledge about urban foraging and opens up important new topics. Like urban foraging itself, *Urban Foraging in the Changing world* connects many dots, revealing a more complete picture of this fundamental relationship between human beings and nature than we have had to date. It marks and propels the study of urban foraging from its infancy to maturity.

As *Urban Foraging in the Changing world* demonstrates, urban foraging is not simply a curiosity. Rather, it has direct implications for policy at scales from the local to the global. Enhanced understanding of urban foraging can and, I would argue, should inform the design of our cities and the ways we govern greenspaces in them. It also has much to say to international efforts to conserve biodiversity and support human well-being. As the IPBES *Assessment of Sustainable Use of Wild Species*, which I co-chaired, notes "dozens to hundreds of wild plant and fungi species [are] gathered for food, medicine, firewood, decoration and cultural practices in urban ecosystems worldwide" (2022:12). These uses can make largely unacknowledged contributions to realizing the United Nations 2030 Agenda, notably Sustainable Development Goals 2. Zero Hunger, 3. Good Health and Well-being, and 11. Sustainable Cities and Communities. As it turns out, the brambles of Edinburgh, like the case studies throughout this book, offer us windows into existing and potential strategies for realizing those goals.

Burlington, VT, USA
28 August 2023

Marla R. Emery

Acknowledgments

The editors would like to express heartfelt gratitude to all involved, directly and indirectly, for their constant support and contributions, that have helped compile this book on the futuristic subject of urban foraging. We have been successful in putting together this important scientific contribution that we are sure will be crucial to understanding and acknowledging the growing relevance of this topic in both the Global North and South. A topic that is relevant in the present context and will be equally important in the future, not only for providing nutritional security in fast growing urban areas but also to improve sense of belongingness, nature connectedness, and place-based attachment to improve urban resilience.

We are immensely grateful to numerous subject matter experts, researchers, and practitioners from different corners of the world who invested significant amounts of time beyond their professional obligations to contribute to this volume. Their devoted and sustained efforts have yielded this rich and diverse compilation in which each contribution develops the nuances of urban foraging as a highly prevalent practice. We would also like to thank Mr. Manu Thomas (CSIR-NEERI, Nagpur) for his assistance in formatting the content. We would also like to express our gratitude to the Springer Production Team, in particular Mrs. Aakanksha Tyagi and Ms. Muthuneela Muthukumar, technical assistance, and for generously keeping time while we ensured that we collectively submit a well-rounded and comprehensive book.

We acknowledge that some geographies and aspects are still under-represented in this volume, and hope that this publication stirs interest and action towards greater inclusivity. Despite our best efforts to ensure accuracy, it is still possible that a few errors may have crept in because of the pace of production and the diversity of formats. The editors have read through all of the chapters and personally reviewed them in accordance with international standards, including publication ethics that specifically aim to reduce plagiarism and resemblance from already existing publications. If readers of this book have any relevant feedback to help us improve our future releases, we would be ever grateful to have it.

We express our gratitude to our families, especially our parents, siblings, spouses, and children, for their support, understanding, and encouragement as we engaged in this mammoth task.

We conclude by expressing our deep gratitude to Dr. Marla Emery, subject authority on the subject, Co-Chair of the Sustainable Use Assessment (2022) in

Co-Chair of the Assessment of Sustainable Use of Wild Species, a report of the Intergovernmental Science-Policy Platform on Biodiversity and Ecosystem Services (IPBES), and former employee of the U.S. Department of Agriculture (USDA), Forest Service, for kindly authoring the foreword for this book.

-Shalini Dhyani and Mallika Sardeshpande

Contents

Editors and Contributors

About the Editors

Shalini Dhyani (PhD) is a Principal Scientist at the National Environment Engineering Research Institute, a constituent laboratory under the Council for Science and Industrial Research, India. She is Asia Vice Chair and Steering Committee member in the Commission on Ecosystems Management of the International Union for Conservation of Nature (IUCN). She is a UNESCO-TWAS (The World Academy of Science) Associate and visiting scientist at the Center for International Forestry Research (CIFOR), Bogor. She has been the lead author in the IPBES global thematic assessment on the sustainable use of wild species and the Asia-Pacific regional assessment of biodiversity and ecosystem services. Shalini works on interlinkages between socio-ecological systems, following diverse knowledge systems and sustainability science approaches for mainstreaming the relevance of ecosystems and biodiversity, multi-sectoral and people centric approaches in decision-making. She is involved in the editorial boards of Sustainability Science, PLOS One, and Anthropocene Science journals.

Mallika Sardeshpande (PhD) is a researcher affiliated with the Centre for Transformative Food Systems at the University of KwaZulu-Natal, South Africa, and the Ashoka Trust for Research in Ecology and the Environment, India. She works on natural resource systems with a focus on agroecology, biodiversity, forestry, and urban systems. Her primary work involves the design and development of productive spaces with indigenous species in collaboration with communities, local governments, and non-government organizations. She also works with various global consortia to synthesize research and policy to make food systems ecologically, nutritionally, and socially sustainable and to increase urban resilience through green infrastructure management. She is an associate editor at Cities and the Environment Journal.

Contributors

Janaína Deane de Abreu Sá Diniz University of Brasília, Brasilia, Brazil

Opeyemi Adeyemi Department of Environmental Science, Rhodes University, Makhanda (Grahamstown), South Africa

Sarah Becker Ursinus College, Collegeville, PA, USA

Dhanya Bhaskar Indian Institute of Forest Management, Bhopal, India

Tuybia Bilal Faculty of Forestry, Sher-e-Kashmir University of Agricultural Sciences & Technology, Wadoora, Jammu & Kashmir, India

Barbara Loureiro Borges Department of Forest Sciences, University of Brasília, Brasilia, Brazil

Sadia Butt Daniels Institute of Frestry and Conservation, University of Toronto, Toronto, ON, Canada

Joana Chan University of Vermont, Burlington, VT, USA

Lihua Cui Graduate School of Global Environmental Studies, Kyoto University, Kyoto, Japan

Oinam Sunanda Devi Assam State Biodiversity Board, Aranya Bhawan, Guwahati, Assam, India

Shalini Dhyani CSIR-National Environmental Engineering Research Institute (CSIR-NEERI), Nehru Marg, Nagpur, Maharashtra, India

Commission on Ecosystem Management (CEM), International Union for the Conservation of Nature (IUCN), Gland, Switzerland

Academy of Scientific and Innovative Research (AcSIR), Ghaziabad, Uttar Pradesh, India

Marla R. Emery Norwegian Institute for Nature Research and Rubenstein School of Environment and Natural Resources, University of Vermont, Burlington, VT, USA

S. A. Gangoo Faculty of Forestry, Sher-e-Kashmir University of Agricultural Sciences & Technology, Wadoora, Jammu & Kashmir, India

Jessica Pereira Garcia Center for Sustainable Development, University of Brasília, Brasilia, Brazil

Sonali Ghosh Chief Conservator of Forests (REWP), Aranya Bhawan, Guwahati, Assam, India

Mikelis Grivins Riga Stradins University, Riga, Latvia

Stéphane Guéneau Cirad – Moisa (French Agricultural Research Centre for International Development – Montpellier Interdisciplinary Center on Sustainable Agri-Food Systems), Montpellier, France

Humara Gulzar Faculty of Forestry, Sher-e-Kashmir University of Agricultural Sciences & Technology, Wadoora, Jammu & Kashmir, India

Md Abdul Halim Daniels Institute of Forestry and Conservation, University of Toronto, Toronto, ON, Canada

Holly Harris School of Anthropology and Conservation, University of Kent, Canterbury, UK

Patrick T. Hurley Ursinus College, Collegeville, PA, USA

M. A. Islam Faculty of Forestry, Sher-e-Kashmir University of Agricultural Sciences & Technology, Wadoora, Jammu & Kashmir, India

Imrul Kayes Daniels Institute of Forestry and Conservation, University of Toronto, Toronto, ON, Canada

Ashish Lahoti Institute for Global Environmental Strategies (IGES), Tokyo, Japan

Shruti Lahoti Institute for Global Environmental Strategies (IGES), Tokyo, Japan

Brian J. Leffler Lewis and Clark National Historic Trail, Omaha, NE, USA

A. R. Malik Faculty of Forestry, Sher-e-Kashmir University of Agricultural Sciences & Technology, Wadoora, Jammu & Kashmir, India

N. Manika FloraFauna Science Foundation, Lucknow, Uttar Pradesh, India

Commission on Ecosystem Management (CEM), International Union for the Conservation of Nature (IUCN), Gland, Switzerland

T. H. Masoodi Faculty of Forestry, Sher-e-Kashmir University of Agricultural Sciences & Technology, Wadoora, Jammu & Kashmir, India

Laura McCarthy School of Arts, University of Gloucestershire, Gloucestershire, UK

Kristin McGillis Department of Environmental Studies, Ursinus College, Collegeville, PA, USA

Tahir Mushtaq Faculty of Forestry, Sher-e-Kashmir University of Agricultural Sciences & Technology, Wadoora, Jammu & Kashmir, India

Mir Muskan Un Nisa Faculty of Forestry, Sher-e-Kashmir University of Agricultural Sciences & Technology, Wadoora, Jammu & Kashmir, India

N. A. Pala Faculty of Forestry, Sher-e-Kashmir University of Agricultural Sciences & Technology, Wadoora, Jammu & Kashmir, India

Ishtiyak Ahmad Peerzada Faculty of Forestry, Sher-e-Kashmir University of Agricultural Sciences & Technology, Wadoora, Jammu & Kashmir, India

Dhanya Rajaram Bangalore, India

Samudyatha Ramananda Indian Institute of Forest Management, Bhopal, India

Sharmistha Rej Indian Institute of Forest Management, Bhopal, India

Christoph D. D. Rupprecht Department of Environmental Design, Faculty for Collaborative Regional Innovation, Ehime University, Ehime, Japan

Alessio Russo School of Arts, University of Gloucestershire, Gloucestershire, UK

Osamu Saito Institute for Global Environmental Strategies (IGES), Tokyo, Japan

Mallika Sardeshpande Centre for Transformative Agriculture and Food Systems, University of KwaZulu-Natal, Scottsville, South Africa

Ashoka Trust for Research in Ecology and the Environment, Bengaluru, India

Mallika Sardeshpande Centre for Transformative Agriculture and Food Systems, University of KwaZulu-Natal, Scottsville, South Africa

Ashoka Trust for Research in Ecology and the Environment, Bengaluru, India

Charlie Shackleton Department of Environmental Science, Rhodes University, Makhanda (Grahamstown), South Africa

Rika Shinkai Research Institute for Humanity and Nature, FEAST NPO, Kyoto, Japan

Jayshree Shukla National Environmental Engineering Research Institute (CSIR-NEERI), Nagpur, Maharashtra, India

Abhijit Sinha Delhi, India

Sharmila Sinha Delhi, India

Sivajanani Sivarajah Département des sciences du bois et de la forêt, Université Laval, Québec, QC, Canada

P. A. Sofi Faculty of Forestry, Sher-e-Kashmir University of Agricultural Sciences & Technology, Wadoora, Jammu & Kashmir, India

Radhika Sood CSIR-National Environmental Engineering Research Institute, Nagpur, Maharashtra, India

Academy of Scientific and Innovative Research (AcSIR), Ghaziabad, Uttar Pradesh, India

Maximilian Spiegelberg FEAST NPO, Kyoto, Japan

Aoi Yoshida AOI Landscape Design LLC, Tokyo, Japan

Abbreviations

ACS	American Community Survey
AMRUT	Atal Mission for Rejuvenation and Urban Transformation
AoF	Association of Foragers
ASL	Above Sea Level
AYUSH	Ayurveda, Yoga and Naturopathy, Unani, Siddha and Homeopathy
BMC	Biodiversity Management Committees
BNP	Bannerghatta National Park
CBD	Convention on Biological Diversity
CHF	Congestive Heart Failure
CLU	Crown Land Use
CPR	Common Pool Resource
CSCAF	Climate Smart Cities Assessment Framework
DAFF	Department of Agriculture, Fisheries, and Forests
EDTEA	Economic Development, Tourism, and Environmental Affairs
EoN	Experiences of Nature
ES	Ecosystem Services
ESZ	Eco-Sensitive Zone
FFC	Foraging Field Courses
GIS	Geographic Information System
HNC	Human-Nature Connections
IGP	Indo Gangetic Plain
IVI	Importance Value Index
JFMC	Joint Forest Management Committee
LGAs	Local Government Areas
LPG	Liquid Petroleum Gas
MECP	Ministry of Environment, Conservation and Parks
MoHUA	Ministry of Housing and Urban Affairs
NbS	Nature-based Solutions
NE	Natural England
NMC	Nagpur Municipal Corporation
NTAs	Neighborhood Tabulation Areas
OMNRF	Ontario Ministry of Natural Resources and Forestry
PADCNR	Pennsylvania Department of Conservation and Natural Resources
PAs	Protected Areas

PBR	People's Biodiversity Register
RDP	Reconstruction and Development Programme
RMP	Revised Master Plan
SARA	Species at Risk Act
SEM	Social-Ecological Memory
SPSS	Statistical Package for Social Sciences
TEK	Traditional Ecological Knowledge
TKDL	Traditional Knowledge Digital Library
UF	Urban Forests
UFMP	Urban Forestry Management Plan
UGS	Urban Green Spaces

Part I

Why Do Foragers Forage

Urban Foraging in a Changing World: An Introduction

1

Mallika Sardeshpande and Shalini Dhyani

Abstract

Foraging by humans is an activity that is increasingly being observed in urban areas. This book presents a collection of scientific research and practitioner perspectives on urban foraging from different parts of the world. The sections in the book document what urban humans forage for, why, where, and how, and what urban foraging could look like in the emerging future. Acknowledging that these aspects are inherently interlinked, this chapter iterates the relationships between urban foraging and experiential learning, human well-being, and environmental stewardship. We invite readers to delve into the diversity and possibilities of urban foraging.

Keywords

Ethnobiology · Local ecological knowledge · Socio-economics · Social-ecological systems · Urbanization

Humans forage for numerous things over the course of the day, not just juicy fruit on the farm or fresh produce at the shops. When we look for information in the news or entertainment on-screen, we continue to sift through our environment for meaning, inspiration, and connection. The act of foraging has intrigued researchers for decades, leading to a veritable body of work on the behavior of bees, birds, bears, and so on (Cestari et al. 2020; Fehlmann et al. 2021; Colin et al. 2022). There are

M. Sardeshpande (✉)
Centre for Transformative Agriculture and Food Systems, University of KwaZulu-Natal, Scottsville, South Africa

Ashoka Trust for Research in Ecology and the Environment, Bengaluru, India

S. Dhyani
CSIR-National Environmental Engineering Research Institute, Nagpur, India
e-mail: s_dhyani@neeri.res.in

S. Dhyani, M. Sardeshpande (eds.), *Urban Foraging in the Changing World*, https://doi.org/10.1007/978-981-97-0345-6_1

also several studies on human societies that live non-agrarian, non-sedentary lifestyles, and how this compares to the paradigm of present industrial societies (House et al. 2013; Gupta et al. 2017; Reyes-García et al. 2018). At the turn of the century came the recognition that humans also forage in urban environments, and it is not necessarily for food (Leckie and Given 2005; McCullough 2013), nor necessarily from economic deprivation (Cucullu 2006; Bates 2013). Since then, there has been a steady increase in the interest in urban foraging by humans, with the seminal work of Shackleton and colleagues (2017) adding momentum to the research.

In this book, we invite readers to explore the world of urban foraging through the lenses of diverse practitioners and researchers. The mechanistic definition of urban foraging is the practice of collecting naturally occurring resources from urban environments (Shackleton et al. 2017). However, much of the research on urban foraging goes beyond *what* is collected or foraged and investigates *why* people forage in urban environments. The tacit assumption that people forage to fill gaps in their household economies has been long overturned in the Global North (e.g., Poe et al. 2013), and more recently so in the Global South (e.g., Sardeshpande and Shackleton 2023). This book aims to provide a current overview of the knowledge on urban foraging for everyone from the uninitiated to the well-versed.

Cities are places of rapid sociodemographic and structural change. For example, urban influx can result in social densification, informality, and mobility (Iveson et al. 2019; Hamilton et al. 2020). These conditions often lead to increased pressures on physical resources such as infrastructure and energy (Petrescu et al. 2021; Freeman et al. 2023). Furthermore, urbanization can have significant impacts on ecological processes, altering the interactions between humans, nonhumans, and their environment (Teng et al. 2020). Urbanization continues to be an important driver of social-ecological change and a priority for more than half of the world's population (UNDESA 2018). It is believed that urban foraging can help people to adapt to these changes and complexities at different levels (Sardeshpande et al. 2021).

In this chapter, we outline the conceptual framings that have been linked to urban foraging that appear in the chapters that follow. At a personal level, urban foraging can help people experience their surroundings in a way that enriches tourists (Vaittinen and McGookin 2016) or enables new urban migrants to feel at home (Marquina et al. 2022). These personal experiences may also have psychological or spiritual benefits (Stöckelová et al. 2023; Ware 2022). Interpersonally within the wider community, urban foraging can help people build social ties across generations (Garekae and Shackleton 2020; Poe et al. 2013) as well as across different cultural backgrounds (Johnson Gaither et al. 2020). Intergenerational knowledge transfer related to urban foraging helps improve social cohesion and human well-being in the long run (Guenat et al. 2023).

In household economies, foraging can enable savings on expenditure and cash income from foraged goods and also be a source of food and financial security in times of shock (Clouse 2022; Sachdeva et al. 2018). Gender has an important role to play in food sovereignty, and typically more women are involved in foraging wild plants than men (Palliwoda et al. 2017; Howard 2003). Urban foraging can also contribute to dietary diversity, which is an important component of nutrition and health (Garekae and Shackleton 2020). Foraging for resources in urban environments allows people to develop knowledge about their urban environments and their nonhuman cohabitants, fostering environmental learning and stewardship (Fischer and Kowarik 2020). Urban foraging improves place-based attachment, and sense of belonging, that can provide opportunities for co-governance of urban spaces by residents, local governments, and nongovernmental entities. Restoration of degraded urban green spaces for foraging purposes can develop resilience in urban spaces and among urban residents, as a co-benefit (Guenat et al. 2023; Arrington et al. 2017; McLain et al. 2014; Poe et al. 2014).

The book is divided into six sections; the first four sections respond to the basic questions, namely, why people forage, what is foraged, where, and how. Many of the chapters also characterize who the foragers are, based on their motivations, sociodemographics, or locations. Indeed, in the first section on why people forage, Grivins engages foragers (and the readers) in an exploration of forager identity—Is it about what you forage for, what you do with your foraged goods, or what personally motivates you to forage? Harris delves into the experience of foraging as a new way of (re)connecting with nature, fellow human beings, and food. Adeyemi and Shackleton explore foragers' own perceptions of their foraging activity, including previously undocumented pros and cons such as safety, shame, and gentrification. Dhyani describes the role of urban foraging in enabling cultural continuity of centuries-old traditions.

The second section on what is foraged documents ethnobiological aspects of resources foraged in different landscapes. Manika and Dhyani richly elaborate on the species, spaces, seasons, and festivities related to foraging in multiple cities from the populous Indian state of Uttar Pradesh. Peerzada et al. present a detailed economic valuation of the contribution of foraged species in urban areas of the relatively understudied Indian state of Kashmir. Lahoti et al. establish a baseline for forageable plant species across different urban green spaces in the growing Indian city of Nagpur.

The third section on where urban foraging occurs describes the spatial aspects that are key to understanding patterns, which can aid urban planning and policy. Becker et al. use a geographic information systems approach to map some culturally important plant species in New York, and their relative accessibility by public transport—a very relevant analysis, considering much of urban foraging occurs when people are in transit across town. Russo and McCarthy trace the evolution of green spaces where foraging occurs and offer a glimpse into what foragers would want these green spaces to look like, providing actionable suggestions to urban planners. In their rigorous review of Canadian legislation interfacing with urban foraging, Sivarajah et al. find many restrictive clauses and recommend further social-ecological evidence be gathered to help tailor foraging-friendly policies.

The fourth section on how foraging occurs includes examples of the pathways through which foraging is enabled or inhibited. Hurley et al. describe the role of social media in mobilizing the knowledge of sustainable foraging to enable wider utility. Bhaskar et al. look at some emerging initiatives in the Global South that promote urban foraging and discuss how urbanization can also reduce the feasibility of foraging in some cases. Sood and Dhyani document diminishing urban foraging practices in mountainous regions and suggest that growing tourism pressures can be cautiously converted to win-win solutions for sustainable development.

The fifth section presents some spontaneous perspectives from practitioners that speak to a confluence of all the previous themes. Rajaram presents an evocative and vibrant picture of fast-urbanizing areas where local foragers hold their own by creating awareness among urbanites. Sinha and Sinha walk readers through some cherished urban spaces with their passionate co-citizens and a sense of wonder. Devi and Ghosh walk readers through a traditional market, sharing local knowledge about edible and medicinal foraged species. Shukla presents the relevance of urban foraging in cultivating sustainable urban environments for social cohesion by balancing the risks and benefits associated.

The sixth section holds examples of how foraging could look going forward into a changing future. Diniz et al. describe three inspiring emerging initiatives linked to urban foraging, linked to improving child nutrition through school gardens, promoting sustainable urban foraging through a mobile application, and an urban greening movement in Brazil. Rupprecht et al. look at examples of community gardens, vacant lot restoration, and rewilded rooftops in Japan to start thinking of sharing our urban spaces with more-than-human foragers. Sardeshpande summarizes this journey with a light theoretical touch and offers a quick guide to safe foraging.

We hope that readers will enjoy this journey across cities, spaces, and cultures. We hope to contribute to the very urgent agenda of making cities more livable for all and to stimulate constructive dialogue on nature-based solutions. We welcome you to a world of possibilities and invite your thoughts, ideas, and dreams of a world where we forage for a better future.

References

Arrington AB, Diemont Stewart AW, Phillips CT, Welty EZ (2017) Demographic and landscape-level urban foraging trends in the USA derived from web and mobile app usage. J Urban Ecol 3(1):jux006. https://doi.org/10.1093/jue/jux006

Bates C (2013) Sustainable urban foragings in the Canadian metropolis: rummaging through Rita Wong's forage and Nicholas Dickner's Nikolski. Br J Can Stud 26(2):191–212

Clouse C (2022) The resurgence of urban foraging under COVID-19. Landsc Res 47(3):285–299. https://doi.org/10.1080/01426397.2022.2047911

Cestari C, da Silva Gonçalves C, de Melo C (2020) Keeping safe and fed: large heterospecific shorebird flocks to decrease intraspecific competition. J Avian Biol 51(1)

Colin T, Warren RJ, Quarrell SR, Allen GR, Barron AB (2022) Evaluating the foraging performance of individual honey bees in different environments with automated field RFID systems. Ecosphere 13(5):e4088

Cucullu L (2006) Over-eating: Pilgrimage's food mania and the Flânerie of public foraging. Modernist Cultures 2(1):42–57

Fehlmann G, O'riain MJ, FÜrtbauer I, King AJ (2021) Behavioral causes, ecological consequences, and management challenges associated with wildlife foraging in human-modified landscapes. Bioscience 71(1):40–54

Fischer LK, Kowarik I (2020) Connecting people to biodiversity in cities of tomorrow: is urban foraging a powerful tool? Ecol Indic 112:106087

Freeman J, Baggio JA, Miranda L, Anderies JM (2023) Infrastructure and the energy use of human polities. Cross-Cult Res 57(2–3):294–322
Garekae H, Shackleton CM (2020) Foraging wild food in urban spaces: the contribution of wild foods to urban dietary diversity in South Africa. Sustainability 12(2):678
Guenat S, Bailey-Athias JP, Fischer LK (2023) Urban foraging in Brazilian public greenspaces. Ambio 52:1248–1261. https://doi.org/10.1007/s13280-023-01847-y
Gupta VK, Paul S, Dutta C (2017) Geography, ethnicity or subsistence-specific variations in human microbiome composition and diversity. Front Microbiol 8:1162
Hamilton MJ, Walker RS, Buchanan B, Sandeford DS (2020) Scaling human sociopolitical complexity. PloS one 15(7):e0234615
House BR, Silk JB, Henrich J, Barrett HC, Scelza BA, Boyette AH et al (2013) Ontogeny of prosocial behavior across diverse societies. Proc Natl Acad Sci 110(36):14586–14591
Howard PL (2003) Women and the plant world: an exploration. In: Howard PL (ed) Women & plants. Case studies on gender relations in biodiversity management & conservation, London, pp 1–31
Iveson K, Lyons C, Clark S, Weir S (2019) The informal Australian city. Aust Geogr 50(1):11–27
Johnson Gaither C, Aragon A, Madden M, Alford S, Wynn M, Emery M (2020) Black folks do forage: examining wild food gathering in Southeast Atlanta communities. Urban For Urban Green 56:126860. 9. https://doi.org/10.1016/j.ufug.2020.126860
Leckie GJ, Given LM (2005) Understanding information-seeking: the public library context. In: Advances in librarianship, vol 29. Emerald Group Publishing Limited, pp 1–72
Marquina T, Emery M, Hurley P, Gould RK (2022) The 'quiet hunt': the significance of mushroom foraging among Russian-speaking immigrants in new York City. Ecosystems and People 18(1):226–240
McCullough M (2013) Attention in urban foraging. IxD&A (Interact Des Archit(s)) 16:27–36
McLain RJ, Hurley PT, Emery MR, Poe MR (2014) Gathering "wild" food in the city: rethinking the role of foraging in urban ecosystem planning and management. Local environment 19(2):220–240
Palliwoda J, Kowarik I, Lippe M (2017) Human-biodiversity interactions in urban parks: the species level matters. Landsc Urban Plan 157:394–406
Petrescu D, Petcou C, Safri M, Gibson K (2021) Calculating the value of the commons: generating resilient urban futures. Environ Policy Gov 31(3):159–174
Poe MR, McLain RJ, Emery M, Hurley PT (2013) Urban forest justice and the rights to wild foods, medicines, and materials in the city. Hum Ecol 41:409–422
Poe MR, LeCompte J, McLain R, Hurley P (2014) Urban foraging and the relational ecologies of belonging. Soc Cult Geogr 15(8):901–919. https://doi.org/10.1080/14649365.2014.908232
Reyes-García V, Fernández-Llamazares Á, Guèze M, Gallois S (2018) Does weather forecasting relate to foraging productivity? An empirical test among three hunter-gatherer societies. Weather Clim Soc 10(1):163–177
Sardeshpande M, Hurley PT, Mollee E, Garekae H, Dahlberg AC, Emery MR, Shackleton C (2021) How people foraging in urban greenspace can mobilize social–ecological resilience during Covid-19 and beyond. Frontiers in Sustainable Cities 3:686254
Sardeshpande M, Shackleton C (2023) Fruits of the city: the nature, nurture and future of urban foraging. People and Nature 5(1):213–227. 5. https://doi.org/10.1002/pan3.10428
Sachdeva S, Emery MR, Hurley PT (2018) Depiction of wild food foraging practices in the media: impact of the great recession. Soc Nat Resour 31(8):977–993. https://doi.org/10.1080/08941920.2018.1450914
Shackleton CM, Hurley PT, Dahlberg AC, Emery MR, Nagendra H (2017) Urban foraging: A ubiquitous human practice overlooked by urban planners, policy, and research. Sustainability 9(10):1884
Stöckelová T, Senft L, Kolářová K (2023) Sympoietic growth: living and producing with fungi in times of ecological distress. Agric Hum Values 40(1):359–371
Teng SN, Svenning JC, Santana J, Reino L, Abades S, Xu C (2020) Linking landscape ecology and macroecology by scaling biodiversity in space and time. Curr Landsc Ecol Rep 5:25–34

UNDESA. (2018). United Nations, Department of Economic and Social Affairs, Population Division. World Urbanization Prospects: The 2018 Revision, Online Edition. Accessed 03 June 2019

Vaittinen T, McGookin D (2016, June). Phases of urban tourists' exploratory navigation: A field study. In *proceedings of the 2016 ACM conference on designing interactive systems* (pp 1111–1122)

Ware SM (2022) Foraging the future: Forest baths, engaged pedagogy, and planting ourselves into the future. Qual Inq 28(2):236–243

Mallika Sardeshpande obtained her PhD at Rhodes University, South Africa, and is now a research associate at University of KwaZulu-Natal, Scottsville, South Africa, and Ashoka Trust for Research in Ecology and the Environment, Bangalore, India. Her work spans the continuum between biodiversity conservation, ecological restoration, agroecology and agroforestry, environmental and food policy, human wellbeing, and urban ecology.

Shalini Dhyani (PhD) is a Principal Scientist at the CSIR-National Environment Engineering Research Institute, India. She is Asia Vice Chair and Steering Committee member in the Commission on Ecosystems Management of IUCN and Visiting Scientist at CIFOR-ICRAF.

Making Sense of Diversity of Foragers

2

Mikelis Grivins

Abstract

If foraging is considered only as a practice of engaging with wild products, then all foragers might look the same. However, if diverse knowledge claims, diverse motivations, and diverse uses of wild products are considered, then the people behind the practice start to appear very different. This chapter engages with this diversity by reflecting on my attempts to introduce a classification of foragers that would allow me to structure this diversity. To do this, the chapter raises two questions: (1) What are the categories that can be used to classify foragers? (2) What can we learn from discussing various possible classifications – first of all about foraging and secondly about ways in which scientists can work with the unknown and invisible differences within target groups. The chapter answers these questions by discussing the strengths and weaknesses of the forager classifications I have tried to apply to my work with foragers.

Keywords

Foraging · Urban · Wild resources · Consumption · Sharing · Selling

2.1 Introduction

There is a story that I first heard while interviewing foragers from Latvia that was later also repeated by respondents from Estonia (two countries sharing somewhat similar history). The story respondents shared with me was concerning the encounters they have had with part of the population that was resettled to the two countries during the

M. Grivins (✉)
Riga Stradins University, Riga, Latvia
e-mail: mikelis.grivins@rsu.lv

S. Dhyani, M. Sardeshpande (eds.), *Urban Foraging in the Changing World*,
https://doi.org/10.1007/978-981-97-0345-6_2

period when these countries were occupied by the Soviet Union. The story was trying to make sense of different foraging approaches/preferences between them and the newcomers. A substantial number of people were moved from other parts of the Soviet Union to the two Baltic states during the period while these countries were occupied, and those who have been following the history of the Baltic states will know that topics related to politics of the Soviet Union and memories of the period are highly sensitive in the region till this day (see Vullnetari and King 2008; Agarin and Grivins 2016). In these stories, the mushroom picking newcomers from unknown corners of the Soviet Union were presented as completely ignorant and whimsical – collecting inedible mushrooms that were full of worms and carrying them in a way that mashed their appearance, thus damaging their aesthetics. As a researcher well-read in discourse theory and power relations, I interpreted these remarks as an extension of thinking about the historical events as well as a possibility for interviewed persons to stress the outstanding individual picking knowledge and uniqueness of local foraging tradition. Also, the fact that the same story was repeated in more than one interview made me cautious that the story might be based on nothing more than willingness to smear particular groups. I never really thought that these stories should be examined more closely. After all, they resembled so many other similar stories trying to present the differences between the native communities and groups of people resettled by the Soviet Union. As such, to my mind, these stories were just one more way in which people were expressing their discontent with the historical injustices.

While conducting these interviews, I was also engaging in literature review. During this inquiry, I stumbled upon an article by Peintner et al. (2013) which, among other things, suggested that there is what could be called Slavic and Romanic foraging cultures in Europe. The article elaborated that the Romanic mushroom culture is more oriented to just some specific mushroom types (more specifically chanterelles and boletes) while Slavic tradition recognizes significantly more mushroom types. This discovery made me get back in the data I gathered and reassess what I had heard. The story discussing foraging differences suddenly presented itself in a completely different light. Now the story appeared to me as a clear miscommunication between two foraging approaches. It also occurred to me that the ignorance of respondents that served as a soil for these stories was not just an ideological statement but represents much deeper differences in knowledge and completely different attitudes towards the wild products between the two groups. While respondents telling these stories were mainly looking for the aesthetically pleasing mushrooms and were not really creating real interlinkages between wild resources and diets that would incorporate wild products, people being in the center of these stories were significantly more engaged in everything wild product related. They knew more ways in which mushrooms can be prepared, were eating much broader spectrum of mushroom varieties, and in general did not care so much about how these mushrooms look.

This finding forced me to recognize two claims. Firstly, foragers differ, and these differences can be quite substantial. Secondly, the practice of foraging, regarded from a distance, might seem alike no matter who forages. Thus, we would not be able to understand these differences without really looking deeper into their knowledge and interpretations. Consequently, it might be a mistake to analyze foragers, as it is often done, as a unified category. Underneath the practice that looks similar everywhere, we can identify diverse knowledge claims, and knowing these claims makes the practice people are engaged in quite different. An illustration of this final point is a question I have asked all respondents during my interviews I conducted between 2018 and 2019: Are there any ethical aspects of foraging that are important to them? Many of the respondents without real hesitation were saying that they are aware that they could be overharvesting the plants they are looking for and this might have a long-term impact on local ecosystems. On the other hand, there were many respondents, who even when directly asked did not identify that there could be a possibility of overharvesting or that there are any ethical issues tied to the practice at all. Not making any claims as to who is right, I just want to use this example to underline that this leads to two very different ways in which foragers can engage with nature and the factors they will take into consideration while picking wild products. Thus, if knowledge is scrutinized, we will suddenly notice differences in practice as well.

The goal of this chapter is to explore the dimensions that allow us to distinguish between various groups of foragers of edible products. I do this by discussing in this chapter two issues: (1) what are the categories that can be used to classify foragers and (2) what can we learn from discussing various possible classifications – first of all about foraging and secondly about ways in which scientists can work with the unknown and invisible differences within target groups. For this purpose, in the chapter, three different classification approaches I have consequently worked with while studying foraging and foragers are presented and discussed. This chapter reflects on the findings I have made while conducting in-depth interviews with foragers in the years 2018 and 2019. In total, 44 interviews with foragers from Latvia, Estonia, the Netherlands, and the UK were conducted. Although the ideas presented here could be attributed to other parts of the world, the chapter is mainly looking at examples from Europe. The chapter uses words forager and picker interchangeably. It also consciously uses the notion "wild product" to describe wild booty. Wild products as a notion cover all products that can be picked but are not cultivated now for picking (thus, foraging can take place in forests, but it could also take place in parks).

The chapter starts by providing a short theoretical background – it discusses the classifications of foragers that are already presented in the academic literature. It continues by providing a short overview of the data behind the presented classifications. This is followed by three propositions on how wild product pickers can be classified: based on the product they prefer to forage; based on the intended use of product; and based on their motivations and knowledge. Finally, discussion remarks on the overall challenges that must be faced when classifying foragers.

2.2 Diversity of Foragers Across Europe

While most articles discussing foraging and foragers are in fact looking at various subgroups of people engaging with wild products, very few have chosen to reflect on the differences between the chosen subgroup and broader practice of foraging (see Landor-Yamagata et al. 2018; Paddeu 2019; Nyman 2019). Despite the scarcity of literature addressing these issues, some of the existing articles offer a good entry point to the understanding of diversity among foragers.

The already mentioned Peintner et al. (2013) suggest that there are cultural differences that allow identifying two large groups of foragers in Europe. Peintner and his colleagues' work assumes that in countries where wild products are popular, this popularity will be reflected in legal documents. This allows the researchers to suggest that the state of mushroom picking across Europe can be derived from a targeted analysis of documents regulating mushroom trade in each Europe's country. Findings from this analysis allow authors' suggesting that there are two groups of countries in Europe – mycophilic and mycophobic. The mycophobic countries are in general Romanic-speaking countries or countries with substantial Romanic-speaking minorities. In the opposite side, there are the mycophilic countries that are predominantly Slavic-speaking countries. The reference to a particular Slavic foraging tradition is reiterated by quite a few studies (see Łuczaj and Nieroda 2011). Some studies have been reiterating the same distinction in designation between the West and the East of the EU (Svanberg et al. 2012; Schulp et al. 2014; Lovrić et al. 2020).

Returning to Peintner et al. (2013), it is clear that the distinction is very important as it helps to grasp the differences in diversity of wild products consumed between various cultures. Still the assumption that the relevance of the wild products will be reflected in the trade regulation could and should be approached cautiously. Firstly, clearly regulations will not always reflect the public appeal of a phenomenon. Second, because of the focus on regulations, the abovementioned study was forced to use borders of countries to designate foraging cultures. This approach, although it seems like the most obvious way to divide different cultures, misses the nuances that can be observed on the ground level. Even more importantly, there are a great number of articles (see Landor-Yamagata et al. 2018; Paddeu 2019; Nyman 2019) describing smaller groups of foragers (communities of pickers) representing a region of the country or sometimes even crossing the borders of countries that have highly intensive engagement with wild products. Focus on countries, although helpful, makes us not to notice these groups that are exploring wild products. Foraging is not a phenomenon that aligns itself with national borders.

Wiersum et al. (2018) use results of the work conducted by Cesaro et al. (1995) and propose a more detailed distinction of diversity of European regions. Wiersum et al. (2018) identify five main orientations of non-timber forest product (NTFP) potential: (1) production of NTFP as cultural resources, (2) forest enterprise development, (3) regional development, (4) development of common pool resources, and (5) production relevant to bioeconomy. After assessing the importance of the orientations in the five regions (Northern Europe, Atlantic Europe, Central Europe, Southeast Europe, Mediterranean), authors conclude that concepts like culture, identity, and self-actualization are crucial to understand the role these products have in all five of the regions. It is worth pointing out that the five orientations clearly designate the distinction between commercial potential of wild products and other possible motivations foragers might have. Although Wiersum et al. (2018) do not delve too much in addressing the differences between commercial and noncommercial foraging, it is crucial that this distinction is made.

Foraging has always been strongly related to tradition and values (Grivins 2021). However, if the focus is shifted from a spatial distinction of foraging cultures to engaging with the change the tradition has had over time, one most likely will notice that the meaning of the orientations suggested by Wiersum et al. (2018) is changing. Łuczaj et al. (2012), in an article "Wild food plant use in 21st century Europe," distinguish between what could be described as the fading heritage foraging and currently emerging new foraging traditions. Engagement we have with nature – and consequently with wild products – tends to change over time. In the case of wild products, in general, we can observe across various cultures a drop in the diversity of products that are used by local communities (Łuczaj et al. 2012; Grivins 2021). Several reasons can be presented for this trend. However, the overarching argument seems to be that shift from traditional and rural to modern and urban has caused a shift in significance and role these products have – they are not so much seen as products anymore as they are tied with experiences or services (Reyes-Garcia et al. 2015). This finding has been rephrased by Wiersum (2017) who suggests that the new foraging tradition cannot be regarded as an extension of the old product-oriented approach as it is based on a completely new way of rethinking engagement with nature. Although this observation is hard to argue with, it sits uneasily with evidence from countries (for example, Latvia, Estonia, Finland) that still have a high share of foragers who jointly collect huge quantities of wild products (for examples, see Grivins 2016; Vidale et al. 2015). In many of these cases, it is clearly a historical tradition that has helped maintain the close relations between communities and wild products (as has been illustrated by Bardone and Pungas-Kohv 2015).

These theoretical considerations offer strong starting point to engage with ways to classify foragers. The literature illustrates that when it comes to foraging, there are differences between countries. However, links between society and the wild are going through constant transformation, and we are witnessing new foraging traditions emerging (Łuczaj et al. 2012; Grivins 2021). This process is taking place differently across Europe. Finally, on top of these considerations, it should be kept in mind that wild products are also part of lucrative market, and this cannot be overlooked when thinking about differences between foragers.

2.3 The Data Used in the Chapter

This chapter follows my attempts to develop a better way to structure the diversity of foragers. It focuses on two tasks simultaneously – it asks how to structure diversity and tries to capture conceptual challenges that arise while working on a comprehensive classification of foragers. Because of this, this chapter takes somewhat an unconventional structure. At its core – it follows and discusses the findings of my research at different stages of my work.

The chapter is based on in-depth interviews conducted in 2018 and 2019. The interviews were conducted in Latvia (30 interviews), Estonia (3 interviews), the UK (5 interviews), and the Netherlands (6 interviews). All 44 respondents were occasional and professional foragers. However, they have been selected so that they would capture as much diversity between foragers as possible. The respondents were selected so that they would capture the diversity of products that are foraged, considering their engagement level in foraging, and looking at the intended use of booty.

It was not the initial aim of the data gathering to develop a classification of foragers. Instead, the data gathering was focusing primarily on two aspects of foraging – how communities are formed around foraging and ways in which foragers commercialize wild products. Nevertheless, it became apparent relatively early that any meaningful claims regarding the two issues can be made only if subgroups of foragers are considered. Because of this, a third target was added to the study – to identify subgroups of foragers allowing to provide more elaborated answers to the initial two questions. Because the classification of foragers was not the initial goal, its development did not start with a theoretical inquiry. Instead, the process could be described as iterative – it started with data analysis and initial proposition for classification, which then led to delving into theoretical literature, adding additional questions in interview guidelines and eventually new propositions for classification.

Three classifications were developed, tested, and either rejected or improved during the data-gathering process. The first version was based on the type of wild products respondents were foraging. Although it allowed making new conclusions regarding the foraging process and the knowledge foragers need, it had a substantial number of overlapping groups. It was also unable to reflect upon various applications of wild products. Finally, the approach was toothless when it came to offering insights regarding broader foraging-related dynamics that can be observed in Europe. The second version primarily focused on the application of wild products interpreting application from the perspective of ways in which wild products reach the final consumer. The classification allowed to have an in-depth engagement with notions like commercial picking and subsistence picking in the classification. However, the approach fell short in presenting the differences in diversity of products foraged. Additionally, the classification was inefficient in explaining the practices of various lifestyle groups of foragers. As a result, a third classification was developed built around the motivation of forager and rootedness of practice. All three classifications are expanded on in the results section of the chapter.

For data analysis, a two-step process has been used. During the first step, the information provided by the respondents was entered in an Excel sheet combining information about interviewed forager. Some examples of information collected include gender, products foraged, distance travelled to foraging spots, use of wild products, particular plants collected, and estimations regarding the amount collected. The author developed the initial matrix to be filled before fieldwork started. It was then supplemented with additional criteria all through the study. This table was used as the basis to develop the classification. During the second stage of data analysis, additional in-depth analysis of interviews was conducted that allowed to tease out from the data gathered the characteristics of the identified groups. The used approach could be characterized as theoretically guided open coding.

2.4 The Three Ways to Classify Foragers

The following sections will illustrate the three ways to classify foragers discussing the characteristics of the groups presented as well as reflecting upon the strong and the weak elements of the proposed approaches. The three suggested classifications should be perceived as ideal types. They represent an image of reality that can rarely be observed completely as described – for each respondent, there are some deviations from these absolute groups. It is also worth noting that in all three cases, there are foragers that would technically fit in more than one group. In these cases, based on respondents' answer, they were classified in the group that corresponded their main foraging activities (for example, for the first classification, this meant considering how much time they spent engaging with each group of wild products). However, it is also worth noting that in most cases, the answers respondents gave made it very obvious in which group they fit in.

2.4.1 Tell Me What You Forage, and I Will Tell Who You Are

My first attempt to make some sense of the diversity in the data led me into a direction that was very simple yet relatively unexplored in academic literature – the interviews led me to think that there are differences in groups of foragers that focus on different products. Various groups of wild products are quite different in their properties. These differences affect the way in which people engage with the wild. Think about berries – knowing the preferred environment of various berry species, one can easily predict the location of the best berry spots, and each spot can contain substantially more booty than a forager might need. Mushrooms, on the other hand, can be difficult to find even for an experienced forager. Herbal teas as a category are again a completely different experience of foraging – these can be easily spotted, and one can easily collect substantial amounts of product in just one spot. Same goes for the edible leaves and other similar products. The interviews illustrate that people picking wild products are aware of the peculiarities associated with various product groups, thus illustrating that there is a certain reflection among the foragers on why and how they are engaging in the practice. This helps ensuring that foragers choose the engagement with wild products (possibly unconsciously) and consequently with the environment that is most suitable to them.

2.4.1.1 Berries

The interviewed foragers focusing on berries stressed the importance of repetitive nature of berry picking. Respondents also indicated that one can easily predict where one will find the best spots for each type of berries, and the best spots do not usually move from year to year. Furthermore, most of the berries have very clear visual characteristics that allow them being recognized by even a first-time forager. Because of these factors, looking for wild berries does not require too much specific knowledge. Furthermore, there is practically no risk in collecting the wrong species. However, it does require willingness to engage with repetitive movements – it is a repetition of moving berries from a branch to foragers' foraging container, which is also often done while being in an uncomfortable position, sometimes squatting for hours. Most of the interviewed foragers (or at least those foragers who are not engaging in the practice for commercial purposes) claimed that they are not using any instruments that would allow them to pick berries faster. Instead, they were relying on their hands, making the process slower, yet also creating a more personal bond with the product picked while also ensuring that the final booty is of high quality. Foragers focusing on berry picking often chose to talk about the mental state that they were in while foraging. A number of respondents described the repetitive nature of berry picking using the word "meditation" (the same word yet in a different meaning has often been used by foragers focusing on plant parts as well). Respondents were explaining that while result is important for them, it was rather the slow and repetitive process that attracted them to berry picking. Engaging in this process allowed them to run away from their daily worries and focus on a practice that is very clear and requires very little decision-making. In this sense, it was their meditation. Berry picking is a practice for somebody who is willing to put in work in the activity.

2.4.1.2 Mushrooms

Wild products that require a completely different approach are mushrooms. There are significantly more various mushrooms than berries, and quite a few of the mushrooms that can be found in forests are not edible. Thus, when it comes to mushroom picking, knowledge plays a significantly more prominent role. Furthermore, the knowledge regarding the properties of mushrooms is something that can be constantly improved. Typically, a mushroom forager starts with just a couple of mushrooms one feels safe to pick. Even recognizing a couple of edible mushrooms might be challenging for a novice forager. The struggle of a mushroom picker just starting to engage in mushroom picking is nicely captured by Pollan (2006). However, with time, this list can become substantially longer. This is due to the possibility to improve knowledge of what is edible and what is not. This is also due to an improved ability to identify nuances of visual properties of mushrooms that allow distinguishing various varieties of mushrooms. Finally, this is also because foragers learn new ways to prepare mushrooms. Just as a side note, it is worth noting here that not all mushroom foragers are looking for this broader knowledge in mushroom picking. In a number of occasions, precisely the opposite was true – people were losing their knowledge and reducing the number of mushrooms they were picking. Still, even in these cases, there was apparent pride when respondents were talking about the biggest mushrooms or their best spots. So even if knowledge is not important, there is still the dimension of competition.

The other important aspect that distinguishes mushrooms from berries is the hunt and the discovery. The hunt for mushrooms demands a trained eye. Mushrooms are difficult to spot. Only practice allows one to become effective in this activity. Thus, there is a clear dimension of experience that changes the performance of mushroom picker. Taking these two dimensions together – experience and knowledge – offers mushroom pickers a clear trajectory for skill development, something that is not possible if berries are considered. So, compared to berry pickers, mushroom pickers are focusing on a process that does not require repetitive activity. This is a process that requires mastery and as such offers a possibility for a competition. Consequently, it is possible to suggest that mushroom picking can be associated with very different expectations from berry picking. Additionally, a thing that foragers have made fun of in some interviews is the difference that mushroom foragers must carry a knife with them. This automatically makes mushroom picking much more like hunting – after all, one must carry a weapon and must be skillful to find the prey (some respondents tied these claims to gender saying that berry picking is a more feminine activity, while mushroom picking is a more masculine activity). This description of experience of mushroom picking could be summarized stating that mushrooms offer foragers a space to grow and to become better, and thus, it can be easier transformed into a competition or a hobby or can be easier linked to self-actualization. Thus, mushroom picking is for somebody who is ready to invest time in a hunt and might be looking for an activity where he or she can grow.

2.4.1.3 Roots and Leaves

Finally, there are edible roots and leaves and other parts of plants one can forage. This is a broad category, and thus, there are exceptions – plants that can be difficult to recognize or to find. However, in broad terms, these plants do not require the same labor that is associated with berries, nor are they as challenging to find or identify as mushrooms. Also, on most occasions, if commercial foraging is excluded, the amounts that picker tries to gather can be picked relatively easily. Consequently, this is a group of wild products that can help people to maintain relations with the wild yet does not require substantial investments from the person. This is also a simple way to establish new relations with the wild.

This is a rather diverse group. However, one aspect that was often present in interviews and that distinguishes this group of foragers from the other two is the way in which these foragers link themselves to the particular products. While the nature of engagement with the foraged product remains like in other groups of products, there is often an additional link created between these products and broader knowledge. It does not tie into knowledge regarding the wild, but to broader claims about the world. During the interviews, people who were foraging these products were more informed about the environmental issues and were more willing to discuss these issues as well as see their practice in the context of environmental challenges. People from this group were also talking about ethical issues they faced while foraged (this could be maybe explained by the fact that unlike fruits and mushrooms, overharvesting of leaves and roots is more likely to be destructive if overdone or done carelessly). Finally, quite a few representatives of this group linked foraging to personal spiritual development. This last claim is one of the most interesting – foragers from this group were often using the whole foraging process as the way to reconnect with nature. In several cases, this search for a way to connect with the environment incorporates esoteric claims about nature. Respondents discuss issues like feelings of plants, consciousness of forest, and energies of various areas. For this group, it was more evident that foraging is also a means to illustrate their worldviews.

Classifying foragers depending on their wild product preferences offers a promising perspective on the diversity of foragers. To better understand the feasibility of the proposed classification, a quantitative study needs to be conducted. Yet, even without a quantitative analysis, it is possible to claim that this approach reveals new perspectives on what various foragers are looking for in the practice. It also helps us to develop understanding on how foragers engage with these products. Furthermore, as I have indicated, the differences of engagement with various wild products are well known to foragers – it is common that during interviews, claims about groups of foragers are made based on what they are foraging. Still, although there is a clear potential in this relatively simple approach, there are also limitations that must be resolved to say that this approach could be optimal to explain differences among foragers.

2.4.2 What Do You Do with the Product You Forage?

The classification stressing the properties of various wild plants is helpful, yet it becomes obsolete once we start to debate how the foraging practices are affected by the intended use of booty. The classification lacks a dimension presenting the use of the products. To some, this might not seem important; however, the understanding of the intended use of the wild booty allows illuminating an additional dimension of motivations and offers some insights as to how much of wild product people are collecting, thus offering a better understanding of potential impacts of the practice. Also, once foraging is not meant only for satisfying own personal needs, it needs to take into consideration the expectations of the consumer (no matter who the consumer is). The second approach presented here presupposes that the foragers can be placed on an axis classifying the envisioned use of these products. Using this approach presupposes two main ways in which wild products can be used: foragers can collect them either for self-consumption or with an intention to sell them. In between these two oppositions, there are several additional strategies. Movement along this scale presupposes a gradual shift from a more personal (foraging for self-consumption) to purely pragmatic (foraging just for trade) relations with the wild products and, by extension, nature. If compared to the first proposed classification, this is a much more structured approach based on the idea that wild product foraging is automatically linked to some sort of supply chains and assumes that the supply chain properties the forager is linked to will affect the forager.

2.4.2.1 Self-Consumption

Foragers of this group are foraging to satisfy their own needs. The product on its own is an important motivator for many of the representatives of this group. However, self-consumption can also be centered around foraging experience. Representatives of this group have a good understanding of the amount of product they need to forage and have a very clear vision as to which products they are looking for. The products this group is looking for are strongly linked to lifestyle and culture of the forager, which predefine considerations the forager has regarding what is edible and what is good food. This group has an in-depth knowledge related to possible uses of wild products.

2.4.2.2 Sharing

Sharing foraged products is the simplest way in which wild products can penetrate food supply chains. Foragers representing this group also consume the wild booty; however, they also share the products with people from their social network. The motivation to do so differs. For example, some respondents were telling that this is a way in which they maintain relations with people surrounding them – they were sharing jams, pickled mushrooms, herbal teas, or other goods with people they felt close with. In some other cases, it seemed that networks had a designated representative who collected wild products for all of them. Herbal teas are a good example of this – it was quite common that in interviews, it was explained that a broader circle of people had one unofficial key person who had assumed the responsibility to ensure that everybody has the teas that they might need. Meanwhile, in other interviews, sharing seemed to be a way in which foragers ensured that they could pick more. Especially among mushroom pickers, it seemed the case – it was common that during interviews, stories were told of finding spots that had mushrooms significantly more than one could consume. Respondents were explaining that they shared the booty with people around them. Unfortunately, it was not clear whether the people receiving the gift of mushrooms were really excited about this present.

In the case of sharing, foraging is very much a social activity that presupposes mutual recognition of what products are edible and what is a good food. Because of this, the diversity of products picked by foragers sharing the booty is lower – they must align what they pick with what is accepted by the people who are not foraging. This also presupposes social trust – people engaged presuppose that all of them can distinguish between various edible and potentially harmful products. A good example of this once more is mushrooms. This is a product that requires deep knowledge of various subspecies, and thus, we can assume that people receiving the product must hold the belief that the person providing the product has a firm grasp on what is it that he or she is giving to others (this is even more important when the product is preserved and has lost its visual properties). In Latvia, there have been loud cases when well-intended sharing of wild products has seriously endangered people receiving these products.

2.4.2.3 Selling Experience

The next level of use of (benefitting from) wild products is a form of relations that can be placed in between personal and commercial engagement with nature. In this case, foragers are not selling wild products. Instead, consumers are offered foraging-related experiences which are packed as practical engagement episodes focusing on either foraging-related knowledge, skills, or sensations. It means that the forager is maintaining educational or other types of events that allow groups of people to engage and learn. From the interviews conducted, it was evident that a few people who were heavily engaged with foraging chose at least on some occasions to organize events for people not familiar with the practice (some do it regularly). Also, many foragers with less experience in foraging were saying that they have participated in educational events to improve their foraging skills.

The form and the focus this engagement with wild products takes are strongly linked to the expectations of participants of these events (those paying the bill). This means that these events can differ from one another quite substantially, and this is true for both the form they can take and the broader knowledge fields these activities can be linked to. During interviews, respondents were talking about events that were teaching to recognize various mushrooms, focusing on the environmental challenges of foraging, linking wild products to cooking and to alternative medicine, linking wild products to claims related to spirituality, etc. Some themes of these events were questionable – like an event dedicated to creating tea mandalas that would represent the herbal tea most suitable for the person creating the mandala. Still, foragers who had participated in these activities as consumers reported that they broadened their knowledge and widened their understanding of the surrounding environment. Furthermore, this engagement presupposes that everybody is spending more time in the nature developing deeper appreciation and developing more personal relations with it.

2.4.2.4 Selling in Short Supply Chains

Foragers operating in short supply chains are mainly focusing on selling the collected products to the community of people appreciating certain properties that are associated with the notion of "wild." In this case, forager or a person close to forager is also selling the booty. The motivation of people engaging in this activity is clearly commercial. Still, these foragers maintain a strong link to a particular territory, thus developing a deep understanding of the environment they are foraging in and consequently having a deep understanding of the impact they might be causing. However, it is also worth noting that this group is not making so much profits out of the amount of the product sold as it is focused on asking for a higher price (a price that satisfies them) for the product. This does not mean that they have substantial income from the activity (although this might be the case). Instead, it means that in most cases, they pick only a limited amount of product.

In most cases, these supply chains deal with a limited number of wild products – those that are well known and appreciated by consumers. For example, in Latvia, this means that there is a predominant focus on some berries and few species of best known mushrooms – the products that can be easily transported, do not lose their aesthetic properties, and can be easily visually recognized. On a couple of occasions, the respondents mixed the offer from wild products with products grown in their gardens in this way diversifying their offer.

2.4.2.5 Selling to Mainstream Chains

The product is sold in quantities through conventional supply chains. Foragers focus on just some products that are in high demand (in Northern Europe, this mainly means focusing on blueberries, lingonberries, chanterelles, boletes). Once there is a demand for the product, it is heavily foraged by this group of foragers. Consequently, the groups' activity might leave a substantial environmental impact. In this case, the foraging practice is not related to any personal considerations yet is motivated by practical considerations – the fact that wild products are easily accessible and often on public land that allows foragers to easily access them without additional fees. Because of this, foraging these products is an appealing niche for those that do not have the capital to start other activities and in general have few other opportunities. They are highly mobile and are willing to move closer to the best new spots of products that have the high-income promise. Foragers representing this group claim that they enjoy foraging and being a part of these supply chains. However, the physical nature of the work and lack of social warranties make this type of income making a risky and challenging endeavor.

As it is shown, linking foraging practice with the intended use of product helps explain where exactly one sees the value of foraging and what is foragers' potential impact. In this regard, the strength of this classification is its ability to link claims regarding environmental impacts with social considerations and social structures. In this case, the borders of the practice are substantially widened – the picking experience is seen in the context of a much broader perspective on various social institutions. Any practice undoubtedly is to be perceived as inevitably social, and thus, in this regard, this classification provides a much more realistic account of reality than the first classification. However, the weakness of the classification emerges from the same factor that is also its strength, namely, the claim that just the use can sufficiently explain the diversity. It certainly offers one more valuable starting point, yet its focus on the link with consumption limits the classification's ability to talk about the ideological differences of foragers and the properties of the product and even more importantly makes it say little about the process itself.

2.4.3 Linking Motivation and Use

The final classification to be presented here was recently introduced in an article published by Sociologia Ruralis (Grivins 2021), so I will not go into detail explaining the theoretical assumptions behind the classification. This classification was developed based on the findings from the first and the second classification. From the second classification, the notion of wild product use was borrowed. Yet in this case, the notion of use was replaced with the notion of motivation, thus incorporating a possibility that pickers are not just focusing on the product, but might be foraging because they enjoy the process. Consequently, this version distinguishes two different motivations foragers might have – they can be motivated by product or by the process. Furthermore, the classification distinguishes two knowledge frames – based on traditional knowledge and based on newly established lifestyles. The dominance of the two knowledge frames has been stressed by the existing literature on the matter. However, it was also an aspect that was not captured by the previous classifications. This allows the article to identify four groups of foragers: (1) rooted foragers – motivated by the process and operating with traditional knowledge; (2)

lifestyle foragers – motivated by the process and operating with newly established knowledge frames; (3) subsistence foragers – motivated by the product and benefiting from traditional knowledge; and (4) commercial foragers – motivated by the product and benefiting from the novel knowledge.

2.4.3.1 Rooted Foragers

This is a group that has learned to forage in their childhood and continues to do so. They have learned the products that can be foraged and the way in which these products can be prepared when they are young. This gives them a sense of security when they engage with the wilderness and the wild products. Also, it gives them a perspective on how to interpret and to engage with nature. Engaging with the wild comes natural to them. When it comes to foraging, it is not the product that is important to them – instead, it is the process that allows them to maintain relations with their previous experiences and consequently identity. During the interviews, it was quite typical that the representatives of this group shared information regarding the best find and they were keen to share the photos of perfect specimens online, thus illustrating the bounding significance of foraging. The other surprising fact that can be observed is that at least in the case of rooted foragers I have interviewed, I observe that they tend to lose the diversity of the products they once foraged. It is not a necessity for them to ensure that they have a particular amount of product, and thus, they do not feel obliged to ensure that they constantly have a particular amount of booty. On the one hand, this leads to the fact that people maintain interest in just the most recognizable and most common products. This results in people focusing on products that are most aesthetically pleasing easily discarding products that are not as pleasant to the eye, leading to a situation where most of the interviewed rooted foragers focused on chantarelles, boletes, wild strawberries, and maybe a handful of other products. On the other hand, this leads to a situation, where foragers go into a forest even if there are few possibilities that they will find anything.

The final point to make here is regarding what this group knows and how it perceives what it knows. It mainly relies on the claims that they have learned from the predecessors and do not really look for new perspectives on how to regard this process. As a result, this group is well informed about the local insights into local ecosystems. Yet they lack more recent elements to confront this knowledge with – like claims regarding sustainability or ethical aspects of foraging.

2.4.3.2 Lifestyle Foragers

A group of foragers have decided on their own to learn to forage. These are foragers that before their decision to engage with wild products had only limited knowledge regarding this activity. Because of this, the foragers are much more open to various foraging-related information that has resulted in a fact that they are better read in various perspectives of how wild products could be interpreted. Consequently, the lifestyle foragers are potentially much more responsible and conscious regarding the impacts of their activities on the wild than the rooted foragers. Lack of traditional knowledge is also the reason why this group is prone to exploring the diversity of wild products and more open to try new things. They are also more willing to explore various foraging locations.

Similarly, as it is for rooted foragers, this group is not so much concerned with the product and rather sees the value in practice. However, again, since the practice is not tied in the historical experiences but in the willingness to search for the new or to revitalize existing identities, the lifestyle foragers are not even tied to particular foraging spaces – places where there is a particular flora. The link with identities is important because it helps substantiating what aspects of foraging lifestyle foragers will bring forward. Consequently, there are those that are looking for a healthier way of living, those trying to be more environmentally conscious, those looking for new flavors, and many other groups among lifestyle foragers.

2.4.3.3 Subsistence Foragers

This is the group that maintains the cultural knowledge and is also recognizing the value of the wild product. Among the foragers interviewed, none were dependent on wild products – they could have bought everything for their meals if they wanted. However, they chose to maintain strong relations with the wild constantly ensuring that some part of their daily meals was coming from the nature. Foragers representing this group know which products can be foraged at what time of the year and have a very clear idea as to how much product they should gather and where they can find it. For this group, foraging is more work-like – they are setting a goal, and it is important to reach the set amount of product. Also, for this group, the number of different species foraged is typically higher and is tied to biodiversity of the surrounding areas. This is strongly related to the fact that subsistence foragers are mostly engaging with the areas in their direct proximity. Most of them were very aware of the peculiarities of the surrounding flora. This awareness, of course, was the result of lifelong engagement with the landscape. Interestingly, while most of the respondents from this group had a deep understanding of the surrounding environment, they had very little to say when it came to broader foraging-related sustainability claims.

Because subsistence foragers appreciate the product, they are less likely to lose foraging knowledge they have. Yet, on the other hand, from the foragers I have interviewed, there were very little evidence that foragers representing this group would have been motivated to take up new products. Instead, they stick with what they know. Still, what they know is enough to become a steward of local surroundings.

2.4.3.4 Commercial Foragers

The nature of commercial foraging presupposes that the focus of foragers is on the product. Quite a few commercial foragers I have encountered have a long experience with foraging. Still, when it comes to trade, the focus on a particular product is not motivated by the tradition but by the demand. This is especially true once global supply chains are regarded. This group of foragers is extremely diverse. For example, those selling product in local markets often forage in local spots. If they are selling their product through local short supply chains, then the amount they forage is rather limited.

Meanwhile, those operating in global supply chains are quite different. Global supply chains have almost unlimited capacity to absorb products, allowing foragers to sell substantial amounts of picked goods. Many of these products promise high added value, and thus, foragers are highly motivated to forage. Consequently, many of them are highly mobile constantly looking for best foraging spots and moving from one spot to another and willing to invest (in transport, foraging equipment) to increase their profit.

2.5 Discussion

In this chapter, I offer three perspectives on how foragers can be classified and a reflection of the process explaining how these classifications have been building on the results of the previous one. All three of them offer some new aspects as to what characteristics of foraging can be used to build a classification. However, all of them also illustrate at least some shortcoming of the whole endeavor of trying to classify foragers, and a number of these shortcomings are shared by all three classifications, suggesting that these are broader issues to be resolved, yet not emerging from the task at hand. Here, I suggest that three of these overarching issues should be discussed.

The first challenge to raise here is also probably the most obvious – all three leave some aspects of foraging uncovered (this is less of an issue for the third classification). This is related to the fact that each of the three classifications represents a very different perspective on how one can look at foraging as a practice. So, the real challenge is not related to the classifications but to the question – What exactly do we expect from wild products? Is it just one more commercial opportunity? Is it a way to engage with sustainability crisis of agro-food systems? Is it an activity to help us to reestablish stronger links between people and nature? Probably, the truth is somewhere in between these claims. Answer to these claims might be the basis for the fourth classification focusing on social-ecological motivations. However, the fact that there is no definite answer to what this practice is all about is the reason both why we need a more nuanced perspective to scientifically engage with it and why it is difficult to develop this more attuned perspective. The classifications offered here are a starting point to illustrate how the various objectives link to various subgroups of foragers. However, clearly, much more work must be done in this regard. Yet, before we have this very clear focus, the third classification could be used as an entry point to develop a deeper understanding of diversity among foragers.

This leads me to the second challenge – How capable the offered instruments are to help develop meaningful explanations capturing the nuances of various subgroups of foragers? For each of the described categories, there is a diversity of subgroups underneath that cannot be properly captured without introducing additional analytical elements. For capturing these, there would be a need for a significantly larger set of subgroups. This would clearly improve the precision of description yet would also reduce the analytical applicability of the classification. A balance needs to be found between the willingness to capture details and maintaining the analytical instrument sharp.

On the other hand, the complexity that is emerging from intertwining diverse motivations, knowledge, skill levels, intended use, surrounding flora, etc. most likely cannot be fully captured by any classification. This is because this is the intended role of this instrument – stressing the crucial categories creating groups of people while paying less attention to minor differences. For classification like this, being analytically sharp means that some nuances can be glossed over and generalized. Furthermore, the proposed categories are what are called ideal types – an aggregation of the results which most likely cannot be found in reality in pure form. The question then is – Have we stressed and generalized the right nuances? To answer this question, a quantitative assessment of the classifications would be needed.

Finally, the third challenge is the question as to where to draw a border for the explored practice. The first borders that are difficult to draw are the ones concerning the notion of wild products. Most literature would prefer to use the concept of NTFP. However, this stresses the location where these products grow and tie foraging to forests. Regarding the practice as a way people can engage with forests is limiting because in practice there are people who forage in meadows, in parks, and even in abandoned gardens. So, it is an engagement with the nature that is not cultivated at the moment for the particular purpose. Thus, the focus should be on the practice. However, then the question is – where this practice ends – should we consider only what is happening when a person picks a product? This chapter has shown that foragers' motivation can change how exactly this is done. This would mean that some broader considerations should be included.

References

Agarin T, Grivins M (2016) Chasing the green buck? Environmental activism in post-communist Baltic States. *Communist and Post-Communist Studies* 49(1):243–254. In Press Available online 17 June 2016

Bardone E, Pungas-Kohv P (2015) Changing values of wild berries in Estonian households: recollections from an ethnographic archive. J Balt Stud 46(3):319–336

Cesaro L, Linddahl M, Pettenella D (1995) The economic role of non-wood forest products and services in rural development. Medit 2:28–34

Grivins M (2016) A comparative study of the legal and grey wild product supply chains. J Rural Stud 45:66–75

Grivins M (2021) Are all foragers the same? Towards a classification of foragers. Sociol Rural 61(2):518–539. https://doi.org/10.1111/soru.12335

Landor-Yamagata JL, Kowarik I, Fischer LK (2018) Urban foraging in Berlin: people, plants and practices within the metropolitan green infrastructure. *Sustainability 10*(6):1873. https://doi.org/10.3390/su10061873

Lovric M, Da Re R, Vidale E, Prokofieva I, Wong J, Pettenella D et al (2020) Non-wood forest products in Europe – a quantitative overview. Forest Policy Econ 116:102175. https://doi.org/10.1016/j.forpol.2020.102175

Łuczaj Ł, Nieroda Z (2011) Collecting and learning to identify edible fungi in southeastern Poland: age and gender differences. Ecol Food Nutr 50(4):319–336. https://doi.org/10.1080/03670244.2011.586314

Łuczaj L, Pieroni A, Tardío J, Pardo-de-Santayana M, Sõukand R, Svanberg I, Kalle R (2012) Wild food plant use in 21st century Europe: the disappearance of old traditions and the search for new cuisines involving wild edibles. Acta Soc Bot Pol 81(4):359–370. https://doi.org/10.5586/asbp.2012.031

Nyman M (2019) Food, meaning-making and ontological uncertainty: exploring 'urban foraging' and productive landscapes in London. Geoforum 99:170–180. https://doi.org/10.1016/j.geoforum.2018.10.009

Paddeu F (2019) Waste, weeds, and wild food: a critical geography of urban food collecting. EchoGeo 47(47). https://doi.org/10.4000/echogeo.16623

Peintner U, Schwarz S, Mešić A, Moreau PA, Moreno G, Saviuc P (2013) Mycophilic or Mycophobic? Legislation and guidelines on wild mushroom commerce reveal different consumption behaviour in European Countries. PLoS ONE 8(5):e63926

Pollan M (2006) The Omnivore's dilemma. A natural history of four meals. Penguin Press, Bloomsbury, London

Reyes-Garcia V, Menendez-Baceta G, Aceituno-Mata L, Acosta-Naranjo R, Calvet-Mir L, Domínguez P, Garnatje T, Gómez-Baggethun E et al (2015) From famine foods to delicatessen: interpreting trends in the use of wild edible plants through cultural ecosystem services. Ecol Econ 120:303–311. http://icta.uab.cat/Etnoecologia/Docs/[495]-delicatessen.pdf

Schulp CJE, Thuiller W, Verburg PH (2014) Wild food in Europe: a synthesis of knowledge and data of terrestrial wild food as an ecosystem service. Ecol Econ 105:292–305. https://doi.org/10.1016/j.ecolecon.2014.06.018

Svanberg I, Sõukand R, Łuczaj Ł, Kalle R, Zyryanova O, Dénes A et al (2012) Uses of tree saps in northern and eastern parts of Europe. Acta Soc Bot Pol 81(4):343–357. https://doi.org/10.5586/asbp.2012.036

Vidale E, Da Re R Pettenella D (2015) Trends, rural impacts and future developments of regional WFP market. Project deliverable D3.2. StarTree project (EU project 311919)

Vullnetari J, King R (2008) 'Does your granny eat grass?' On mass migration, care drain and the fate of older people in rural Albania. Global Netw 8(2):139–171. https://doi.org/10.1111/j.1471-0374.2008.00189.x

Wiersum KF (2017) New interest in Wild Forest Products in Europe as an Expression of Biocultural Dynamics. Hum Ecol 45(6):787–794. https://link.springer.com/article/10.1007/s10745-017-9949-7

Wiersum KF, Wong JLG, Vacik H (2018) Perspectives on non-wood forest product development in Europe. Int For Rev 20(2):250–262. http://www.bioone.org/doi/abs/10.1505/146554818823767546

Mikelis Grivins (PhD) is a Senior Researcher at the Baltic Studies Centre, a research institute focused on studying sustainable rural and regional development, agro-food systems, farming, and innovations.

Experience Foraging: Connecting with Nature and Each Other Through Foraging Field Courses in Kent, England

3

Holly Harris

Abstract

Foraging field courses, run by wild food experts, have become a popular way to learn about wild edible species in urban societies. These courses are facilitating greater access to nature by members of the general public and can unsettle conservation's accepted wisdoms of people's appropriate interactions with nature where foodie fashion is concerned. This chapter explores the experiences and motivations of a particular group of urban consumers motivated by novel foods and nature experiences. Through an ethnography of foraging field courses, the chapter suggests that these courses align with 'new experiences of nature', enhancing human-nature connections as well as facilitating human-to-human connections through this shared experience. The wider relational values evident in participants' narratives open up new spaces for considering what connection to nature looks like in all its diversity and can contribute to more inclusive conservation planning.

Keywords

Nature experiences · Foraging field courses · Wild food education · Relational values · Sustainable foraging

3.1 Introduction

One of the most tangible effects of the contemporary foraging movement can be seen in the development of foraging field courses (FFCs, hereafter) where members of the public, guided by an expert forager, participate in formalised walks or trails,

---

H. Harris (✉)
School of Anthropology and Conservation, University of Kent, Canterbury, UK
e-mail: H.Harris@kent.ac.uk

S. Dhyani, M. Sardeshpande (eds.), *Urban Foraging in the Changing World*,
https://doi.org/10.1007/978-981-97-0345-6_3

learning how to forage for wild edible species. FFCs are run in a variety of settings ranging from the more 'wilderness' areas of woodland, coasts and nature reserves to public parks, car parks, churchyards and edges of urban centres such as riverbanks or industrial wastelands. Unlike reading guidebooks and participating in Internet forums, this form of learning takes place in situ, in a variety of spaces and formats, from short, guided walks to longer events that include a meal made up of wild foods. As such, courses offer people opportunities to both learn about wild foods and experience nature directly. Field courses are increasingly popular in the UK, advertised through a range of media, and vary in content, locations and prices.[1] This chapter examines the experiences of participants attending FFCs at the higher end of the FFC price offer, drawing on ethnographic fieldwork with foragers and natural resource managers undertaken as part of the author's PhD research in the southeast of England. In this chapter, I consider how FFCs may be viewed as an example of the 'new experiences of nature' (EoN) proposed by Clayton et al. (2017). As a form of 'nature experience', FFCs offer immersive experiences with wild foods, connecting people to nature and to each other but which can unsettle conservation's accepted wisdoms of people's appropriate interactions with nature (McLain et al. 2014). The discussion responds to developments in conservation that increasingly call attention to the need for strengthening 'human-nature connection' (HNC) as an urgent and prerequisite condition for biodiversity conservation (Ives et al. 2017). It examines the significance of consumers' experiences of FFCs to our understanding of the different ways in which people experience and connect *to* and *through* nature. The experiences of participants described here provide a more complex view of the stereotypical affluent, urban dweller, pursuing wild food fashions at the expense of nature—a conservation and public authority narrative found to be prevalent in the study area. By illuminating how participants experience the field courses and engage with nature through the medium of wild foods, the discussion highlights other value systems of significance, opening up our theoretical literature on what connection with nature looks like.

3.2 Method

During October 2017–April 2019, I conducted ethnographic fieldwork with foragers, natural resource managers and conservation authorities in East Kent. Participant observation was undertaken with forager-educators running courses in the study region, one of whom was a registered forager-teacher with whom I spent a season, assisting in the preparation and running of her field course events. This forager is a registered member of a professional association that facilitates peoples' learning about wild foods through their teaching practices.[2] The association promotes principles of sustainable practice and ecological stewardship through their work (Łuczaj

[1] 1–2-h foraging walks cost from £10 per person, community talks are usually free, while 5–8-h events can cost between £150 and £190 per person.

[2] The Association of Foragers: https://foragers-association.org/ (Accessed: 29/06/2023).

Table 3.1 Participant survey (n = 69) of reasons for attending the foraging course

Reasons for attending	Not important (%)	Neutral (%)	Important (%)
To learn about wild plants that are edible	6	10	84
To try the taste of wild plant foods	7	13	80
To connect with the natural environment	10	12	78
To experience something new	13	18	69
To learn how to cook with wild plants	18	16	66
To socialise in a natural environment	22	28	50
To connect to our past cultural heritage	28	28	44
To increase knowledge of wild plant foods for potential times of emergency/crisis	44	25	31

et al. 2021), of which teaching field and online courses are a core activity for its members (Townsend 2020). The spaces in which the courses occurred were predominantly coastal (seashore), but also included urban parks and cemeteries, nature reserves on the edges of towns and open meadows and grasslands. The FFCs I describe here were attended by about 250 members of the public over the course of one season. During the first quarter of the season, I ran an online survey for people who had attended an event to gain a preliminary profile of participants, including questions on where they had come from, their age and educational backgrounds, and their motivations for attending the course. Motivations were surveyed by asking participants to rank a series of statements about reasons for attending a course (Table 3.1). Sixty-nine respondents completed the survey, and the results are triangulated with the observations and interviews made during participant observation on the courses. The narratives of participants attending courses were recorded into a field journal after each event and coded thematically for salient themes (Braun and Clarke 2012) using NVivo 12 qualitative analysis software. Open-ended interviews were also conducted with natural resource managers, conservation officers and public authorities. Interviews with these stakeholders focused on their attitudes towards foraging and foragers operating in the region and perceptions of the environmental impacts of foraging in landscapes reserved for biodiversity conservation.

3.3 Repurposing Human-Nature Connections for Sustainability

The concept that HNC is vital for the sustainability of all life has been largely determined by Kellert and Wilson (1993) who proposed that humans are intrinsically motivated to seek connections with other forms of life due to our interdependent and co-evolutionary histories with nature. As a consequence of this *biophilia hypothesis*, a related set of theories posit that as the *social* world has developed further and further away from the natural one, so too have humans become increasingly alienated from nature. Pyle (1993) describes this phenomenon as the ‘extinction of experience’, and since then the concept has been developed and expanded, particularly in relation to the way the increasing urbanisation of societies is leading to ever greater reductions in opportunities for nature contact among urban populations (Pyle 2003; Miller 2005; Soga and Gaston 2016). The importance of HNC has gained momentum (Ives et al. 2017) and is becoming established within public health policies (McHale et al. 2020), with a growing body of evidence reporting on the benefits to human well-being of increased exposure to and activity in nature (Mitchell and Popham 2008; Hartig et al. 2014).

In the UK, the government's statutory advisory agency for the natural environment, Natural England (NE), published its 5-year plan in 2020, signalling a shift away from 'conservation' to 'nature recovery'. At the heart of the 'Building Partnerships for Nature's Recovery' strategy is [a] 'greater connection between people and planet' (Natural England 2020:3). The importance of nature-connected citizens to the natural resource management, biodiversity conservation, public health and economic development of the UK is emphasised in all of the programmes, which form the action plans identified to deliver on 'Nature's Recovery'. Although embedded across all the programmes, 'Connecting People to Nature' is also a separate programme within this strategy, focused on '[scaling up] our work in connecting the public to Nature' (Natural England 2020:20). This programme places particular emphasis on improving access to nature and works across the health, education and planning sectors to deliver on this ambition. Initiatives include, for example, green social prescribing, urban green infrastructure design, nature-friendly schools and greater public access to coastal areas through the development of a national public trail around all of England's coast. It may be argued that underpinning these policy and HNC narratives is a value system based on the instrumental value of nature to satisfy the intrinsic needs of society. In this regard, Chan et al. (2016) introduce the concept of 'relational values' arguing that for environmental policy to be effective, our conceptualisations of nature must move beyond logic motivated by intrinsic and instrumental values to recognising the value of 'the relationship as an end in itself' (Chan et al. 2016:1463). Their analysis extends thinking about human-nature relationships that are oriented towards individual and ecosystem benefits to an understanding of the wider values that impact and shape our relationships with nature. These values are derived from and through relationships and practices that can hold personal, social, symbolic and material meanings at the level of interpersonal relationships but also as Chan et al. (2016:1462) observe, 'as articulated by policies and social norms'. This seminal contribution to conservation science has brought focus to the social contexts that affect and are affected by and between society and nature. The significance of social and cultural diversity as constitutive of biodiversity has also gained prominence within this field, calling for a greater understanding of biocultural approaches to biodiversity conservation (Hanspach et al. 2020; Elands et al. 2019). It is within this critical terrain that Clayton and colleagues developed their concept of EoN, countering Pyle, that far from being 'extinct', there are in fact new experiences of nature that are mediated, shaped and produced in the changing social and cultural conditions of society. They observe that the use of technology, for example, is transforming the ways in which we connect to nature and propose a new conceptualisation of experiences of nature that challenge the dominant approach within conservation discourse; in their view, this dominant paradigm 'largely maintains a disjunction between humans and nature that ultimately inhibits successful conservation efforts' (Clayton et al. 2017:650). Clayton and colleagues offer a typology of 'dimensions of nature experience'. Central to their argument is the idea that the social contexts of these experiences are an important, but often neglected, component but which must be considered in order to effectively address the biodiversity crisis. Their typology offers a useful starting point to illuminate the experiences of participants of FFCs, which resonate with many of the dimensions they propose.

3.4 'New' Nature Relationships: The Dimensions of Human-Nature Experiences

Clayton et al.'s (2017) framework categorises the diverse kinds of EoN and how these relate to conservation outcomes. They outline five different dimensions of these experiences presenting each on a continuum of experience that has implications for conservation dependent on the type of behaviours elicited by a particular experience. These are 'observing vs. interacting', such as the observations by a bird watcher compared with someone who is directly engaged in feeding the bird; 'consumptive vs. appreciative', where 'consumptive' may be regarded as actively modifying nature as a 'resource', while 'appreciative' experiences are predicated on simply appreciating nature without seeing it as an exploitable resource; 'self-directed vs. other-directed' contrasts experiences that are independently sought by an individual versus those that are determined by others, such as an educational school programme; 'separate vs. integrated' refers to experiences that are part of an individual's daily routine compared with an experience in which an individual has to depart from their routine to make a special effort to experience nature; 'solitary vs. shared' compares a solitary experience to one shared with others; and 'positive vs. negative' refers to the emotional responses elicited during an experience. Four of these dimensions, 'interactive', 'consumptive', 'integrated' and 'shared', are applied to this analysis of participants' experiences of FFCs in the study region.

3.5 FFCs, Foraging and Sustainability Narratives

Interest in FFCs is evident by the results of an online search for 'foraging workshops UK', which results in thousands of hits, ranging from specialist species workshops such as 'fungi workshops' or 'seaweed courses' to more general seasonally based courses. The search results signpost to professional foragers' websites where courses can be booked or bought as gift experiences, as well as to a range of 'lifestyle' media such as magazines,[3] news reports and television programmes reflecting the food media and public interest in foraged foods and foraging lifestyles. For example, from the mid-2000s, foraged foods became popularised on food and

[3] *Countryfile Magazine* April 14th, 2023. Best foraging courses in the UK (April). Available at https://www.countryfile.com/.

cooking television shows such as *Masterchef*[4] and *Jamie and Jimmy's Friday Night Feast,*[5] featuring chefs and aspiring cooks using a range of wild edibles from the British coast and countryside. These programmes often include short features with a wild food expert showcasing the different wild foods available in their area with suggestions on how they can be incorporated into domestic cooking practices. During this time, affluent art[6] and lifestyle magazines[7] also featured stories about entrepreneurial foragers, including details of the courses available in different locations throughout the UK. These more fashionable associations with foraging are associated with an epicurean heritage (Mabey 2006) and 'classed consumerism' (de Jong and Varley 2018) that can attract sharp criticism on the sustainability and ethics of foraging from conservation groups and public authorities in England when viewed as a fashionable, *foodie* vogue (Barrie 2015; Horton 2023). The concern over the impact of gastronomic trends on wild food species is not new in Britain (Rotheroe 1998); however, the proliferation of FFCs has created greater access for the general public to engage in foraging and greater visibility.

Despite the high online profile and proliferation of FFC in the UK, there is limited knowledge about the specific roles they play in human-nature and human-human relationships. Field courses have been mentioned in the context of new ways of learning about wild foods and the motivations driving participation by members of the public, for example, by people interested in 'food independence, survival or a healthy lifestyle' (Łuczaj et al. 2012:84). They are recognised for their value in reproducing 'intangible cultural heritage traditions' in regions where gathering wild plants used to be part of agricultural and pastoral identities (Grasser et al. 2012), and more generally, as sustaining cultural connections in societies no longer dependent on wild foods for subsistence (Reyes-García et al. 2015). An exception in this literature is de Jong and Varley's (2018) study that examines the experiences of foraging course leaders and attendees in England and Scotland. This study analysed the content of participants' blogs in order to understand how participants made sense of foraging after the experience. It concluded that attendees' understanding of foraging was not to see it as a solution to unsustainable food systems, but rather, as a response to 'perceived human/non-human disconnections' (de Jong and Varley 2018:698). This suggested that attendees were aware of the discourses surrounding HNC and the ways in which they had become disconnected from individual everyday experiences with nature. In contrast, the link between foraging, public interest and sustainable food systems is also made in a study on truffle hunting in northern Italy, which examined the impacts of changing social and environmental dynamics on the traditional ethnoecological knowledge and practices associated with truffle hunting (Pieroni 2016). Here, workshops offering tourists the chance to be involved in the 'spectacularization' of the search for truffles are implicated in the pressures

[4] BBC, 2005–present.

[5] Channel 4. Available at https://www.channel4.com/programmes/jamie-and-jimmys-friday-night-feast.

[6] *Country Living* 2016. Superfood from the Shore. (July) 58–64.

[7] *Perfect Bound*, Call of the Wild, Vol 1, 2017. The magazine is also online via Facebook.

seen to be facing the uncertain future of truffle gathering. This market focus on the 'magic' truffle is problematised, vis-à-vis 'sustainable gastronomy', which, the author argues, would be facilitated by a focus on '*sense of place*' so that consumers become aware of the environmental and social pressures surrounding the truffle and its community of gatherers (Pieroni 2016:10, *original emphasis*).

Notwithstanding these tensions in academic debate that surround foraging and gastronomy, scholarship also demonstrates the diverse potential in foraging to contribute to sustainability agendas: for example, as an alternative sustainable food system (Molina et al. 2014); to address food waste (Nyman 2019); to promote food sovereignty and food security (Parada et al. 2011; Nolan and Pieroni 2014); to encourage ecological stewardship (Grivins 2021); and to facilitate deeper connections between people and nature (Simmonds 2021; Poe et al. 2014). Connecting to nature is cited as a key motivation to forage in English studies that have explored professional foragers' motivations (Fenton 2008; Lee 2012; Hawkes 2019), but as an experience of nature by members of the public accessing these courses, the significance to individual-nature relationships remains obscured by generalisation. The remainder of this chapter explores peoples' varied experiences of the FFC, applying the four dimensions of Clayton et al.'s framework to illustrate how FFCs can be understood as an EoN.

3.6 Foraging in East Kent

When I met Elle, a registered forager-teacher, she had been running courses along a stretch of the East Kent coast for over 10 years. Attracting an average of 18 participants to each event, she attributed part of her success to advertising through a well-known online retailer of novelty gifts and experiences. Like all professional foragers, Elle also advertises through her social media profile on Instagram, personal website and word of mouth, but the addition of the online retail platform had considerably boosted her profile and course subscriptions. Elle's courses run on weekends from March to December and are repeated twice or more in a month depending on demand and on the species in season. For example, wild garlic has a short season in March so forays occur only once in this month, while shellfish forays might occur as many as four or five times in the year as the harvest season is more prolonged. Her courses are divided into two parts: a guided field excursion to forage for wild plants, seaweeds and shellfish, and a meal prepared by Elle and her team consisting of a variety of wild food dishes and drinks served *al fresco* style on the beach, at her house and, on a few occasions, as a car-boot picnic on the side of the road adjacent to the sea.

3.7 Participants

Participants' ages ranged from 6 years old (the youngest child accompanying parents) to 65+ years with a majority of participants aged between 34 and 54 years old. The socio-demographic survey data also showed that 70% of people were educated to degree level (i.e. an undergraduate or postgraduate university degree). The majority of people (82%)

were from London and the remaining from other regional towns and cities; a few groups of families and couples had come from other European cities such as Paris and Amsterdam. This demographic data correlates with previous studies that have found the majority of participants to be younger to middle-aged, educated professionals living in predominantly urban places (Carvalho and Morales 2010; Fenton 2008).

3.8 An Interactive Experience

In this study, people expressed a variety of motivations for attending the courses, with the two most important reasons described in the survey reported as 'to learn about wild edible plants' and 'to try the taste of wild foods'. 'Connecting to nature' and 'experiencing something new' also scored highly in the survey and in participant narratives, described hereafter.

These results suggest that people are interested in an interactive, multisensory experience, which the format of the field courses provides through direct and sensory encounters with wild foods. This interactive experience of learning and recreating while walking, identifying, picking and tasting wild foods resonates with the 'interactive' dimension of nature experiences proposed in Clayton et al.'s (2017) framework. They contrast the effects of an interactive experience with 'mere observation', noting that the former is likely to be 'more vivid and multisensory, engaging more emotions and creating a more lasting memory' (Clayton et al. 2017:647). They also note that interactive experiences are more likely to cause concern from a conservation perspective due to the perceived risks of human impact on nature. This concern was expressed by local conservation officers whom I interviewed, who reported that groups of people foraging near to nature reserves might disturb nesting birds or trample fragile species underfoot in their efforts to secure a particular species.

Their roles in managing protected areas and nature reserves involved keeping human activity away from areas reserved for wildlife, thus enforcing a separation between people and nature. Participants did not necessarily view nature as separate from themselves however, or from their lives, and expressed a variety of motivations for their interest in foraging; the responses from participants in the survey suggested that participants are seeking an interactive experience that involves learning, recreating and consuming in nature and where their interests in food, leisure experiences and outdoors intersect:

> *'We enjoy cooking and we were keen to have a consumable gift rather than an object as we are trying to reduce the amount of stuff in our lives. I'm a landscape architect and am mindful about environmental changes. We both enjoy reading about the food industry and natural history. We grow herbs and edible flowers in our garden, and we wanted the course to educate us about the world of wild food'.*
>
> *'Because I love wildlife, the outdoors, food and getting muddy!'*
>
> *'I am interested in environmental conservation and locally sourced food generally, so foraging closely aligns with my ideals in terms of living in closer harmony with the environment'.*
>
> *'I am interested in healthy locally sourced and organic foods'.*

Wanting to learn about wild foods is articulated through the activities and behaviours individuals enact as broader lifestyle choices and preferences such as cooking, gardening and outdoor activities; this interactive experience with wild foods reflected their choices and identities in relation to nature, which in this context was articulated as a direct and interactive relationship to food produced from the 'wild':

During an early summer foray, while the group enjoyed a picnic lunch on the beach following several hours of foraging for samphire, sea aster and sea blite among the marshes, I made my way around the group, offering to refill drinks. A retired couple, in their late 60s, were discussing a fungi foraging course that they had attended on Hampstead Heath close to where they lived, recalling how they had been walking the heath for years but it was not until they had been on the foraging course that they had become aware of 'fungi everywhere'. I stopped to listen as the woman remarked how wild food seemed to be coming back into fashion, popularised by chefs. 'I think people are just sick of all this supermarket food', she said, 'it's been on the shelves for days.' A man agreed, saying that he thought people were more aware of the effects of 'industry on food and its impact on health', adding, 'people want something else'. The woman agreed, 'that's why it's good to come here and pick food you know is fresh'. There was a murmur of agreement from the others. Simon, sitting opposite, added that he had been introduced to samphire in a restaurant he had worked at in London, 'but this is something else, actually seeing it in the wild like this'. Laughing, he went on, 'I always thought I'd want to be a hunter in another kind of society, but maybe now I think a forager would be good'. Everyone laughed, and there was a pause in the conversation, before Bill who had come with his two daughters for his 60th birthday volunteered, 'when I retire, I'd like to buy a narrow boat and go down the canals and learn about foods for free' [referring to Richard Mabey's classic foraging guide[8]. His daughters burst out laughing when he said, 'foods for free', and one of them affectionately patted his shoulder. Earlier in the day, one of Bill's daughters had asked me what ethnobotany was and, in response to my explanation, responded, 'it must be disappointing to learn that most (English) people's knowledge of plants comes from a supermarket bag!'

The comment by Bill's daughter was flippant, but the sentiments it expressed and those of other participants in this conversation resonate with current research that highlights how this consumer interest in 'wild' cuisines as healthier alternatives to industrial food systems has become established among urban professionals (Carvalho and Morales 2010; Whitney et al. 2012). This idea has a longer legacy than the last two decades however, having been described by Forbes (1976) in her study of foraging in Greece in which she notes that the symbolic value of wild foods as cleaner and purer than domesticated was a view held by both rural and city dwellers alike. The observation by Bill's daughter hinted at a disconnected relationship to food, contained by the supermarket shop and the plastic bag holding salad. In contrast, the act of foraging provides participants with a direct experience that is meaningful and appreciated for bringing people closer to food sources; for people like

[8] *Food for Free* (1972).

Bill, this interaction prompted reflections on his future that involved closer living with nature, but for others, it stimulated thinking about their own behaviours and relationships with the environment, as demonstrated in the following:

Two people were talking about wild harvesting and harm to plants. Brenden, a keen fungi harvester, was saying that he had been back to an area where he frequently finds *boletus*, to find them all gone. 'People are getting there before me'. Charlotte, in her early 20s, had joined him standing by the water watching the tide slowly begin to swell and submerge the rocks. As we reflected on the idea that once you tell someone [about a patch or species], then everyone is going to go there and harvest, causing a potential over-harvesting of species, Charlotte remarked on the unlikelihood of this happening due to the time it takes to harvest. Brenden agreed: 'I can spend a whole day looking for mushrooms and only find one, but then sometimes, in a few hours, lots'. He agreed with Charlotte that with society being time poor, it is unlikely that the countryside will be overrun with people collecting wild plants or fungi. 'Yet', he added, 'commercial mushroom harvesting has been banned by National trusts and other forests. It's like egg collecting, I used to collect eggs as a child, then everyone was doing it, but no harm, we were just children and you got to know your trees too'. Charlotte replied saying, 'a lot of conservation started like that, with egg collecting. Bill Oddie for example started that way'.

This exchange struck me as being one of the few explicit discussions I heard during fieldwork about the sustainability of foraging, and it raises interesting points about how people have interacted with nature in ways that have been considered harmful in the past, yet these direct encounters have resulted in transformations in their relationships and behaviours towards nature later on. The example mentioned above, of the childhood interest of the conservationist, Bill Oddie, had also been recounted to me by a senior conservation officer during an interview. It had been a passing observation, in the context of our discussion. He was resolute in his opinion that foraging was the same as previous collecting activities 'by nature enthusiasts and OCD types', and Bill Oddie's egg collecting had been mentioned. Learning by 'doing', or 'experiential learning', has been discussed as an alternative approach to sustainability education that could promote environmental sustainability through transformative learning encounters in nature (Beckwith et al. 2016). While FFCs are not promoted as part of a sustainability education, the courses do provide direct contact with nature where participants are learning *by doing, sensing and socialising*. Consuming wild foods is part of this interaction, as they are eaten as a wild food meal and tasted during the course, and participants also usually collected a small amount to take home to cook or experiment with.

3.9 Not for Human Consumption

The following takes up Clayton et al.'s (2017) second dimension of experience, 'consumptive vs. appreciative', and explores what is arguably the most significant barrier to foraging in East Kent—resistance to the idea that wild foods are a legitimate choice for human consumption. The perception of foraging as a fashionable but unnecessary 'foodie' activity in modern times and therefore wilfully exploitative of nature is particularly strong among the natural resource managers and conservation officers interviewed for this research. This attitude is a perplexing one when one considers that the two founding leaders of the British foraging movement were arguably as much motivated by the wild, as with the products that could be made with these species. Richard Mabey's classic, *Food for Free* (1972), and Roger Phillips' *Wild Food* (1983) are celebrated cultural texts (Moran 2014), which are also testament to the culinary influences of this movement in Britain. However, media coverage of the foraging revival has been mixed, and sensational straplines accusing foragers of destructive and irresponsible behaviour towards nature have not helped. In East Kent, local authorities and conservationists are especially concerned with commercial wild harvesting due to the location of a commercial enterprise in the area that for many years has served the needs of the London gastronomic market (Fenton 2008; Jones 2016). Professional foragers teaching the public are also the objects of surveillance and prohibitions, which I personally witnessed on several occasions during fieldwork. This approach has been observed in other parts of the world where conservation practitioners are protective of green spaces, viewing foraging as an 'inappropriate and incompatible' use of wild resources (Poe et al. 2014:235). Sardeshpande and Shackleton's (2020) research, however, found that land managers are broadly in favour of foraging in urban green spaces provided that sustainable harvesting principles are observed and transparent supply chains implemented for commercial collecting. In an interview with the head of a prominent national environmental charity responsible for the management of much of the coast and meadowland where FFCs are run in East Kent, the contested domain of 'wild food' for human consumption was evident:

We were discussing the problem of sea buckthorn, a native species that is spreading rapidly across the dunes of the reserve causing enormous efforts by the reserve team to get it under control. Its abundance along the dunes both within and without the reserve makes it a popular foraging spot for locals and foragers' field site courses, and despite concerted attempts over the last few years to remove it, large stands still dominate the area. In response to my suggestion that foragers might be included in managing its spread, and permitted to collect the berries, Tim was not convinced. He explained their position:

> *'When we see people collecting these berries, we say this is a nature reserve, one of the few areas which is protected for wildlife. "We're only collecting berries" [they say]. Yeah, but this is food for field mice, wood mice and a raft of birds. Are you really dependent on these berries? The truth is they're not dependent on any of these things'.*

A few months later, I attended a local parish council meeting with Elle. She had been invited to attend because a council volunteer had found one of her wild food information cards on the common where she had taken a group. Elle had received an email from the parish clerk asking her to stop running foraging courses on the site, and several emails had gone between the two parties before an invitation had been sent to her to meet in person. Elle was pleased to accept the invitation and viewed this as an opportunity to open a dialogue with the council about her educational courses. She had asked if I would accompany her, and I had agreed. After a brief introduction, the chair of the meeting had got straight to the point, explaining that the card they had found had caused 'alarm', explaining: 'What we don't want is people marauding and collecting in order to augment their dishes. It has made us nervous. What we want is assurance that this is not going to continue'.

These views dismiss wild foods as a valid human food source and go even further by questioning the moral basis for people's actions by pitting these against the needs of wildlife. This perception of people profiteering from nature to 'augment their dishes', however, does not recognise the significant social values inherent in these events and what consumption, as an expression of value, represents in human-nature relationships. Evidence of long-standing relationships with nature, for example, was present in participants' recollections during the course. Many people shared personal stories of being engaged in outdoor activities when they were children, recalling shared experiences with family members or friends suggestive of deeper nature connections than critics would recognise:

I walked back over the rocks to the shore alongside a couple in their late 50s. They had been telling me how much they love the coast, spending their summer holidays in Dorset every year. The tide was coming in and the event officially over. 'We're keen beach people', Laura tells me. Anthony, her husband, grew up in a seaside town just a few kilometres from where we were. They live in a commuter belt town on the outskirts of west London, and both their children are at university. Anthony told me that when he was growing up, his family would regularly forage for mussels and cockles. 'As kids we were always down there', he says, pointing to the sea. Although he has not foraged locally since moving away, he says that there is a place, 'a bay in Dorset, where we go with the kids, even now, where you can find great shellfish'. Anthony works in investment and observed how he was seeing a trend towards environmentally sustainable investments. It is 'big', he says.

Anthony's recollections of foraging as a child were common to many participants, although most who shared stories of foraging in their youth described only collecting blackberries and other familiar hedgerow fruits such as plums, apples and wild cherries that are common to urban and rural habitats across England. For Anthony and many others on the course, gathering wild foods might only be an occasional practice, but it was something that had been a part of their experiences growing up, and one way of connecting to nature in meaningful and material ways. An appreciation of wild food, or nature more generally, is manifested through the acts of consumption described by participants and enacted during the wild food meals. In this context, 'consumptive' experiences such as this have created long-term attachments that are revisited as adults through experiences such as participation in these events.

3.10 Making Time with Nature

The effects of urbanisation on peoples' well-being as well as on biodiversity have led to an awareness of the need to design urban spaces that facilitate *daily* contact with nature (Beatley and Newman 2013), a dimension that is raised by Clayton et al. (2017:647) who discuss the 'separate vs. integrated' dimensions of nature experiences. They observe that experiences that are integrated into *daily* life have a greater chance of influencing people's routines and pro-environmental habits, although they also note that separate nature experiences can produce 'profound cognitive impacts ... leading people to a new perspective'. The people who attended the FFCs as a weekend excursion, making the trip down from London and surrounding towns, might be considered to fall into the latter category. However, it was evident that they also regularly participated in and sought nature-based activities at weekends as part of their leisure time. This suggests that these experiences are an integral part of urban lifestyles albeit not on a daily basis. In Miller's paper which addressed the issues of urban design and planning arrangements necessary for greater human-nature interaction, the particular demands of urban life are expressed as a relation of time: 'As the pace of life accelerates and time becomes commoditized, the rhythms of the human enterprise grow more and more distinct from those of the natural world' (2005:431). Making time for nature in the working lives of participants is important, particularly for younger working professionals with limited access to nearby green space. The examples below illustrate how nature is integrated into urban living and actively pursued and prioritised as leisure time:

Stephanie, in her early 30s, had bought this experience for her boyfriend's birthday. They are physiotherapists working at a London hospital, and were making a weekend of it, camping at a nearby campsite. They had cycled from the train station and had a cycle route planned for the following day. 'What prompted you to find this gift for your boyfriend?' I had asked. Stephanie had been looking for an experience gift for her boyfriend, she told me, and knowing how much he likes the outdoors, 'he grew up near his uncle who has a farm in Oregon', when she saw the event advertised online, she knew it was for them. 'Our flat in London has no greenspace', she explained, 'so we try to get away when we can'.

Similarly, Rosa and Paul, a couple in their late 20s, had travelled from North London. Rosa is Swedish and told me that she foraged as a child 'for mushrooms and berries'. She added, 'the locations of the mushrooms are kept top secret and everyone tries to get you to say where you find them!' Her boyfriend bought her the experience as well as an experience as a zookeeper at a nearby Wildlife Park for the following day. They told me that they are keen walkers and spend their free time 'in the outdoors' as much as possible.

While this contact with nature is not part of daily life for participants living and working in cities, it might be considered as forming part of the broader 'rhythms' of their working lives in which recreational time is punctuated by wider time frames linked to their social and economic contexts. In this study, it was also the case that participants used the wild foods they gathered during the course to create meals and products on their return home, bringing nature 'in' to the domestic and social spaces of urban living, further softening the boundaries of 'separate vs. integrative' experiences.

3.11 Shared Experience: Social Values and Exchange

The 'solitary vs. shared' nature experience is the fourth dimension I apply to participant experiences and reveals the extent to which social values are deeply and inextricably entwined with nature. Research on HNC is often framed in terms of how *individuals* relate to and are transformed by encounter with nature, while the *social* dimensions of these experiences are largely overlooked (Ives et al. 2017; Chan et al. 2016; Clayton et al. 2017). FFCs offer people ways to learn about wild species through the forager instructor, but knowledge exchange also occurs between participants highlighting an important dimension of FFCs: social exchange. The following vignette describes the ways in which this exchange unfolds and how people's interests in wild food and foraging extend beyond the FFC to wider social realms:

As we walked in small groups, singly and in pairs, people offered information about recipes and plants they had collected. It was the first week in June, and the event was organised around the town, at a large cemetery on its periphery and in a smaller park in the town's centre. Although this was not the season for berries, May was telling a couple who walked beside us how she had made sea buckthorn berry and plum jam following an autumn foray with Elle in October. May, a local resident, had attended a previous foray with Elle. Someone had asked her if she had foraged before (a common 'opening' question among group participants), and she was telling them about the autumn event, which she said, is 'highly recommended'. She said that she had looked online for a recipe for sea buckthorn jam, and it had been hard to find but eventually she found one on a blog written by two sisters, 'from Russia, I think'. She added, 'it turned out really well – a brilliant orange colour, the plums were from my garden'. We reached the park, and Elle stopped by a great ash tree producing a jar of ash key pickle for tasting. Someone asked for the recipe, and Elle promised to write it down for them before the end of the day. Neil, a man in his late 20s, said that there were lots of ash trees near his allotment in London, and this was 'definitely something I will try'. Frieda took out her phone and snapped some pictures of the tree, 'so I can identify it later'. We moved on to a bay tree, planted at the back of one of the flower beds, and people started picking the leaves, 'this [plant] needs little introduction', Elle observed, but continued with a description of it and its uses to 'flavour soups and stews'. Rosie, originally from Northern Cyprus, was picking the leaf buds. 'These you can use too', she volunteered, 'you can pickle them'. She tells us that she has a recipe for pickling them, and suddenly people are exchanging numbers with her so she can send them her recipe. Anna and Mark are part of a cooking club and were observing Rosie's picking technique closely; they produced a small plastic container, which they started to fill with buds. 'I can send you the recipe now?' says Rosie, 'on WhatsApp'. Mark gave her the 'thumbs up', 'great', he says, and then turns to me, 'we're cooking for friends next weekend, and we're always keen to impress them with something new'. At the end of the day, people were saying their goodbyes, and Anna was thanking Elle: 'We had a brilliant time meeting everyone, being out in the lovely sea air, sharing a great learning experience and with a wonderful meal at the end'. Elle beamed at her, then pulled out her notepad and pen and began jotting down the ash key pickle recipe.

This type of exchange happened while walking from place to place, prompted by encounters with familiar and new plants. For cooking enthusiasts, it was a chance to share some creative ideas with others and for people who had never foraged before it was an opportunity to learn or to be inspired. What was evident in these exchanges was how cooking with wild foods connected to peoples' wider social and familial networks extending their social value beyond the course and individual participants.

Rosie's pickles for instance would be served at extended family meals. May's sea buckthorn berry and plum jam was produced 'for Christmas presents', and new dishes were made to share with friends over a weekend.

Relational values were also expressed through the 'gifting' of these experiences by friends and family members to participants on the course. A small number of participants had received the course for its novelty value: *'Looking for something unusual as a gift'; 'They knew a new adventure would suit us'*. However, for the majority of gift recipients, its value held personal significance represented in the relationship between the giver and the recipient:

> *'My husband knew that having a Sicilian father I'd grown up foraging food'.*
>
> *'My boyfriend gave me the event for an anniversary gift. He knows how much I like learning about the foods that are available to us in the wild. Foraging is also a favourite past time of mine e.g., Blackberries, elderberries, nettle etc. Something very nostalgic about it now :)'.*
>
> *'My children know that I have had an interest in foraging mainly foraging for fungi. This was a retirement/60th gift with the idea of giving me another interest to develop'.*

Relational value is present in these encounters in a number of ways: as memories of time spent in nature during childhood and youth, as anticipated future meals and gifts for friends and family, and in the pleasure of receiving the experience as a gift, reinforcing social ties and expressions of social identity. Hall (2013:228) notes how foraging invokes feelings of nostalgia in immigrants to New Zealand, who remember the activities as 'something they grew up with'; foraging here connects people to cultural memories, but also people saw it as an opportunity to meet other 'like-minded' people, in a time where the traditions of recreational gathering in family groups have diminished. Foraging fosters not only HNC but also human-human connections becoming a 'deeply relational practice' (Poe et al. 2014:231).

3.12 Conclusion: The Foraging Field Course—A 'New Experience of Nature'

In this chapter, I have described the motivations and experiences of urban residents attending FFCs along the East Kent coast during a year of seasonal activity. Their stories suggest that the majority of participants on these courses are already connected to nature, and that FFCs offer multidimensional experiences that serve to maintain or enhance these connections. For people attending FFC as a novel nature-tourist experience, the FFC can inspire new practices and activities prompting reflections on disconnections to nature inherent in everyday practices (de Jong and Varley 2018). As the concept of HNC has gained prominence in the field of biodiversity conservation (Folke et al. 2011; Ives et al. 2017), the importance of understanding the diversity of cultural interactions with nature has been recognised (Elands et al. 2019). As self-professed *foodies* and *nature lovers*, this group of consumers of FFCs are aware of broader debates about societies' impacts on ecosystems, especially in relation to industrial food systems and food miles. Their practices expressed through FFC thus challenge the narratives shared by conservation practitioners and public authorities over the appropriate use of natural resources and the ethics of *fashionable* consumption, which could be harnessed to inform further development towards a 'sustainable gastronomy' (Pieroni 2016). Clayton et al. (2017) emphasise that *new* experiences of nature are arising in the context of societal changes and that particular changes are transforming the ways in which we connect to nature. Rather than viewing 'technology-based interactions' as a barrier to an 'authentic' nature experience, they argue for scholars and practitioners of conservation to consider the role of technology in shaping people's relationships towards nature afresh. Foraging is very much part of these new transformations and has occupied a significant space in information media for at least the last two decades, reflecting and producing points of influence in society through the media domain (Sachdeva et al. 2018). Knowledge about wild foods is transmitted through online media, and social media profiles are essential platforms for the work of professional foragers (Townsend 2020). During field courses, we see how social media facilitates knowledge and social exchange among participants via WhatsApp and Instagram accounts. It is also the case in the example discussed here that the marketing of courses through an online retailer attracts a high number of participants. These examples demonstrate how information technology facilitates access to nature, wild foods and foraging stories and enables the material practice of gathering through the creation of courses and their subscribers. FFCs offer people direct, consumptive, interactive and shared experiences of nature that are personally, socially and culturally meaningful and which resonate with urban transformations and lifestyles. In this chapter, I conclude that FFCs be considered among the diversity of 'experiences of nature' that hold significant potential for current and future sustainable living.

References

Barrie J (2015) Is the foraging craze damaging the environment? *The Telegraph*, 11 September. http://tinyurl.com/hpnoexc. Accessed 26 June 2023

Beatley T, Newman P (2013) Biophilic cities are sustainable, resilient cities. Sustainability 5(8):3328–3345

Beckwith BR, Halber T, Turner NJ (2016) "You have to do it": creating agency for environmental sustainability through experiential education, transformative learning, and kincentricity. In: Routledge handbook of environmental anthropology. Routledge, pp 412–427

Braun V, Clarke V (2012) Thematic analysis. American Psychological Association

Carvalho AM, Morales R (2010) Northeastern region of Portugal. Ethnobotany in the New Europe: people, health, and wild plant resources, 14, p 147.

Chan KM, Balvanera P, Benessaiah K, Chapman M, Díaz S, Gómez-Baggethun E, Gould R, Hannahs N, Jax K, Klain S, Luck GW (2016) Why protect nature? Rethinking values and the environment. Proc Natl Acad Sci 113(6):1462–1465

Clayton S, Colléony A, Conversy P, Maclouf E, Martin L, Torres AC, Truong MX, Prévot AC (2017) Transformation of experience: toward a new relationship with nature. Conserv Lett 10(5):645–651

de Jong A, Varley P (2018) Foraging tourism: critical moments in sustainable consumption. J Sustain Tour 26(4):685–701

Elands BH, Vierikko K, Andersson E, Fischer LK, Goncalves P, Haase D, Kowarik I, Luz AC, Niemelä J, Santos-Reis M, Wiersum KF (2019) Biocultural diversity: a novel concept to assess human-nature interrelations, nature conservation and stewardship in cities. Urban For Urban Green 40:29–34

Fenton L (2008) An appetite for the wild: an investigation into the contemporary practice, use and importance of wild culinary botanicals for private and commercial consumption in Britain. (MSc Dissertation, University of Kent)

Folke C, Jansson Å, Rockström J, Olsson P, Carpenter SR, Chapin FS, Crépin AS, Daily G, Danell K, Ebbesson J, Elmqvist T (2011) Reconnecting to the biosphere. Ambio 40:719–738

Forbes MHC (1976) Gathering in the argolid: a subsistence subsystem in a Greek agricultural community: gathering in the Argolid. Ann N Y Acad Sci 268(1):251–264

Grasser S, Schunko C, Vogl CR (2012) Gathering "tea"–from necessity to connectedness with nature. Local knowledge about wild plant gathering in the biosphere reserve grosses Walsertal (Austria). J Ethnobiol Ethnomed 8:1–24

Grivins M (2021) Are all foragers the same? Towards a classification of foragers. Sociol Rural 61(2):518–539

Hall CM (2013) Why forage when you don't have to? Personal and cultural meaning in recreational foraging: a New Zealand study. J Herit Tour 8(2–3):224–233

Hanspach J, Haider LJ, Oteros-Rozas E, Olafsson AS, Gulsrud NM, Raymond CM, Torralba M, Martín-López B, Bieling C, García-Martín M, Albert C (2020) Biocultural approaches to sustainability: a systematic review of the scientific literature. People Nat 2(1):1–17

Hartig T, Mitchell R, De Vries S, Frumkin H (2014) Nature and health. Annu Rev Public Health 35:207–228

Hawkes T (2019) Characterising British foragers: motivations, barriers and sustainability issues. (BA dissertation, SRUC and University of Glasgow)

Horton H (2023) Excessive foraging for wild garlic and mushrooms in UK 'a risk to wildlife'. Guardian online 13 May. https://www.theguardian.com/lifeandstyle/2023/may/13/excessive-foraging-for-wild-garlic-and-mushrooms-in-uk-a-risk-to-wildlife. Accessed 20 May 2023

Ives CD, Giusti M, Fischer J, Abson DJ, Klaniecki K, Dorninger C, Laudan J, Barthel S, Abernethy P, Martín-López B, Raymond CM (2017) Human–nature connection: a multidisciplinary review. *Curr Opin Environ Sustain 26*:106–113

Jones L (2016) Commercial foraging in Kent, UK: the phenomenology of an ethical practice (MSc dissertation, University of Kent)

Kellert SR, Wilson EO (1993) The biophilia hypothesis

Lee JL (2012) Through a foraging lens: legal, economic and social change in England (Doctoral dissertation, University of Liverpool)

Luczaj L, Pieroni A, Tardío J, Pardo-de-Santayana M, Sõukand R, Svanberg I, Kalle R (2012) Wild food plant use in 21st century Europe, the disappearance of old traditions and the search for new cuisines involving wild edibles. Acta Soc Bot Pol 81(4):359–370

Łuczaj Ł, Wilde M, Townsend L (2021) The ethnobiology of contemporary British foragers: foods they teach, their sources of inspiration and impact. Sustainability 13(6):3478

Mabey R (1972) Food for free. Collins, London

Mabey R (2006) Where the wildings are. The Guardian. https://www.theguardian.com/books/2006/sep/02/featuresreviews.guardianreview4. Accessed 20 May 2023

McHale S, Pearsons A, Neubeck L, Hanson CL (2020) Green health partnerships in Scotland; pathways for social prescribing and physical activity referral. Int J Environ Res Public Health 17(18):6832

McLain RJ, Hurley PT, Emery MR, Poe MR (2014) Gathering "wild" food in the city: rethinking the role of foraging in urban ecosystem planning and management. Local Environ 19(2):220–240

Miller JR (2005) Biodiversity conservation and the extinction of experience. Trends Ecol Evol 20(8):430–434

Mitchell R, Popham F (2008) Effect of exposure to natural environment on health inequalities: an observational population study. Lancet 372(9650):1655–1660

Molina M, Tardío J, Aceituno-Mata L, Morales R, Reyes-García V, Pardo-de-Santayana M (2014) Weeds and food diversity: natural yield assessment and future alternatives for traditionally consumed wild vegetables. J Ethnobiol 34(1):44–67

Moran J (2014) A cultural history of the new nature writing. Lit Hist 23(1):49–63

Natural England (2020) *Building partnerships for nature's recovery*

Nolan JM, Pieroni A (2014) Food security in a changing world. J Ethnobiol 34(1):4–122

Nyman M (2019) Food, meaning-making and ontological uncertainty: exploring 'urban foraging' and productive landscapes in London. Geoforum 99:170–180

Parada M, Carrió E, Vallès J (2011) Ethnobotany of food plants in the Alt Emporda region (Catalonia, Iberian Peninsula). J Appl Bot Food Qual 84(1):11–25

Phillips R (1983) *Wild food: a complete guide for foragers*. Pan Books, London. Reprint London: Macmillan, 2014

Pieroni A (2016) The changing ethnoecological cobweb of white truffle (tuber magnatum Pico) gatherers in South Piedmont, NW Italy. J Ethnobiol Ethnomed 12(1):1–11

Poe MR, LeCompte J, McLain R, Hurley P (2014) Urban foraging and the relational ecologies of belonging. Soc Cult Geogr 15(8):901–919

Pyle RM (1993) The thunder tree: lessons from an urban wildland, vol 220. Houghton Mifflin, Boston

Pyle RM (2003) Nature matrix: reconnecting people and nature. Oryx 37(2):206–214

Reyes-García V, Menendez-Baceta G, Aceituno-Mata L, Acosta-Naranjo R, Calvet-Mir L, Domínguez P, Garnatje T, Gómez-Baggethun E, Molina-Bustamante M, Molina M, Rodríguez-Franco R (2015) From famine foods to delicatessen: interpreting trends in the use of wild edible plants through cultural ecosystem services. Ecol Econ 120:303–311

Rotheroe M (1998) Comment-wild fungi and the controversy over collecting for the pot. Br Wildl 9:349–356

Sachdeva S, Emery MR, Hurley PT (2018) Depiction of wild food foraging practices in the media: impact of the great recession. Soc Nat Resour 31(8):977–993

Sardeshpande M, Shackleton C (2020) Urban foraging: land management policy, perspectives, and potential. PLoS One 15(4):e0230693

Simmonds IRH (2021) Becoming-with the beyond-human: an ethnography of foraging, inter-species relationships and identity in London (Doctoral dissertation, Department of Anthropology and Sociology SOAS, University of London)

Soga M, Gaston KJ (2016) Extinction of experience: the loss of human–nature interactions. Front Ecol Environ 14(2):94–101

Townsend L (2020) The professional wild food community and Covid-19: the use of online platforms in supporting people to access alternative food sources. Local Dev Soc 1(2):160–165

Whitney CW, Gebauer J, Anderson M (2012) A survey of wild collection and cultivation of indigenous species in Iceland. Hum Ecol 40:781–787

Holly Harris is a postgraduate researcher and research associate in the School of Anthropology and Conservation, University of Kent, Canterbury, United Kingdom. Trained in social anthropology with a long-term interest in indigenous environmental knowledge systems, her work is currently focused on the ethnobotany of foraging in contemporary urban spaces.

4 Practices and Perceptions of Foragers of Wild Resources in Lagos Metropolis, Nigeria

Opeyemi Adeyemi and Charlie M. Shackleton

Abstract

During the past decades, a growing body of literature has recognized the benefits (including ecological, social, economic, cultural, and psychological benefits) of foraging in urban and peri-urban environments. Particularly in the cities of sub-Saharan Africa, the collection and use of raw biological resources play an important role in people's livelihood and well-being. However, perceptions of city residents regarding the collection of foraged resources within the African context and in Nigeria are rarely considered. Understanding how people perceive this practice can offer valuable insights into their attitudes and behaviors towards adhering to policies and management recommendations. This study reports the practices and perceptions of foraging in Lagos metropolis, Nigeria, using a random sample of 276 respondents. Data on foragers' sociodemographic profiles, practices, and perceptions were collected using a semi-structured questionnaire and analyzed using descriptive and inferential statistics. The findings of this study showed that more than half (51%) of the respondents collected biological resources for food and medicine, mostly gathered along the streets. Further analysis showed that males (52%), the self-employed, older people (45–65 years), and people with tertiary education were more likely to forage. Overall, these results indicate that Lagos residents forage despite having both negative and positive feelings and beliefs about foraging. This information can be used to develop initiatives, programs, and policies to enhance foragers' experiences and perceptions of foraging in Lagos and other African cities.

O. Adeyemi (✉) · C. M. Shackleton
Department of Environmental Science, Rhodes University, Makhanda (Grahamstown), South Africa
e-mail: o.adeyemi@ru.ac.za

S. Dhyani, M. Sardeshpande (eds.), *Urban Foraging in the Changing World*,
https://doi.org/10.1007/978-981-97-0345-6_4

Keywords

Livelihoods · Megacity · Perceptions · Urban foraging · Wild foods

4.1 Introduction

The rate of urbanization is rapidly growing globally. The United Nations estimates that 68% of the global population will reside in urban areas by 2050 (UNDESA 2018). However, this rapid growth presents challenges in society, the economy, and the environment. Although the expansion of cities has negatively affected nature and biodiversity, research has shown that interest in using raw biological resources does not diminish as people move from rural to urban areas (Barirega et al. 2012; Sneyd 2013). This loss of biodiversity and natural degradation could result in fewer urban populations having regular and direct experience with nature (Cox et al. 2017), which may consequently reduce residents' interest in nature conservation. Interacting with nature has numerous advantages, including ecological, social, economic, cultural, physical, and mental benefits (Shanahan et al. 2015; Schunko and Brandner 2022), as well as the willingness to take care of the environment (Soga and Gaston 2016).

Foraging in urban areas has been acknowledged as a means of reconnecting urban dwellers with the natural environment (Fischer and Kowarik 2020). According to Shackleton et al. (2017), urban foraging is defined as "the practice of harvesting or gathering raw biological resources (fungi, plants, parts of plants, invertebrates, and vertebrate animals, including fish) within urban and peri-urban settings, primarily for direct consumption, decoration, craft, barter, or small-scale sale." These resources can be domesticated or wild from private or public land, either managed or unmanaged (Poe et al. 2013; Svizzero 2016). Foraging activities in cities help establish links to places (Poe et al. 2014) and provide food, healthcare, income (Svizzero, 2016), ceremonial and spiritual observations, a sense of place, and well-being (Shackleton et al. 2017; Somesh et al. 2021). Additionally, this practice preserves cultural identity and provides products that could be used as a safety net during pandemics (Sardeshpande et al. 2021), crises (Shackleton 2021), or recessions (Sachdeva et al. 2018). It can also improve health in different environments (McMullin et al. 2012), strengthen social bonds within the community (McLain et al. 2014), play an important role in maintaining good mental health (Van den Berg et al. 2010), and supply micronutrient-rich foods (Sogbohossou et al. 2015).

Previous research has shown that foraging is a widespread practice among many people living in cities. For example, 33% of the people under study in Berlin are foragers (Fischer and Kowarik 2020), 47% in Kampala (Mollee et al. 2017), 26% in Atlanta (Gaither et al. 2020), 64% in Vienna (Schunko and Brandner 2022), and 16% in Bengaluru (Somesh et al. 2021). Although there is an increasing number of studies on urban foraging, the focus of most studies has been on understanding the demographic characteristics of people who forage, where they forage, what they forage, and why they forage (Mollee et al. 2017; Garekae et al. 2016; Somesh et al. 2021; Adeyemi and Shackleton 2023), while scant attention has been given to other aspects of foraging practices such as the perception of foraging activity. While understanding the biophysical characteristics of urban foraging (actors, sites, and species) is important, knowledge of what foragers think of the practice can help inform effective policy design. How people perceive the gathering of raw materials from the environment influences their participation and frequency (Garekae et al. 2016). Furthermore, understanding how people perceive this practice can offer valuable insights into their attitudes and behaviors towards adhering to policies and management recommendations.

The rate of urbanization is the highest in African countries, with an estimated 50% of Africans living in cities by 2030 (World Bank 2015). It is expected that most of this growth will occur in metropolitan cities; however, very few studies on foraging have been conducted in the metropolitan cities of Africa. Previous studies on the use of wild natural resources have focused on medium-sized towns and cities (Davenport et al. 2011; Kaoma and Shackleton 2015; Schlesinger et al. 2015; Sardeshpande et al. 2021). Others, for example, in Yaoundé, considered the importance of collecting wild food plants (Sneyd 2013). A few African studies (Mollee et al. 2017; Garekae et al. 2016) have reported that foragers generally have positive attitudes towards foraging. Very few published studies have explicitly examined the perceptions of foragers regarding foraging in an African metropolitan city. Adeyemi and Shackleton (2023) conducted a study on foraging in Lagos, identifying foragers, forage species, forage sites, experiences, and regulations. Although this study provides a comprehensive understanding of foraging practices in Lagos, it overlooks people's perceptions of foraging practices.

Lagos metropolis was selected for this study as it is one of the fastest growing megacities in the world and Nigeria's largest city. Lagos is set to be Africa's largest megacity with potentially 30 million inhabitants by 2035 (Uduku et al. 2021). This expected population growth requires urgent redesigning of Lagos landscapes to accommodate adequate nature interaction experiences. Within this context, this research aimed to understand the practices and perceptions of gathering raw biological resources in Lagos metropolis. Adeyemi and Shackleton (2023) uncovered a wide range of foraging sites, ranging from private to public urban greenspaces. This chapter focuses on foraging sites outside of conventional homegrown or domestic garden resources to understand how foragers feel when they engage in the activity in open or public spaces. This study aimed to address the following research questions: (1) Who are the people who forage? (2) What do they forage for, for what use, and where do they forage? (3) How do they perceive foraging activity?

4.2 Methods

4.2.1 Study Area

The Lagos metropolitan area (6°24′–6°31′ N; 3°16′–3°27′ E) is the largest city in Nigeria and one of the largest megacities in Africa. It is located in the southwestern geopolitical zone of Nigeria (Ajewole et al. 2019). Lagos State's metropolis, which makes up approximately two-fifths of the state's land area, is home to over 85% of its population. In 2018, the estimated population of Lagos State was 25.6 million, with an annual population growth rate of 3.2% (Lagos Bureau of Statistics 2019). Lagos accounts for more than 60% of Nigeria's commercial and industrial activities, contributing 33% of the country's GDP. The major economic activities in Lagos metropolis and the employers of its residents include manufacturing (pharmaceutical, chemical, food, beverage, automobile assembly, etc.), construction, transport, wholesale, and retail trade industries. However, many residents make a living by selling goods and services in the informal sector. There are two climatic seasons: a cold dry season between November and March, and a humid wet season from April to October. The city has a mean annual temperature of 27 °C and mean annual precipitation of approximately 1693 mm (Eresanya et al. 2018). The vegetation of Lagos is dominated by freshwater and mangrove swamp forests, which have a wet equatorial climate owing to their proximity to the Gulf of Guinea along the Atlantic Ocean (Lagos State Government 2021). The Lagos metropolis is divided into 16 local government areas (LGAs) and the metropolis with the other four LGAs that make up Lagos State. This study was conducted in all 16 LGAs in the Lagos metropolitan area.

4.2.2 Data Collection

A semi-structured questionnaire survey was conducted among people aged 18 years and older. The survey was administered to all LGAs to ensure diverse social, cultural, and economic settings that formed the metropolis. The information collected included (1) identifying foragers and their practices, (2) collecting their sociodemographic profiles and resources collected, and (3) the perception of foragers in Lagos metropolis. Sixteen respondents were targeted for each LGA because of limited funds and time. Opportunistic sampling was used to recruit willing and available respondents from diverse locations such as green spaces, social gatherings/events, transport terminals, and markets. The questionnaire was administered in English and translated into Yoruba for a few respondents. The total number of valid surveys was 275, although only foragers were asked further questions regarding their practices and perceptions. Common names of the foraged species were collected from the respondents. The data used in this study is a subset of dataset collected for a larger research project in Lagos.

4.2.3 Data Analysis

Statistical Package for Social Sciences (SPSS) version 28 was used to manage and analyze the data. Descriptive statistics, including measures of central tendency, frequencies, and proportions, were used to summarize the descriptive data.

4.3 Results

4.3.1 Respondent Profile

The study comprised 275 valid respondents, with a disproportionate number of males (61%) compared with females (39%). The majority of the respondents fell within the age range of 25–44 years (57%), were self-employed (61%), and had a middle income (55%). The education level of the participants was mostly tertiary (40%) or secondary (38%) (Fig. 4.1).

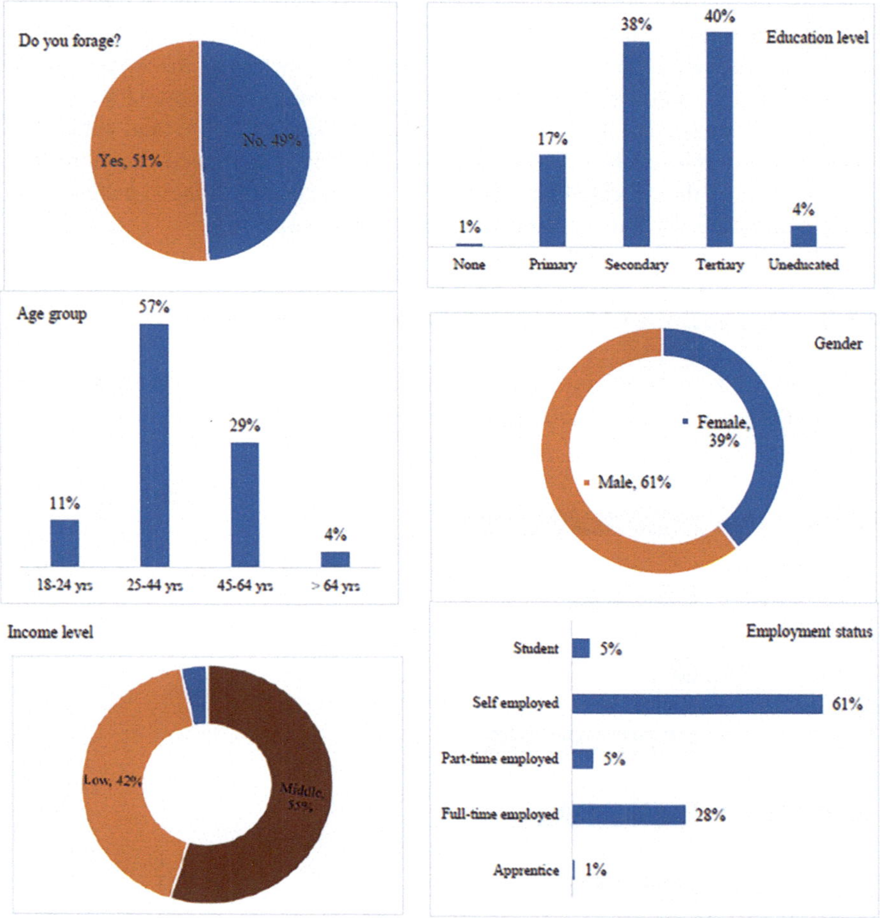

Fig. 4.1 Socio-economic characteristics of the respondents (N = 275)

4.3.2 Who Forages, Where Do They Forage, and What Do They Forage for?

In Lagos, slightly more than half (51%) of the respondents collected free biological materials from public sites. The majority of these collectors were male (52%), particularly those aged 45–65 years. Those with tertiary education tended to forage more frequently, and self-employed individuals and those with low to medium income levels participated more in foraging.

Wild foragers surveyed in this study collected more frequently from the streets (41%), unused lands (14%), and lakes (14%) than from other green spaces (Fig. 4.2). The least visited foraging space was parks (2%). A total of 27 species were mentioned by the foragers in Lagos (Table 4.1). The most common species were mango (*Mangifera indica*) collected by 20.5% of foragers, pawpaw (*Carica papaya*) by 14.3%, almond (*Prunus dulcis*) by 12.7%, fish (various species) by 6.6%, spinach (*Spinacia oleracea*) by 4.1%, snail (*Achatina fulica*) by 3.7%, and bitter leaf (*Vernonia amygdalina*) by 3.7%. The leaves and fruits accounted for more than three-quarters of the collected plant parts. Most foragers seek foraged materials for food (including fruits and vegetables) and medicinal purposes.

4.3.3 Perceptions of Foragers Towards Foraging Activities

In Lagos, 22% of foragers viewed foraging practices as shameful and 42% viewed them as tiresome activities. Even though 35% thought that foraging

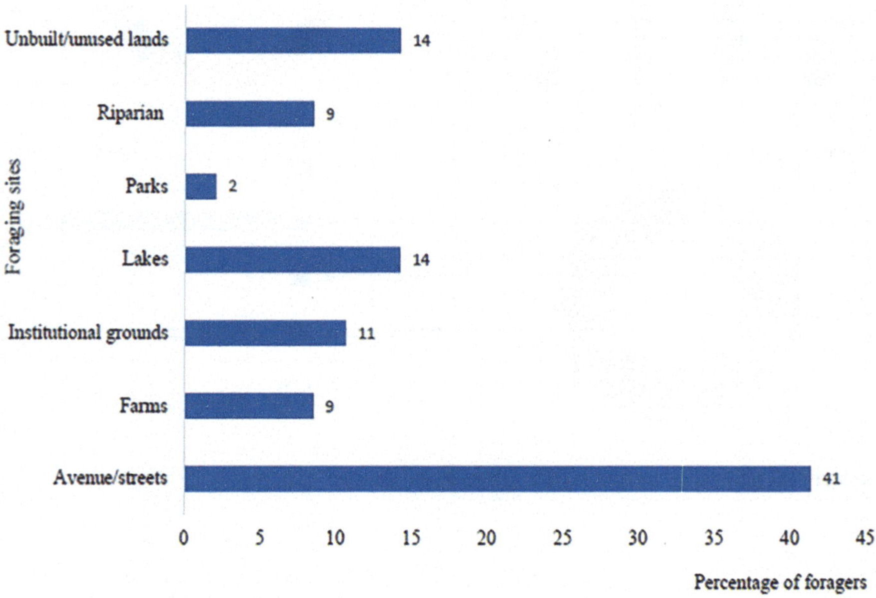

Fig.4.2 Percentage of foragers using different foraging sites (N = 140)

Table 4.1 Foraging species collected

Scientific name	Common name	Parts collected	Uses	Frequency
Cocos nucifera	Coconut	F	Food	2
Citrus aurantiifolia	Lime	L, F, S	Food, medicinal	2
Cymbopogon citratus	Lemon grass	L	Medicinal	3
Manihot esculenta	Cassava	T, R	Food, medicinal	3
Saccharum officinarum	Sugarcane	S	Food	3
Corchorus olitorius	Jute tree	L, F	Food, medicinal	3
Talinum fruticosum	Water leaf	L	Food, medicinal	3
Musa acuminata	Banana	F	Food	3
Dioscorea spp.	Yam	R	Food	4
Musa paradisiaca	Plantain	F	Food	4
Capsicum sp.	Pepper	F, S	Food, medicinal	4
Elaeis guineensis	Oil palm	F, S	Food, medicinal	4
Moringa oleifera	Moringa	L, S	Food, medicinal	4
Telfairia occidentalis	Fluted pumpkin	O	Food	4
Solanum lycopersicum	Tomatoes	F	Food	4
Azadirachta indica	Neem tree	L, B	Food, medicinal	5
Ocimum gratissimum	Scent leaf	L	Food, medicinal	6
Psidium guajava	Guava	F, L	Food, medicinal	7
Brachyura sp.*	Crab	O	Food, decoration	8
Citrus sinensis	Orange	F	Food	8
Vernonia amygdalina	Bitter leaf	L, B	Food, medicinal	9
*Achatina fulica**	Snail	O	Food	9
Spinacia oleracea	Spinach	L	Food	10
Fishes (various species)*	Fish	O	Food	16
Terminalia catappa	Tropical almond	L, F, S	Food, medicinal	31
Carica papaya	Pawpaw	F, S	Food, medicinal	35
Mangifera indica	Mango	L, F, S, B	Food, medicinal	50

*F = fruits, L = leaves, S = stems, B = barks, O = others | * indicates animal species*

species were not easily or freely accessible, 74% believed that it should be an activity open to both natives and foreigners. Forty-three percent disagreed that foraging should take place in rural instead of urban areas, while 76% acknowledged that foraging strengthens social connections among city dwellers. In total, one-third of the respondents (33%) had a favorable view of urban foraging, the majority (58%) had an unfavorable opinion, and a small number (9%) had a neutral stance (see Fig. 4.3).

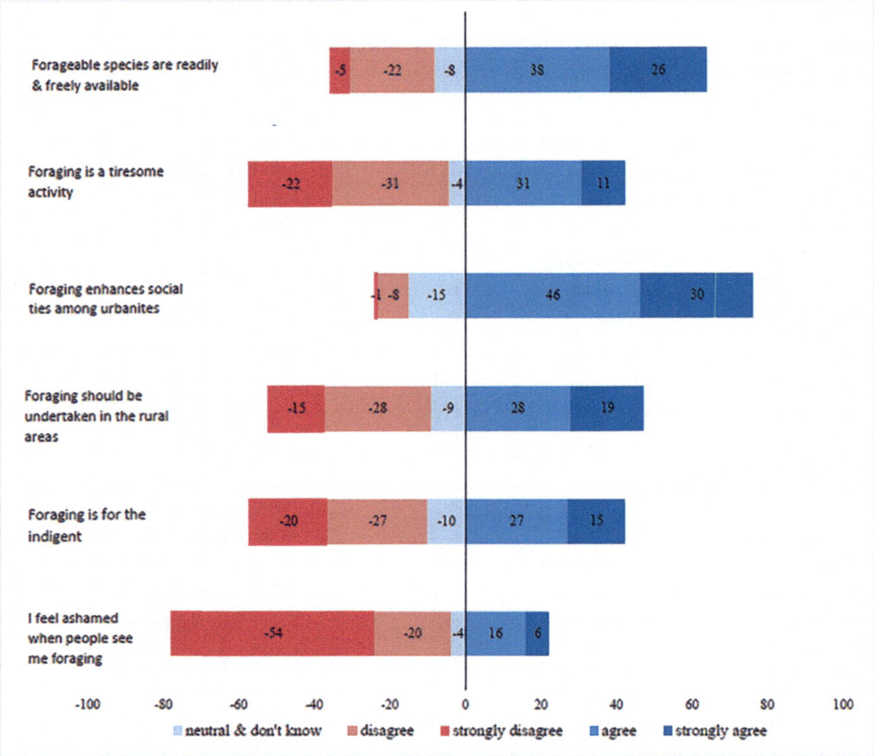

Fig. 4.3 Foragers' perception of foraging (N = 140)

4.4 Discussion

Wild biological resources are essential for human well-being as they are sources of food, medicine, and raw materials. More than half (51%) of the people sampled in Lagos participated in gathering wild biological resources to some extent. This is higher than that reported in corresponding studies in the literature. For example, only about half of wild foragers (47%) were recorded by Mollee et al. (2017) in Kampala, while Fischer and Kowarik (2020) documented one-third in Berlin. However, the percentage of foragers is higher elsewhere; for instance, in two medium-sized towns in South Africa, Garekae et al. (2016) reported that 68% of the total respondents were foragers. The high food insecurity (with or without hunger) of 66% of households in Lagos (Roberts et al. 2019) could explain why a high proportion of people living in Lagos engage in the collection of edible materials. Additionally, more than 15 million people live in Lagos metropolitan areas, and this figure is expected to double by 2050. The large population of Lagos, with a 37.14% unemployment rate as of 2020 (NBS 2020), could also contribute to the high proportion of people collecting raw biological resources from the environment.

While several previous studies (Mollee et al. 2017; Fischer and Kowarik 2020; Somesh et al. 2021; Adeyemi and Shackleton 2023) have reported that females engage more in foraging than males, this result shows that more males collect resources than females in Lagos. This result may be attributed to the fact that safer spaces, such as home gardens, were not included in this study, and the high crime rates in Lagos, including kidnapping, serious assault, sexual violence, and rape, may make many females feel unsafe when collecting biological resources from foraging spaces. Alabi and Balogun (2023) reported an increasing crime rate in Lagos and suggested ways in which such activities could be curbed. This includes functioning streetlights, crime spot detector drones, installation of closed-circuit television, and other civic infrastructure or programs that can reduce poverty and unemployment. The potential relationship between security/safety in cities and foraging demands that the government put more effort into ensuring a peaceful and secure environment for foraging. Further studies are needed to examine the relationship between crime and foraging in urban areas. Public lands such as streets, unused lands, and lakes were visited the most by foragers. This result differs from that of other studies conducted in the Global South. For instance, in Bengaluru, 32% of foragers gather in urban parks (Somesh et al. 2021), which may be linked to the high number of parks in the city (1105 developed parks) (BBMP 2022). On the other hand, there are relatively few parks in Lagos (327 parks), with a population of at least 2 million larger than that in Bengaluru. In addition to the limited number of parks, other factors that could be responsible include fees levied for entrance, formal or informal prohibition of foraging, and absence of species sought by foragers (Adeyemi and Shackleton 2023).

Previous studies have reported that urban dwellers often collect medicine and food in Kampala (Mollee et al. 2017), whereas others prioritize the collection of food, such as in Seattle (Poe et al. 2013), Berlin (Landor-Yamagata et al. 2018), and Bengaluru (Somesh et al. 2021). This study corroborated those studies with collections mostly done for food and medicine. With most people (70%) in Nigeria living below the poverty line (WFP 2023), 57% of the total household expenditure on food (48% for Lagos) (National Bureau of Statistics 2019) could explain this finding. Because medicinal plants are affordable, available, and effective, they are highly dependent on these plants to address several health challenges, such as hypertension, dysentery, low sperm count and weak erection, coated tongue, piles, menstrual disorders, leukorrhea, and fever in Lagos (Olowokudejo et al. 2008). *Mangifera indica* and *Carica papaya* were the two most collected species. These species are both fruit trees and possess numerous medicinal properties, which explains why most foragers collect for food and medicine. What people forage on and for depends on the available forageable resources, and future research should evaluate the supply of these resources within the green spaces in Lagos.

Despite the proportion of foragers and the benefits they derive from their practice, most respondents had negative perceptions of urban foraging. These negative perceptions can be attributed to negative experiences or constraints faced during foraging, as highlighted by Adeyemi and Shackleton (2023). They reported that foragers in Lagos often travel an average of 5.4 km (equivalent to a 50–60-min one-way walk at a normal pace) to gather forageable resources, which could explain why foragers perceive that the activity is tiresome and that gatherable resources are not readily available. This suggests that foraging is a crucial aspect of their livelihoods, and efforts to reduce stress by providing more foraging space and species would contribute to their well-being.

4.5 Conclusion

Foraging can serve as an important approach to supplement food and medical requirements, especially as a safety net in times of economic difficulty. This study reports on Lagos residents' practices and perceptions of foraging to identify attitudes and beliefs, and more than half of the respondents reported gathering wild biological resources in the Lagos metropolitan area. Street was the most visited, with fruit trees (*Mangifera indica, Carica papaya, Terminalia catappa*, etc.) being the most collected resources. Most foragers had a negative perception of foraging, believing that it was a tiresome and shameful activity with the unavailability of forageable species. Conversely, some perceive it as enhancing social ties among urban dwellers and should not be limited to rural areas or indigents.

With a high proportion of people under study interested in foraging activities and foraging resources not always available, there could be an increase in foragers with the establishment of more edible public green spaces within the city. Given the scarcity of spaces for expanding edible green areas for both active and potential foragers, local government bodies should create food forests/community orchards in collaboration with the community and active foragers to ensure sustainable management and harvesting practices (McLain et al. 2014). Several initiatives already being used in European countries could be adopted in Lagos to contribute to human-nature interactions while delivering multiple ecological and economic benefits to residents. Examples of such initiatives include the Andernach Initiative (Germany) and Edible Todmorden (Central UK) (Holthaus 2021). In both cases, edible species were cultivated in public green spaces for free collection and cared for by volunteers (Adeyemi and Shackleton 2023). Such an establishment enhances foragers' experiences, perceptions, and participation. Consequently, it will contribute to addressing urban challenges, including health and food insecurity.

Questionnaire Sample

Practices and perceptions of wild foragers in Lagos metropolis

This research aims to identify foragers, forage sites, and perceptions of urban foragers towards foraging in Lagos metropolis.

In this study, urban foraging is defined as the practice of harvesting or gathering raw biological resources or forageable species (fungi, plants, parts of plants, invertebrate and vertebrate animals, and fish) within urban and peri-urban settings primarily for direct consumption, decoration, crafts, barter, or small-scale sale. This gathering or harvesting can include both wild and domesticated biological resources, in either private or public and managed or unmanaged lands.

Taking part in this survey is voluntary. The information you provide will be treated with strict confidentiality and will be used for research purpose only. Should you have any question(s), please feel free to ask the researcher. Thank you for your kind co-operation.

Section A

Foraging practice

1. Did you or any member of your household gather forageable species during the past 3 years in and around the town or your own garden or fields? [] yes I [] no
2. Which spaces do you normally gather from? [] farms I [] institutional grounds I [] parks I [] riparian zones I [] lakes I [] unbuilt/unused lands I [] avenues/streets I [] other:
3. What species/resources do you gather?
4. Which part do you collect and for what use? (use code below—tick as appropriate)

Part	Use
1. Bark	1. Building poles
2. Branch	2. Craft
3. Flower	3. Decoration
4. Fruit	4. Fencing/poles
5. Leaf	5. Food
6. Root	6. Fruits
7. Seed	7. Fuelwood
8. Stem	8. Furniture
9. Tuber	9. Medicinal
10. Other (specify)	10. Thatching
	11. Utensils and tools
	12. Vegetables
	13. Weaving material
	14. Other (specify)

Respondent profile

5. Sex: [] female I [] male
6. Age (years):
7. Highest education level:
8. Employment status: [] full-time employed I [] part-time employed I [] self-employed I [] student I [] unemployed I [] retired I [] other:
9. How would you describe your income level? [] low I [] middle I [] high

Section B

Perceptions Towards Foraging

What are your perceptions towards gathering?

	Level of agreement					
Items	SD	D	N	A	SA	D/K
I feel ashamed when people see me foraging						
Foraging is the practice for the indigent						
Foraging should be undertaken in rural areas						
Foraging enhances social ties among urban citizens						
Foraging is a tiresome activity						
Forageable species are readily and freely available						

References

Adeyemi O, Shackleton CM (2023) Understanding foraging practices in Lagos metropolis to redesign urban greenspaces in support of human-nature interactions. Urban For Urban Green 79:127805

Ajewole OC, Adebo GM, Fadipe BO (2019) Household attitudes towards sustainable management of urban green spaces and parks in Lagos state, Nigeria. Adv soc sci 6(8):58–66

Alabi AM, Balogun FA (2023) Patterns of urban crimes in Lagos State. Nigeria. KIU Journal of Humanities 8(1):183–189

Barirega A, Agea JG, Van Damme P (2012) Prioritizing wild medicinal and foodplants with potential for commercialization and value chain improvement for livelihood enhancement and poverty reduction in Uganda. Res J Environ Earth Sci 4:668–673

BBMP (2022) Bruhat Bengaluru Mahanagara Palike information. https://bbmp.gov.in/home. Accessed 14 June 2023

Cox DTC, Hudson HL, Shanahan DF et al (2017) The rarity of direct experiences of nature in an urban population. Landsc Urban Plan 160:79–84

Davenport NA, Gambiza J, Shackleton CM (2011) Use and users of municipal commonage around three small towns in the eastern cape, South Africa. J Environ Manag 92:449–1460. https://doi.org/10.1016/j.jenvman.2010.11.003

Eresanya EO, Ajayi VO, Daramola MT, Balogun R (2018) Temperature extremes over selected stations in Nigeria. Phys. Sci. Int. J. 20:1–10

Fischer LK, Kowarik I (2020) Connecting people to biodiversity in cities of tomorrow: is urban foraging a powerful tool? Ecol Indic 112:106087

Gaither CJ, Aragón A, Madden M, Alford S, Wynn A, Emery M (2020) "Black folks do forage": examining wild food gathering in Southeast Atlanta communities. Urban For Urban Green 56:126860. https://doi.org/10.1016/j.ufug.2020.126860

Garekae H, Thakadu OT, Lepetu J (2016) Attitudes of local communities towards forest conservation in Botswana: a case study of Chobe Forest reserve. Int For Rev 18(2):180–191

Holthaus A (2021) The edible city: A concept for the sustainable and resilient city during the COVID-19 pandemic?: A case study of Andernach and Todmorden. Master's thesis,. Department of Human Geography, Stockholm University

Kaoma H, Shackleton CM (2015) The direct-use value of urban tree non-timber forest products to household income in poorer suburbs in south African towns. Forest Policy Econ 61:104–112

Lagos Bureau of Statistics (2019) Abstract of local government statistics; Lagos Bureau of Statistics: Lagos, Nigeria, 2019. http://mepb.lagosstate.gov.ng/wp-content/uploads/sites/29/2020/08/Abstract-of-Local-Government-Statistics-Y2019.pdf. Accessed 16 June 2023

Lagos State Government (2021). About Lagos. Available at: https://lagosstate.gov.ng/about-lagos/ (accessed on 16 May 2023).

Landor-Yamagata JL, Kowarik I, Fischer LK (2018) Urban foraging in Berlin: People, plants and practices within the metropolitan green infrastructure. Sustainability 10(6):1873

McLain RJ, Hurley PT, Emery MR, Poe MR (2014) Gathering "wild" food in the city: rethinking the role of foraging in urban ecosystem planning and management. Local Environ. 19:220–240

McMullin S, Phelan J, Jamnadass R, Iiyama M, Franzel S, Nieuwenhuis M (2012) Trade in medicinal tree and shrub products in three urban centres in Kenya. Forests, Trees and livelihoods 21(3):188–206

Mollee E, Pouliot M, McDonald MA (2017) Into the urban wild: collection of wild urban plants for food and medicine in Kampala, Uganda. Land Use Policy 63:67–77

National Bureau of Statistics (2020) Labor force statistics: unemployment and underemployment Report (Q4 2020). https://nigerianstat.gov.ng/elibrary/read/1238. Accessed 14 Oct 2023

Olowokudejo JD, Kadiri AB, Travih VA (2008) An ethnobotanical survey of herbal markets and medicinal plants in Lagos state of Nigeria. Ethnobot Leafl 12:851–865

Poe MR, McLain RJ, Emery M, Hurley PT (2013) Urban forest justice and the rights to wild foods, medicines, and materials in the city. Hum Ecol 41:409–422

Poe MR, LeCompte J, McLain R, Hurley P (2014) Urban foraging and the relational ecologies of belonging. Soc Cult Geogr 15(8):901–919

Roberts AA, Osadare JO, Inem VA (2019) Hunger in the midst of plenty: a survey of household food security among urban families in Lagos State. Nigeria. J. Public Health Afr. 10:1

Sachdeva S, Emery MR, Hurley PT (2018) Depiction of wild food foraging practices in the media: impact of the great recession. Soc Nat Resour 31(8):977–993

Sardeshpande M, Hurley PT, Garekae H, Mollee E, Emery M, Dahlberg A, Shackleton CM (2021) How people foraging in urban greenspaces can mobilise social-ecological resilience during Covid-19 and beyond. Front Sustain 3:686254

Schlesinger J, Drescher A, Shackleton CM (2015) Socio-spatial dynamics in the use of wild natural resources: evidence from six rapidly growing medium-sized cities in Africa. Appl Geogr 56:107–115. https://doi.org/10.1016/j.apgeog.2014.11.013

Schunko C, Brandner A (2022) Urban nature at the fingertips: investigating wild food foraging to enable nature interactions of urban dwellers. Ambio 51(5):1168–1178

Shackleton CM (2021) Provisioning services in Global South cities. In: Shackleton CM, Cilliers SS, Davoren E, du Toit M (eds) Urban ecology in the global south. Springer, Cham, pp 203–226. https://doi.org/10.1007/978-3-030-67650-6_8

Shackleton CM, Hurley PT, Dahlberg AC, Emery MR, Nagendra H (2017) Urban foraging: a ubiquitous human practice overlooked by urban planners, policy, and research. Sustain For 9(10):1884

Shanahan DF, Lin BB, Bush R, Gaston KJ, Dean JH, Barber E, Fuller RA (2015) Toward improved public health outcomes from urban nature. Am J Public Health 105:470–477

Sneyd L (2013) Wild food, prices, diets and development: sustainability and food security in urban Cameroon. Sustain For 5:4728–4759. https://doi.org/10.3390/su5114728

Soga M, Gaston KJ (2016) Extinction of experience: the loss of human–nature interactions. Front Ecol Environ 14:94–101

Sogbohossou OE, Achigan-Dako EG, Komlan FA, Ahanchede A (2015) Diversity and differential utilization of Amaranthus spp. along the urban-rural continuum of southern Benin. Econ Bot 69(1):9–25

Somesh D, Rao R, Murali R, Nagendra H (2021) Patterns of urban foraging in Bengaluru city. Urban For Urban Green 57:126940

Svizzero S (2016) Foraging wild resources: evolving goals of an ubiquitous human behavior. Anthropol 4

Uduku O, Lawanson T, Ogodo O (2021) Lagos: city scoping study. African Cities Research Consortium, The University of Manchester, Manchester, UK. www.african-cities.org

UNDESA (2018) News. https://www.un.org/development/desa/en/news/population/2018-revision-of-world-urbanization-prospects.html. Accessed 10 June 2023

Van den Berg AE, Maas J, Verheij RA, Groenewegen PP (2010) Green space as a buffer between stressful life events and health. Soc Sci Med 70(8):1203–1210

World Bank (2015) Urbanization in Africa: Trends, promises, and challenges. https://www.worldbank.org/en/events/2015/06/01/urbanization-in-africa-trends-promises-and-challenges. Accessed 14 Apr 2023

World Food Programme (2023) Nigeria: country brief. https://www.wfp.org/countries/nigeria. Accessed 14 Apr 2023

Opeyemi Adeyemi is a Postdoctoral Researcher at the Environmental Science Department of Rhodes University, Makhanda, South Africa. Yemi is an emerging researcher dedicated to promoting sustainable livelihoods in both urban and rural areas through the study of human-environment interactions.

Charlie M. Shackleton is a Distinguished Professor and DSI/NRF SARChi Research Chair in "Interdisciplinary Science in Land & Natural Resource Use for Sustainable Livelihoods", Department of Environmental Science, Rhodes University, Makhanda, South Africa. Charlie's research interests centre on the links between landscapes, biological resources and peoples' uses thereof.

Traditional Connect to Urban Foraging: Experiences from Bali, Indonesia

5

Shalini Dhyani

Abstract

Non-timber forest products (NTFPs), especially wild edibles and medicinal plants, have always been of nutritional as well as cultural interest of people since time immemorial. Over the years, increasing urbanisation and migration of people from rural to urban settings have not deterred many in foraging wild plants for various purposes. In many urban settings, foraging is still influenced and inspired by traditions and cultural connection with nature. This chapter provides an overview of urban foraging experiences from Bali, Indonesia. Bali over the years has seen enhanced tourism interest, and global influx of tourists has also led to urbanisation and developmental pressure on the island. The local Balinese people still having rich traditional connect forage on different wild plants especially for traditional offerings, food preparations, medicinal purposes, etc. The chapter explores the impact of urbanisation, loss of green spaces, legal protection to green spaces along with motivation of urban residents for foraging for culture, and food and livelihood needs in fast-developing Bali using mixed methods. Foraging for wild leaves and flowers as a small-scale economic activity has specially supported marginalised women. It is observed that while urbanisation has led to loss of green spaces, depleting intergenerational connect can affect urban foraging in coming years. Social media influence seems to have attracted and generated interest among tourists for local traditional food preparations. This has enhanced curiosity for urban foraging, and the Covid-19 pandemic was important for local Balinese people to explore and realise the benefits of urban foraging. Urban foraging insights from Bali presented in this chapter are relevant for researchers working in the

S. Dhyani (✉)
CSIR-National Environmental Engineering Research Institute, Nehru Marg, Nagpur, Maharashtra, India

S. Dhyani, M. Sardeshpande (eds.), *Urban Foraging in the Changing World*,
https://doi.org/10.1007/978-981-97-0345-6_5

domain of urban foraging in Bali and elsewhere. Study will also be helpful for urban planners to protect remaining urban green spaces for strengthening nature-culture connections in fast-growing Bali.

Keywords

Wild edibles · Urban green spaces · Culture

5.1 Introduction

According to the United Nations Food and Agriculture Organization (FAO), only around 4% of the approximately 250,000–300,000 types of edible plants in the world are consumed by humans on a regular basis (FAO 1996; Thrupp 2003). Many of the plants that are collected from the wild are from the planet's most biodiverse habitats, consumed traditionally by local people for decades. Many of these provisioning goods (wild edibles, medicinal plants used in rituals) have strong linkages to specific cultures and locations even if they have a worldwide market and are important sources of economic benefits. The bio-cultural concept supports urban foraging that is the outcome of the combination and interdependence of biological and cultural factors that affects human behaviour (Poe et al. 2014). This interdependence not only provides and supports nutritional security but also strengthens food sovereignty and place-based attachment among urban residents (Sujarwo et al. 2019). Wild plant ingredients are presently in high demand, because of not only their immense nutritional and medicinal benefits but also the boredom of consuming the same kind of food for years, and hence their global market has increased by more than 75% in the last two decades (Schindler et al. 2022).

Nature has consistently provided abundant resources to support human well-being beyond what is on display in grocery store shelves. Foraging has traditional connections and is considered a means of nutritional support for marginalised rural as well as growing poor migrant population in cities. However, the practice has slowly grown as a source of recreation, nature-enhancing place-based attachment and nature connectedness for well-to-do urban residents in the fast-growing urban sprawls in Global South (Dhyani and Kadaverugu 2020). Urban foraging has been widely examined and reported largely from Global North; however, it is slowly gaining interest among researchers in Global South too (Shackleton et al. 2017). While foraging always had its roots in cultures and traditions practiced in rural areas, rapid urbanisation has resulted in a rural practice gaining ground in urban areas for cultural continuity.

Indonesia is a tropical Asian nation having a strong and vast array of distinct cultural expressions. Indonesia is home to a wide variety of cultures, customs, communities, traditional wisdom, ancient technology and indigenous languages, including localised medical practices (Jati et al. 2018). Many of these alluring and also highly valuable assets of the country have not been fully studied, recognised or harnessed. The pace of urbanisation in Indonesia, like any other developing country in Asia, continues to be fast and has been exponential in the last few decades. The population of the country is around 263 million with about 55% of population living in urban centres. By 2040, it is anticipated that the urban population of Indonesia will reach approximately 70% of its total population, with annual growth of urban residents expected to be 2.5% (Putu Indra Christiawan 2019).[1]

Among Indonesia's biodiversity-rich states, Bali is one of the richest (Sujarwo et al. 2014). Bali is also among the world's most frequently visited tourist attractions. However, the island has undergone exceptional development as a result of fresh initiatives boosting tourism sector, which have taken top priority. In order to meet the need for services and infrastructure, the city is continuously undergoing transformation (Prajnawrdhi et al. 2012). While Indonesia is known to have the largest Muslim population (86%), the island of Bali exceptionally has about 87% of its population identifying themselves as Hindus. The Hindu population finds its connection with nature since time immemorial and, even after urbanisation and influx of Western tourists, has kept their ties with religion, culture and nature intact.

Considering the growing relevance of urban foraging across the world, the study attempts to provide forging experiences and lessons from foragers in Bali. The study follows a mixed method approach by (a) reviewing available publications on the topic including grey literature and (b) the author's personal observations as well as exploring the social media pictures and articles from Bali and (c) field surveys and informal discussions with urban residents of Bali, especially in Sanur, Bali, to explore and understand the urban foraging context and its relevance along with the key challenges the practice faces. This study is an outcome and one of the first attempts of bringing instances and reflections of urban foraging from Indonesia with a focus on Bali and is expected to be of interest for researchers working in the urban foraging domain. The author brings experiences of largely Hindu population of Bali Island and their traditional practices that are presently supported by urban foraging. The chapter largely focuses on foraging of plants, fungi and ferns. However, with Bali's considerable coastline, one can also observe foraging of crustaceans, fishes, seaweed, coral and shells from Bali Sea for edible as well as other purposes. However, this has not been covered in this chapter and may be an interesting research area for future researchers who are interested in exploring foraging practices in Bali.

5.2 Relevance of Urban Green Spaces

According to Article 29, paragraphs 1 and 2 of Law of the Republic of Indonesia No. 26, 2007, regarding urban planning, the percentage of green open space in the urban area must be at least 30% and a percentage of public green open space must be at least 20% of the total urban area. However, Indonesian cities fall short of the required amount of green open space. Owing to the extensive infrastructural development over the past 30 years, there are fewer green spaces over the entire country of Indonesia. Only 13 out of 174 Indonesian cities, according to data from the Ministry of Public Works and Public Housing as of mid-2019, have prioritised green open space for regional growth (Sukartini et al. 2021). Bali Island with a surface area of 5577 km^2 has less than 20% forest cover

[1] https://www.paulhypepage.co.id/id/urban-life-in-indonesia/

(7.8% primary forests, 10.1% secondary forests and 0.3% planted forests) (Sujarwo and Zuccarello 2019). The extent of urban green space in Denpasar (the administrative district of Bali) is 28.22 km^2 that is 22.1% of the total area of the administrative region of Denpasar Bali. Denpasar Selatan district has the most urban green space in Indonesia (Wirayuda et al. 2023). Green spaces are an inherent part of tradition and culture in Bali. There is a strong connection of local people with their natural surroundings not only for the provisioning benefits but largely for the cultural ecosystem services and traditional connection with the green spaces.

Urban green space is defined as a part of an urban area's public space that has been planted with vegetation to offer ecological, social, economic, cultural, and aesthetic values in Article 1, Point 2, of Permendagri No. 1 of 2007. According to Permen PU No. 05/PRT/M/2008, urban green open space is an extended area/lane and/or grouped with a more open environment where plants are allowed to grow naturally as well as for purpose. Act Number 26 of 2007 under government regulation mandates a minimum of 30% for each city's total area in Indonesia be set aside for green open spaces. There are several other government regulations to urban green spaces for the city that provide guidelines on having sufficient green space for public access in the city (Fig. 5.1). In most cases, foragers in Bali mostly gather wild edible fruits scattered on the ground or flowers from green spaces, and there are no legal obstacles or restrictions to forage them from public urban green spaces.

One of the basic concepts in urban planning in Bali is the notion of *Tri Hita Karana* (the harmonious interactions between God, Man, and Nature). Green

Fig. 5.1 Common urban green spaces and foraging grounds in Bali

spaces, amenities for the public and entrance points like major roadways, as well as natural areas, served essential roles in everyday life and traditions. This included holy offerings to God, ancestors, humans, animals, supranatural and others as per their *Pancha yadnya* (Dev yadnya, Pitra yadnya, Bhoot yadnya, Manav yadnya) tradition at various public places including roads, corners, beaches, etc. The traditional green connection extends from rural to urban settings, and one can clearly observe this in Bali. Beach and river sides, avenue plantations, city parks, city forests and cemeteries are part of relevant public urban green spaces considered by local people to have cultural ties and considered living spaces by them. In other parts of the world, these places as urban green spaces have value, but in Bali, beyond urban green spaces, they are worshipped and offerings are made every day as well as on special days of full moon and no moon under permanent rituals.

By achieving a harmonious relationship between the built-up area and the natural environment, open green space in urban areas attempts to preserve balance and harmony in the urban ecosystem and to improve the living standards in cities (Indrawati et al. 2019). Balinese Hindus have an inherent connection with nature, and their nature culture ties are part of their lives and lifestyle.

In Bali, there is tradition to manage the landscape when constructing a home. Traditional knowledge-based designs and structures, for example, *telajakan*, serve to preserve ecological harmony. *Telajakan* continues to support groundwater infiltration of water and acts as a water catchment region. Groundwater reserves continue to be recharged because of *telajakan*, which drains rainwater into the soil throughout rainy seasons. *Telajakan* was basically a customary green space between a residential compound's boundary and a drainage hole or footpath in a roadside that once offered a variety of foraging choices. A provincial regulation covering green open spaces, or *telajakan*, was historically developed in the Jasri traditional village with the goal to render *Peraturan Daerah* (Local Regulation) 17 concerning the comprehensive spatial plan for the urban area as a legal requirement. The gradient has a unique capacity to serve and effectively been utilised for creating green spaces. The gradual disappearance of this feature was owed to urban sprawl and unplanned growth in the city. Traditionally, *telajakan* supported a variety of purposes, including providing plant parts in daily Balinese rituals especially floral materials, controlling rainwater runoff from the premises and providing a semi-public area for nearby residents and street sellers. The *telajakan* is a traditional custom maintained by the community at large and is set up right in the front yard of the house (Darmadi et al. 2019). This customary practice of having *telajakan* near the houses is simply referred to as *awig-awig* in Bali and has not been formalised. *Telajakan* offers benefits for society and the economy alongside its environmental benefit of improving groundwater recharge and protecting microclimate and aesthetic significance.

5.3 Foraging for Culture and Traditions

Plants utilised in Hindu rituals and herbal remedies were both maintained by the Balinese people traditionally for cultural botany interests. For the customary use, people had been using plants for several hundred years (Andila et al. 2022). Many tropical fruits and native vegetables that one could possibly consume in one lifetime may be found in Bali. All of them are extremely nourishing and can support the immune system to be more resilient, especially in these modern times.

Balinese Hinduism incorporates native spiritual customs and tributes that are important for the "five holy ceremonies" or "*Pancha yadnya*" religious rituals, and even living in urban settings, they are very much performed with the same interests and devotion as in rural areas. These everyday offerings are very much part and parcel of the lifestyle of Balinese Hindus, and it is still practiced with a lot of devotion and dedication (Fig. 5.2).

Many plants and plant-based offerings are a vital part of predominant Balinese Hindu culture even in urban areas of the Bali Island. It is unlikely to miss the endless vibrant offerings made of various flowers that can be found across the island and especially close to temples and throughout religious rituals. Tropical flowers and abundant greenery abound in the stunning Balinese-style gardens are unmissable. One can easily locate diverse *Plumeria* trees with different coloured flowers in every street and neighbourhood. It is also common to see jackfruit (*Artocarpus heterophyllus*), mango (*Mangifera indica*), banana (*Musa* spp.), neem (*Azadirachta indica*) and many other fruiting and flowering trees across the island.

Hindu offerings can be found all across Bali using a variety of native aromatic Balinese flowers and leaves (Fig. 5.3). Most of the offering plants are grown either in the backyard or in the home garden and are also foraged by locals for personal use or for livelihoods as they are frequently sold in traditional marketplaces. The plant species present in offerings are typically native

Fig. 5.2 Daily offering from foraged plants from urban green spaces of Bali: traditional connect to urban foraging. The pictures show local Balinese offerings on the platter known as canang sari made of coconut leaves; this container includes Wija or rice; Porosan made of betel leaves, gambir and lime; Jajan in the form of sugar cane and bananas; red, yellow and blue flowers; fragrant oil; and an incense stick. A candy and edibles are also kept in the canang

Fig. 5.3 Pictures representing everyday religious offerings on various private and public spaces in Bali ranging across household, roads, beach sides, shops, etc.

and easy for Balinese people to procure. Many of these are also foraged from urban and suburban locations by urban poor who earn livelihood by foraging and selling flowers and other offering requirements. Among many culturally relevant species, *Arenga pinnata* is considered a cultural keystone species as all parts of this wild palm are used for various purposes, viz. food, culture, tradition, etc. (Sujarwo et al. 2014). Many exotic species have also been observed in these offerings and are widely used by the Balinese, notably *Celosia spicata, Eleusine indica, Imperata cylindrica, Colocasia esculenta, Arenga pinnata, Cinnamomum burmanni, Artocarpus heterophyllus, Piper betle, Lablab purpureus, Manihot esculenta, Musa paradisiaca, Blumea balsamifera, Ipomoea batatas* and *Phyllanthus niruri*; some of these are foraged for flowers as well as leaves from public green spaces (Sujarwo et al. 2019).

Studies on plants having high cultural food significance index because of their high cultural value in Balinese food preparation have listed the top 20 plant species (viz. *Arenga pinnata, Artocarpus heterophyllus, Lablab purpureus, Colocasia esculenta, Cinnamomum burmanni, Piper betle, Paederia foetida, Zingiber officinale, Blumea balsamifera, Kaempferia galanga, Syzygium polyanthum, Curcuma longa, Antidesma bunius, Luffa acutangula, Alstonia scholaris, Centella asiatica, Ipomoea batatas, Carica papaya Manihot esculenta* and *Averrhoa bilimbi*) that are mostly foraged from various urban green spaces (Sujarwo and Caneva 2015).

Dioscorea hispida, Alstonia scholaris, Psidium guajava, Carica papaya, Antidesma bunius, Paederia foetida, Cajanus cajan, Diplazium esculentum, Luffa acutangula and *Moringa oleifera* are considered to have high cultural value index and are among the top 20 plant species considered to be holding high cultural value to the Balinese people; many of them are not cultivated but collected from green spaces (Sujarwo and Caneva 2015).

5.4 Foraging for Offerings

In Bali, small palm leaf baskets, filled with rice, colourful flowers of *Plumeria acuminata* and *Bougainvillea* species, food, cigarettes and incense, scattered in front of every entrance door are a common site near the front of shops and houses, on statues, at the foot of temples and many other places. The *canang sari* or offerings are given to the Gods every morning by the Hindu Balinese as a form of thanking for balance and peace in our world that is renewed each day. These small offerings are in small palm leaf, coconut leaf, betel nut and lime leaf basket, which are later sprinkled by holy water, the *canang*, using *jepun* flower (*Plumeria* species) (Sujarwo and Caneva 2016). A collection of 36 types of flowers, including 29 species that are grown along with 7 that are not cultivated but foraged from green spaces, are utilised in traditional Hindu religious events (Darma et al. 2021). Ten plants, including *Cananga odorata, Impatiens balsamina, Hydrangea macrophylla, Plumeria acuminata, Tagetes erecta, Magnolia champaca, Gomphrena globosa*/bunga ratna, *Gardenia jasminoides*/jempiring, *Nymphaea* spp./tunjung and *Hibiscus rosa-sinensis* L., are considered of high value in local Hindu offerings (Darma et al. 2021). In Bali, the streets and compounds of the houses are dominated by various varieties of Frangipani (*Plumeria* species) locally known as *bunga jepun* or *bunga kamboja* that are quintessential part of daily offering and can be seen growing on nearly every street corner, home and region; it is considered by the majority of Balinese as the most prevalent factor and a symbol of their island.

Coconut trees in Bali are everywhere and are regularly foraged for various religious offerings, and hence unique foraged leaf appearances are common across the island (Fig. 5.4). Cut-leaved coconut crowns are very common across the island as the coconut leaves are regularly harvested for religious purposes, which provides a glimpse to relationship between people and the natural environment. Young and old coconut leaves are

Fig. 5.4 Foraged coconut leaves to prepare *canang sari* in one of the Balinese Hindu temples

frequently harvested and used in religious offerings, including *canang, penjor* and *sanggah cucuk* for *Dewi Sri* offerings. Coconut leaves are extensively used in decorative pieces and arrangements along with an array of colourful petals (Miyaura et al. 2015).

5.5 Foraging for Food

5.5.1 Flowers

Flowers have a special place in Balinese culture and traditions, and apart from cultural offerings, they are frequently consumed in local food preparations. Flowers make delicious food sources in Balinese food. Flowers are not only used in Balinese Hindu offerings, but a variety of wild flowers are also used in various Balinese food preparations that are listed in Table 5.1. There is a great deal of knowledge regarding different kinds of flowers as well as how to use them in food and cultural practices, but

Table 5.1 Important flowers foraged and used in traditional Balinese food preparations

Sr. No.	Vernacular name	Botanical name	Use
1	Papaya	*Carica papaya*	Flowers are boiled or mixed with spices to be made edible
2	Kecombrang	*Etlingera elatior*	Flowers used as spice in fish preparations
3	Turi	*Sesbania grandiflora*	Use in vegetables and traditional salad
4	Genjer	*Limnocharis flava*	Flowers are boiled or stir fried
5	Banana	*Musa paradisiaca*	Flowers are boiled or roasted
6	Sedam malam	*Polianthes tuberosa*	In Tekwan soup also consumed fresh and dried
7	Pumpkin	*Cucurbita moschata*	Stir fried and mixed in vegetable preparations
8	Teland	*Clitoria ternate*	In eye drops for treatment, consumed raw or in Telang tea
9	Jasmine	*Jasminum sambac*	Jasmine tea
10	Melinjo	*Gnetum gnemon*	Famous Syur Asam soup
11	Rosella	*Hibiscus sabdariffa*	Consumed raw or in rosella flowers
12	Hibiscus	*Hibiscus species*	Mixed in salads
13	Begonia	*Begonia species*	Consumed fresh
14	Dandelions	*Taraxacum species*	Dandelion tea
15	Kenikir	*Cosmos caudatus*	Stir fried for consumption
16	Yard long bean	*Vigna unguiculata*	Stir fried and mixed with other vegetables
18	Ketul	*Biden pilosa*	Fresh as well as cooked
19	Belimbingan	*Oxalis barrelieri*	Consumed fresh
20	Calincing	*Oxalis corniculata*	Flowers, stem, leaves all are consumed
21	Basil	*Ocimum basilicum*	Mixed in salads
22	Maman	*Cleome rutidosperma*	Flower consumed along with stem and leaves
23	Bunga Kupu-kupu	*Bauhinia purpurea*	Consumed fresh and cooked

Source: Andila et al. 2022

there is also immense therapeutic significance of flowers used in the Balinese Hindu culture and still used by families migrated in urban areas (Andila et al. 2022).

While *P. acuminata* flowers are an essential part of Balinese Hindu offering rituals, fresh juvenile leaves of the plant are used in vegetable mixture traditionally known as '*Urab Don Jepun*'. Fresh juvenile leaves of *P. acuminata* are boiled, chopped and kneaded with Balinese spices, viz. *Kacang Panjang* and *Toge*, until the seasoning mix is mixed well (Andila et al. 2022).

5.5.2 Ferns

Being a tropical island, one can observe ferns growing across the city in public and personal green spaces. Many of these ferns are used in food preparations too as Balinese people like to relish food preparations made from small fresh fronds of the ferns. Fronds of *Blechnum orientale* and *Diplazium esculentum* ferns are relished as green vegetables in Bali households (Table 5.2), and one can find them selling in the traditional markets as well as supermarkets largely foraged from urban green spaces and peri-urban or rural parts of Bali (Sujarwo et al. 2014).

5.5.2.1 Mushrooms

Foraging mushrooms from urban green space is a rare activity for urban foragers in Bali. There are several mushrooms foraged by rural people living in villages. There are more than 50 species of mushrooms reported to be frequently used for food and traditional medicine in Indonesia (Putra et al. 2022). During the pandemic years, especially 2020 and 2021, when market supplies went irregular, a lot of urban poor started exploring urban green spaces, especially the city forest, for foraging mushroom as nutritional support.

> *Initially we were not sure of finding mushrooms in urban areas but after we started locating them in city forests we really got interested in foraging mushrooms, identification identifying them if they are poisonous or non-poisonous. This interest later grew up and we started posting pictures on different social media platforms.*
> *-Wayan Ratna, a local resident of Sanur Bali*

Table 5.2 Preferred ferns foraged from green spaces for food preparations in traditional Balinese culture

Sr. No.	Vernacular name	Botanical name	Uses
1	Paku jahe	*Blechnum orientale*	Cooked leaves added to vegetable soups
2	Paku lindung	*Pneumatopteris callosa*	Cooked leaves added to vegetable soups; leaf decoction used for fever and hypertension
3	Paku jukut, paku kedis	*Diplazium esculentum*	Cooked young leaves added to vegetable soups
4	Paku udang, paku labuh	*Diplazium repandum*	Cooked young leaves added to vegetable soups

Source: Sujarwo et al. 2014

There are also mushroom foraging groups known as mushroom hunters who regularly share information about locations and safe mushroom foraging ways on close WhatsApp and Facebook groups. Due to mushroom poisoning, mushroom foraging is risky if the forager is not able to distinguish between edible and poisonous mushrooms. Hence, do's and don'ts are regularly shared in these online mushroom hunting groups. *Termitomyces microcarpus*, or termite mushroom as it often grows over the termite nests, is an important edible mushroom because all the termite mushrooms are non-poisonous and are preferred for their nutritional and medicinal properties by foragers in Indonesia. Wood ear mushroom (*Auricularia auricula-judae*) is reported as preferred mushroom foraged as posted on social media by city foragers. One of the urban forgers, Ferry, along with the mycologist, Ivan Permana Putra, also wrote scientific articles *Termitomyces di kawasan urban kecamatan Gunungpati, Kota Semarang, Jawa Tengah* (*Termitomyces* in urban areas, Gunungpati District, Semarang City, Central Java) and *Keragaman dan potensi jamur di hutan Kota Semarang, Jawa Tengah* (Diversity and potency of macrofungi in the city forest of Semarang, Central Java) (Wardhani 2022).

5.6 Foraging for a Livelihood

Balinese people prefer making customary offerings using the entire plant parts, and the last few years have seen an increase in the sale of coconut leaflets also in the urban markets that helps in local livelihood. Every Hindu household in Bali dedicates daily offerings of *canang sari*. Traditionally, women had the role of putting together a *canang sari* everyday. Every family makes offerings ranging from 15 to 50 in numbers, and at least 15 offerings are common across the Balinese Hindu households. It is not unusual to see multiple generations of women sitting side by side over a collection of tiny square baskets, making *canang sari* in either a solitary or a group activity. However, it is increasingly becoming typical for locals in Bali to buy *canang sari* from the local markets as more women in urban settings do formal jobs for income generation. Though both men and women are involved in the small-scale businesses of selling offering material, women dominate it. Traditionally, women have been involved in preparing *canang sari* at household level; hence, more women belonging to poor socio-economic backgrounds are opting it as a potential livelihood alternative. Apart from selling the *canang sari*, they also provide offering requirements that include flowers foraged from urban as well as peri-urban areas. Nowadays, it is almost as simple to find *canang sari* sellers on Bali's streets as it is to spot a grocery store (Fig. 5.5). Street vendors selling canang sari are part of the informal economy, but they also possess assets having monetary, social, cultural and symbolic value (Atmadja et al. 2016).

Fig. 5.5 A street vendor with foraged coconut leaves for making *canang sari*, and a couple making preparations from foraged mango leaves for full moon offerings

5.7 Vintage Yet Relevant

It has been widely observed that urban foraging got a new meaning during the pandemic. For many poor families, foraging helped in fulfilling their food requirements, whereas for many others, it helped in finding resources when market supplies were not organised. Many wild flowers, ferns as well as fungi were foraged for food during this time following traditional knowledge systems that were otherwise forgotten and went waste in urban green spaces (Tables 5.1 and 5.2).

> Foraging keeps them aware about the value of nature in their surroundings and how relevant it is for their existence as well as their daily rituals and traditional offerings. It is especially important for younger generations leading for strengthening their relationship to the place they live in, by giving them a taste of their city very literally and fulfilling many of their everyday requirements including required for daily offerings.
>
> *-Pak Gusti, Kuta Bali*

The pandemic outbreak in Bali between the years 2020 and 2021 also encouraged many people to appreciate the outdoors and public green spaces. Urban foragers in the island used the practice as a way to revive traditional methods that encouraged interpersonal relationships that are not often fostered in urban settings. Many urban people mentioned during the informal discussions how they were first introduced to foraging through the sharing of stories among neighbours and elders or just by observing them.

> *Traditionally, moringa (Moringa oleifera) tree served as a fence around most Balinese family complexes. It served as a talisman as it was considered magical foodstuff that is incredibly nutritious, and medicine all bundled into one. Traditional Balinese cuisine frequently included moringa leaves, which were frequently simmered in a hot stew. However, it became stigmatized in recent decades as a poor man's food and a component of antiquated mystical notions.*
>
> Ketut, Denpasar

However, as the workers of Bali's collapsed tourism industry returned to live off the land during the pandemic outbreak, people started to re-adopt the advantages of the wonderful tree that spreads like a weed. The raw leaves were added to smoothies, omelettes, soups, stews, curries and other dishes, and now moringa has a new meaning in many vegan eateries and cooking workshops as superfood. Similarly, *Centella asiatica* leaves were also used in various food preparations like green juices during the pandemic and are now introduced in Balinese cooking classes and preparations. There are many medicinal and nutraceutical benefits of *Centella* that were valued by Balinese, but these went out of frequent use because of loss of green spaces and unavailability in surrounding areas. However, during the pandemic, locals went back to many such traditional and medicinal plants and revived their use in food and medicines.

5.8 Urban Foraging in Social Media

Foraging to many urban dwellers still seems like a rural-only activity. However, urban foragers are reviving this custom today by posting photos of their excursions on social media (Evan 2021). One can find many interesting sites and pages dedicated to urban foraging.[2,3,4] Chefs and home cooks alike are venturing out into the countryside as well as nearby urban green spaces to pick wild vegetables, herbs and fruits for food as foraging has become one of the biggest culinary trends across the world and is getting popular in Bali, especially in Ubud area of Bali. Urban foragers are using fresh and cut flowers to spice up their homes in addition to using them for meals and beverages. A rising number of foragers in the country are learning new pleasures from the purposeful quest for sustenance from the wilderness as well as

[2] https://holisticbali.com/2015/03/04/foraging-in-bali/.

[3] https://beyondthemapletree.com/foraging-near-ubud.

[4] https://www.thejakartapost.com/life/2021/10/14/going-wild-urban-indonesians-go-back-to-nature-in-search-for-food.html.

green spaces in urban areas. During the pandemic, many urban residents got attracted to foraging because it allowed them to appreciate nature more than earlier.

> *"As I learn more about edible wild plants, I am more aware of how little I know about the planet I call home. In addition, foraging teaches me about the Indonesian biodiversity, which makes me feel proud of my country."*
> *-Dyah, Indah Kumala (a content writer from Bali)*

For many urban foragers, finding food is more than simply a fun activity; it is a way of life. They have been implementing their discoveries into the family's home-cooked meals and beverages. According to Indah, the impact is felt throughout the body.

> *I am less prone to become ill now that I eat wild mushrooms more frequently. There are many anti-inflammatory qualities in wild mushrooms.* However, *drinking herbal tea infused with edible flowers helped mer unwind.*
> *-Wayan Suchi*

One can also find many restaurants in Ubud, Bali, that serve preparations and food from foraged plant parts, especially moringa leaves considered to be a superfood in Bali. Ubud is a vegetarian's paradise in Bali, and there are many cooking workshops now that follow vegan and used foraged ingredients. One can find information about these restaurants and cooking classes not only by visiting them but also through social media where there are many foragers advertising their products and classes. There are several other wild fruits, leaves and mushrooms that have gained popularity as foraged food and are regularly used in food preparations. Some of these diverse foraged wild plants preferred by young foragers and showcased in social media handles as Instagram are provided in Table 5.3.

Table 5.3 Variety of plant parts preferred by young foragers highlighted by social media handles

Plant part	Vernacular name	Botanical name	Use
Leaves	Wild cilantro	*Culantro Silvestre*	Salad and curries
	Curry leaves	*Murraya koenigii* or *Bergera koenigii*	Salad and curries
	Malabar spinach berries	*Basella alba*	Salad
	Katuk or star gooseberry	*Sauropus androgynus*	Salad
	Wild mugwort	*Artemisia vulgaris*	Medicine
	Garlic chives	*Allium tuberosum*	Salad and curries
Berries	Barbados cherry	*Malpighia emarginata*	Desserts
	Limeberry	*Triphasia trifolia*	Desserts
	Mulberries	*Morus alba*	Desserts
	Indian jujube	*Ziziphus mauritiana*	Consumed raw
	Cantigi berries	*Vaccinium varingiaefolium*	Salad
	Surinam cherry	*Eugenia uniflora*	Consumed raw
Fruits	Kusum	*Schleichera oleosa*	Consumed raw
	Ackee	*Blighia sapida*	As a vegetable
	Rum fruit	*Myrciaria floribunda*	Consumed raw

(continued)

Table 5.3 (continued)

Plant part	Vernacular name	Botanical name	Use
	Sapodillas	*Manilkara zapota*	Consumed raw and desserts
	Cherry guava	*Psidium cattleianum*	Consumed raw
	Mangrove apple or Buah Pedada	*Sonneratia caseolaris*	In noodles
	Cluster figs	*Ficus racemosa*	Consumed raw and desserts
	Noni Mengkudu	*Morinda citrifolia*	Consumed raw and juice
Flowers	Hibiscus	*Hibiscus*	Tea and salad
	Sulphur cosmos	*Cosmos sulphureus*	Tea and salad
	Rose balsam	*Impatiens balsamina*	Tea and salad
	Butterfly pea flower	*Clitoria ternate*a	Tea and salad
	Wild dandelion	*Taraxacum*	Tea and salad
	Redbud tree berries	*Cercis canadensis*	Tea & salad
	Nasturtium	*Nasturtium* species	Salad
	Vegetable hummingbird	*Sesbania grandiflora*	Salad and fritters
Seeds	Wild fennel	*Nigella arvensis*	Salad
	Curly dock	*Rumex crispus*	In curries
	Blue tongue berry	*Melastoma malbathricum*	Salad
Mushrooms	Bali Chatu	*Termitomyces*	Curries
	Wood ear mushroom	*Auricularia auricula-judae*	Curries

https://instagram.com/kumykumy08?igshid=MzRIODBiNWFIZA==

5.9 Challenges for Urban Foraging

As the first tourist attraction in Indonesia, Bali has welcomed visitors since the early years of the twentieth century. However, due to intense developmental and tourism pressure and its rapidly expanding urban population, Bali Island has undergone significant changes in its land use. The reduction, degradation and loss of traditional small urban green spaces in Denpasar, Bali, are clear examples of urban sprawl (Kato et al. 2017). Land use and land cover saw significant change as a result of the rapid increase of tourism. From 2011 to 2019, the public urban green space availability in Bali has declined from 18.32% to 12.49%. The publicly accessible open green spaces have frequently declined, which means that it no more fulfils the 30% of the green demand for the city (Indrawati et al. 2019). A study by Rimba et al. (2020) projects land use and land cover change in Bali from 2000 to 2025. There has been clear change observed from agriculture, open space and urban green spaces to built-up expansion. From 2015 to 2025, the built-up area is further projected to rise by up to 43%, and this will reduce access of local foragers to harvest foraging goods from urban green spaces.

Previously, religious ceremonies were performed on nearly every beachside in Bali that involved many foraged plants collected from various places. However, resorts and other businesses associated with tourism have legally and illegally possessed a majority of the beachfronts due to growing economic and tourist activity. The beachfront needs to be conserved in order to be utilised exclusively for use by the public, according to the Bali Government, who described it as a vital component of the island's cultural heritage. Presently, only particular areas could potentially be used for locally relevant sacred ceremonies (Prajnawrdhi et al. 2012).

Foraging in urban green spaces of Bali, however, necessitates extra caution as also expressed during the discussions because of the increased exposure to toxins like vehicular emissions and animal droppings. Growing urbanisation and tourism pressure have resulted in increasing number of contaminated sites even closer to beaches considered to be sacred by local Balinese people. This restricts local interests to forage from these areas.

Kato et al. (2017) reported loss of *telajakan*, which was appreciated in Balinese houses even in urban settings. However, it has been seen that the variety of planted trees is diminishing and they are presently restricted to a few uses, which are focussed on aesthetics and tourism purposes, which are greatly favoured. This indicates that telajakan's semi-public nature is vanishing and is also affecting foraging for even offering floral requirements (Kato et al. 2017).

Loss of intergenerational connection and transfer of knowledge due to migration and urbanisation can be observed in Bali, and that has been one of the crucial aspects of loss of traditional knowledge related to the use of wild species and has also affected foraging from urban green spaces. The unavailability of green spaces also results in the disconnect that affects youth and next generation to have the experience of foraging. Like any other fast-growing city in the world, Bali is also facing the challenge of rapid urbanisation that is taking a toll on people. Youth are involved in studies and jobs with insufficient time to experience nature in their neighbourhood. However, tourists visiting Bali love to experience green spaces and also enjoy Balinese cooking classes. The intergenerational disconnect because of migration from villages to cities is also affecting the culture and tradition of foraging from wild spaces. Lack of documentation will further erode this system that has been the foundation of urban foraging in Bali having strong connections to its traditions and culture (Sujarwo et al. 2014).

5.10 Conclusions

Urban foraging experiences from Bali endorse a strong foundation of culture and tradition that has deep connections with nature. This was clearly observed from the review of publications as well as interactions and discussions with local Balinese people. Nature-culture interlinkages were clear in everyday lives of Balinese people following a traditional way of living. Nature-culture connect was predominant in Hindu Balinese traditional religious and food practices. Locals frequently used cultivated as well as wild plant parts in their offerings, decorations, handicrafts and food preparations. The flexibility of including many exotic flowers and plants will be useful in changing times, and urbanisation will further include as well as exclude more plants in these rituals as per availability and convenience of local people. Growing urbanisation, land use change and increasing tourist influx are affecting nature-people connections, and urban foraging can be a great way to reconnect urban residents and especially youth to their roots. Commercialisation of traditional requirements may also exert pressure of foragers on remaining urban and peri-urban green spaces, e.g. excessive harvesting of coconut leaves from urban and peri-urban areas. It is important that the government considers certain areas to be declared no-go zones or a limit of development is decided, especially near coastal areas where resorts and hotels are mushrooming and are considered sacred grounds in Balinese culture. Interest of tourists in Balinese cooking and foraging in green spaces may be helpful to protect urban and peri-urban green spaces in Bali as cultural landscapes that can be showcased to tourists, and involving them can be a good opportunity to generate livelihood and conserve traditional nature-culture connect.

Acknowledgements Author acknowledges the support from CIFOR-ICRAF and UNESCO-TWAS as a UNESCO-TWAS Associate and CIFOR-ICRAF Visiting Scientist (2022–2025). Author places her gratitude to Director CSIR-NEERI for the support and encouragement. Author also thanks the Knowledge Resource Centre of CSIR-NEERI, Nagpur, for plagiarism check using licensed version of iThenticate.

References

Andila PS, Warseno T, Syafitri W, Tirta IG (2022) Ethnobotanical study of Hindu Society in Tabanan Bali and the Conservation Efforts. In: 7th International Conference on Biological Science (ICBS 2021). Atlantis Press, pp 590–597. https://doi.org/10.2991/absr.k.220406.085

Atmadja AT, Ariyani LPS, Atmadja NB (2016) Women as canang sari street vendors in Bali. Komunitas 8(1):85–93. https://doi.org/10.15294/komunitas.v8i1.5116

Christiawan PI (2019) Designing the mitigation model of urban sprawl potential impact in suburban Denpasar, Bali. J Phys Conf Ser 1363(1):012100. https://doi.org/10.1088/1742-6596/1363/1/012100

Darma IDP, Sutomo S, Hanum SF, Iryadi R, Rahayu A (2021) Flowers and value of conservation in the culture of Hindu Community in Bali. Biosaintifika: J Biol Biol Educ 13(1):34–40. https://doi.org/10.15294/biosaintifika.v13i1.27054

Darmadi AAK, Wirasiti NN, Ginantra IK (2019) Diversity of species and benefits of telajakan plant as a potential tourism attraction in Bali. J Environ Manag Tour 10(1):109–118. https://doi.org/10.14505/jemt.v10.1(33).11

Dhyani S, Kadaverugu R (2020) Food security and cultural benefits from urban green spaces: exploring urban foraging as a silently growing global movement. Clim Chang Environ Sustain 8(2):219–225. https://doi.org/10.5958/2320-642X.2020.00022.8

Evan V (2021, October 14) Going wild: urban Indonesians go back to nature in search for food. The Jakarta Post. https://www.thejakartapost.com/life/2021/10/14/going-wild-urban-indonesians-go-back-to-nature-in-search-for-food.html. Accessed 14 June 2023

Food and Agriculture Organization (FAO) (1996) Global plan of action for the conservation and sustainable utilization of plant genetic resources for food and agriculture and the Leipzig declaration. Foraging in Bali, food for life, great things to do in Bali, Herbal Bali. Holistic Bali. https://holisticbali.com/2015/03/04/foraging-in-bali/. Accessed 14 June 2023

Indrawati M, Sudiana AAK, Sumantra K (2019) Position and management strategy for public green open spaces in the City of Denpasar, Bali Province. Int J Res -GRANTHAALAYAH 9:75–85. https://doi.org/10.29121/granthaalayah.v9.i7.2021.4082

Jati B, Serafinah I, Bagyo Y (2018) Survey and utilization of wild plants (extractivism) in Tengger society East Java, Indonesia. Int J Mod Bot 8(1):8–14. https://doi.org/10.5923/j.ijmb.20180801.02

Kato S, Hishiyama K, Darmadi AAK, Suprapta DN (2017) Changing roles of traditional small urban green spaces (Telajakan) in Bali, Indonesia. Open J Ecol 7(1):1–11. https://doi.org/10.4236/oje.2017.71001

Miyaura R, Ohno T, Maenaka H, Sumiartha K, Yamaguchi H (2015) A particular Silhouette of human-influenced coconut trees in Hindu Bali, Indonesia: an ethnobotanical field note. Ethnobot Res Appl 14:405–421

Poe MR, LeCompte J, McLain R, Hurley P (2014) Urban foraging and the relational ecologies of belonging. Soc Cult Geogr 15(8):901–919. https://doi.org/10.1080/14649365.2014.908232

Prajnawrdhi T, Sivam A, Karuppannan S (2012) Is urbanised city sustainable? A case study of Denpasar, Bali (Doctoral dissertation, ICBEDC). https://simdos.unud.ac.id/uploads/file_penelitian_dir/ee3c99913e126c960ea66324c1f7f74c.pdf

Putra IP, Nurdebyandaru N, Amelya MP, Hermawan R (2022) Current checklist of local names and utilization information of Indonesian wild mushrooms. J Trop Biodivers Biotechnol 7(3):71407. https://doi.org/10.22146/jtbb.71407

Rimba AB, Atmaja T, Mohan G, Chapagain SK, Arumansawang A, Payus C, Fukushi K (2020) Identifying land use and land cover (LULC) change from 2000 to 2025 driven by tourism growth: a study case in Bali. Int Arch Photogramm Remote Sens Spat Inf Sci 43:1621–1627. https://doi.org/10.5194/isprs-archives-XLIII-B3-2020-1621-2020,2020

Schindler C, Heral E, Drinkwater E, Timoshyna A, Muir G, Walter S, Leaman DJ, Schippmann U (2022) Wildcheck: Assessing the risks and opportunities of trade in wild plant ingredients. Food & Agriculture Organization. https://doi.org/10.4060/cb9267en

Shackleton CM, Hurley PT, Dahlberg AC, Emery MR, Nagendra H (2017) Urban foraging: a ubiquitous human practice overlooked by urban planners, policy, and research. Sustainability 9(10):1884. https://doi.org/10.3390/su9101884

Sujarwo W, Caneva G (2015) Ethnobotanical study of cultivated plants in home gardens of traditional villages in Bali (Indonesia). Hum Ecol 43:769–778. https://doi.org/10.1007/s10745-015-9775-8

Sujarwo W, Caneva G (2016) Using quantitative indices to evaluate the cultural importance of food and nutraceutical plants: comparative data from the Island of Bali (Indonesia). J Cult Herit 18:342–348. https://doi.org/10.1016/j.culher.2015.06.006

Sujarwo W, Caneva G, Zuccarello V (2019) Bio-cultural traits and cultural keystone species, a combined approach: an example of application about plants used for food and nutraceutical purposes in Aga villages in Bali, Indonesia. Hum Ecol 47:917–929. https://doi.org/10.1007/s10745-019-00124-5

Sujarwo W, Lugrayasa N, Caneva G (2014) Ethnobotanical study of edible ferns used in Bali Indonesia. Asia Pacific J Sustain Agric Food Energy 2(2):1–4

Sujarwo W, Zuccarello V (2019) Patterns of plant use in religious offerings in Bali (Indonesia). Acta Bot Bras 34:40–53. https://doi.org/10.1590/0102-33062019abb0110

Sukartini NM, Auwalin I, Rumayya R (2021) The impact of urban green spaces on the probability of urban crime in Indonesia. Dev Stud Res 8(1):161–169. https://doi.org/10.1080/21665095.2021.1950019

Thrupp LA (2003) The central role of agricultural biodiversity. Key Readings, p 57

Wardhani WK (2022) The Indonesians who hunt for wild mushrooms. https://www.thejakartapost.com/culture/2022/03/15/the-indonesians-who-hunt-for-wild-mushrooms.html. Accessed 12 Aug 2023

Wirayuda IKAK, Widayani P, Sekaranom AB (2023) Urban green space analysis and its effect on the surface urban Heat Island phenomenon in Denpasar City, Bali. For Soc 7(1):150–168. https://doi.org/10.24259/fs.v7i1.24526

Shalini Dhyani (PhD) is a Principal Scientist at the CSIR-National Environment Engineering Research Institute, India. She is Asia Vice Chair and Steering Committee member in the Commission on Ecosystems Management of IUCN; UNESCO-TWAS Associate and Visiting Scientist at CIFOR-ICRAF.

Part II

What Is Foraged?

Rampant Urbanization, Loss of Green Spaces, Depleting Foraging Wisdom for Nutrition, Health, and Protecting Urban Greenscapes: Lessons from Populous Uttar Pradesh, India

6

N. Manika and Shalini Dhyani

Abstract

Globally, urban, peri-urban, as well as rural communities practice foraging for sustenance, natural medicines, materials, and social cohesion. Uttar Pradesh, India's most populous state, like any other state, is facing dramatic economic transformation and urban sprawling. The urban landscape of the state also represents prevailing foraging wisdom that is often overlooked and ignored. In this chapter, we provide an overview of urban foraging trends and practices in some of the major cities in the state of Uttar Pradesh. Local observations as well as interviews with foragers and street vendors who sell these wild products harvested from urban areas helped us gather data on foraged plant species. The use of wild food was also examined in relation to cultural beliefs that still find its relevance in these cities. We observed traditions still having an impact on the type and frequency of urban foraging of wild plants and their consumption. Foraging of wild plants has the potential to promote conservation of biodiversity and urban green spaces, bio-cultural linkages, and multifunctionality of the green spaces while in certain instances also promoting sustainable nature-based income sources. We observed foraging practice providing access to local, nutritious, and

N. Manika (✉)
FloraFauna Science Foundation, Lucknow, Uttar Pradesh, India

Commission on Ecosystem Management (CEM), International Union for the Conservation of Nature (IUCN), Gland, Switzerland
e-mail: n.manika@outlook.com

S. Dhyani
Commission on Ecosystem Management (CEM), International Union for the Conservation of Nature (IUCN), Gland, Switzerland

CSIR-National Environmental Engineering Research Institute (CSIR-NEERI), Nagpur, Maharashtra, India
e-mail: s_dhyani@neeri.res.in

S. Dhyani, M. Sardeshpande (eds.), *Urban Foraging in the Changing World*,
https://doi.org/10.1007/978-981-97-0345-6_6

culturally relevant food sources while also fostering an understanding of the environment and an appreciation of local biodiversity. In addition, foraging also provided meaningful livelihoods that are nature based, which promote equity and social justice and create multifunctional green spaces in urban areas. The concept of urban foraging in Uttar Pradesh is supported by a sense of belonging, place-based attachment, nature connectedness, and bio-cultural heritage, which has strengthened the dietary diversity of urban dwellers, but threat to urban foraging has been observed from shrinking urban green spaces. The chapter attempts to bring attention to the relevance of green spaces in the context of one of the populated states of the country with strong ties with nature and urban forging relevance. The study stresses the need of promotion of urban foraging to improve place-based attachment for conserving remaining islands of biodiversity and lesser known wild edibles in the cities.

Keywords

Urban foraging · North India · Urban sprawls · Wild edibles · Ecosystem services Bio-cultural diversity · Provisioning · Cultural · Sustainable cities

6.1 Introduction

Sustainability depends increasingly on how well the key elements of any landscape, namely the communities that reside there and the natural resources that support them, are managed and live in harmony with nature. The importance of this increases when it comes to low-income and lower-middle-income underdeveloped and developing countries, as they are projected to have the fastest pace of urbanization (UN 2019; UNPFII 2015). Approximately 56% of the world's population lives in cities. Urban areas are booming and are expected to have 6.3 billion inhabitants by 2050. It would be 2.5 billion more than the present situation, increasing by 65% (Ritchie and Roser 2018; UN Habitat 2022).

Due to accelerated urbanization, India's land cover and land use are changing dramatically. The country's urban population was only 17% just a few decades ago. As of today, 33% of Indians live in urban areas. By 2036, 600 million people will be living in large cities in India, representing 40% of the population (Athar et al. 2022).

Researchers, policymakers, and city residents have long focused on the future of cities, particularly the diverging demographic, economic, social, environmental, and policy paths that will lead to more sustainable outcomes. Providing adequate space for nature in these growing cities is essential to ensuring an unrestricted flow of ecosystem services for human well-being (Dhyani et al. 2018; Dhyani et al. 2021; Das et al. 2022; Behera et al. 2022). Urban dwellers often turn to nature for food security, undernourishment, livelihood support, etc., with activities such as urban farming, terrace gardening, NTFP collection, and foraging for wild edibles. Many people all over the world practice urban foraging, which includes gathering fruits, fungi, herbs, craft materials, and wood from urban green spaces (Shackleton et al.

2017; Dhyani et al. 2019). In the urban environment, foraging can support and provide cultural, economic, and recreational needs (McLain et al. 2014; Dhyani and Kadaverugu 2020).

Foraging is a fairly common practice across the world, and there are increasing studies on urban foraging of wild foods worldwide (Shackleton et al. 2017; Garekae and Shackleton 2020a, b), but there is an under-explored, understudied, and under-represented facet of the alternative food security from wild resources in urban areas (Landor-Yamagata et al. 2018; Roberts and Shackleton 2018). Wild foods are known to contribute towards the food security of an estimated one billion people globally, and they contribute to sustainable and climate-resilient livelihoods specifically with respect to their safety net function during food shortages (Paumgarten et al. 2018; Misra et al. 2008).

Wild edibles growing in urban and peri-urban green spaces are sources of immense provisioning and cultural ecosystem services (Basu et al. 2021). India has more than 53 urban agglomerates, and people have their special interests in wild edibles. Urban foraging provides nutritional support to many poor people with insufficient options for nutritional and livelihood support. So far, the practice has been poorly understood and goes unacknowledged by urban planners, authorities, or even researchers. This chapter is an attempt to bring insights to urban foraging of wild edibles (fungi, leafy edibles, fruits, etc.) that mostly go unnoticed and unacknowledged in Indian context. The knowledge is fast depleting because of busy lifestyles, shrinking urban green spaces, and fast-depleting traditional knowledge due to lagging intergenerational partnerships (Zurba et al. 2020).

There has been a growing interest of researchers in the practice of gathering and harvesting wild edibles from urban green spaces due to the fascinating dimensions of the activity. Foraging helps to foster socio-ecological resilience and a sense of nature connectedness for the community in addition to supporting food security and food sovereignty (Dhyani and Kadaverugu 2020).

In India, foraging has mostly been observed as an individual or group activity, that facilitates community interaction and inclusion, highly recommended for integrated and responsive mental health in people (WHO 2013a, 2013b). It has been observed that people started talking about what was growing nearby during the COVID pandemic in India, leading to a dialogue about how to explore, rediscover, and incorporate local greens into diets, ultimately enhancing their mental stimulation and happiness and shifting their attention away from depressing pandemic news (Athira 2020) (Fig. 6.1).

Ecosystems and species are intrinsically linked to people in Indian ethos, culture, and traditional beliefs (Dasgupta et al. 2021). There is an inextricable connection between the components of biodiversity in this culture and its spiritual and social foundation.

In North Indian states, rural-urban migration is on the rise as a result of rural poverty, land fragmentation, increasing industrialization that promotes strong desire for upward economic mobility, and an attraction to the city (Babu et al. 2017). Since most migrant laborers do not have access to power or fuel when they arrive in cities (Rao et al. 2020), they collect food supplies and fuelwood from urban green spaces

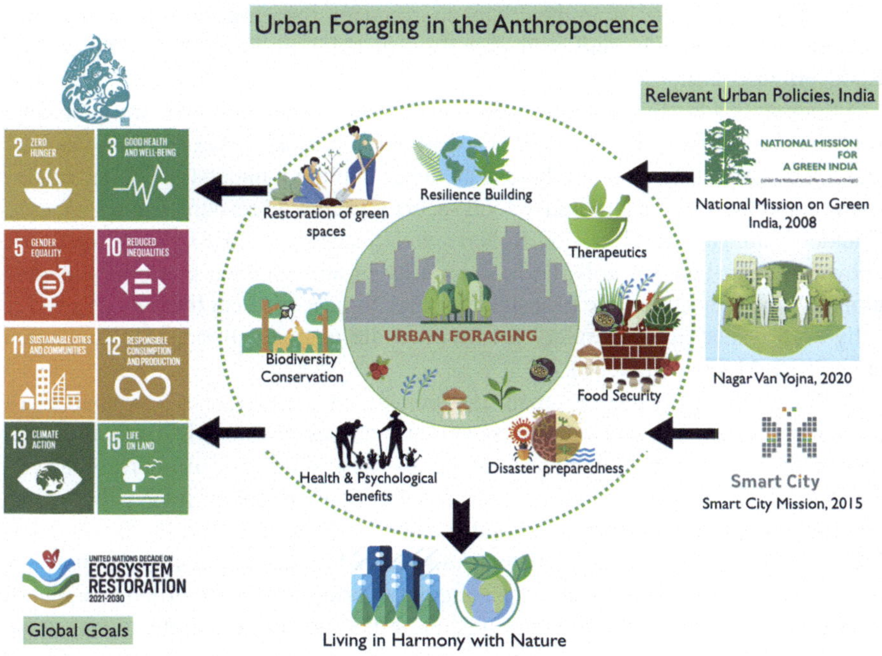

Fig. 6.1 Depicting the importance of urban foraging for fast-expanding urban sprawls of India in the warming Anthropocene

in the early settlement days. In order to meet their daily needs, it becomes vital to design cities that are conducive to foraging (Manika 2022).

The plains of Uttar Pradesh are ideal agricultural zones due to their alluvial soil. Socio-economic development in the region has been characterized by rapid urbanization processes, demographic growth, and also technological as well as industrial advancements, creating a complex mosaic of human-nature relationships (Indian National Commission on Population 2020) that is supported by a drastic change in land use and land cover in the region leading to subsequent transformations.

In light of the rapid expansion of urban sprawls in this landscape, this chapter aims to understand the current practices of urban foraging in Uttar Pradesh, a prominent, large, and highly populated state of India, and further provide insights on how these practices can contribute to a sustainable future.

This chapter brings insights into local attempts to secure bio-cultural diversity, nature connectedness, and sense of belonging while appreciating the dietary benefits of wild edibles. Since there is insufficient information available on urban foraging in India, this chapter attempts to bridge that gap and understand the different socio-ecological contexts that can support conservation of lesser known wild edibles. In addition to providing local evidence for urban foraging in sustainable urban

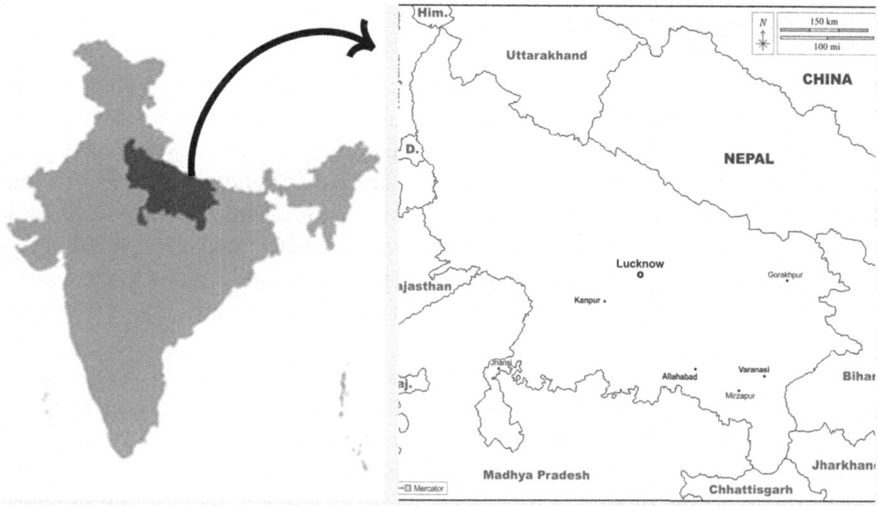

Fig. 6.2 Map showing the locations of the cities in the state of Uttar Pradesh (Map not to scale)

planning, the study also fills the gap by providing ethnographic examples of foraging in India.

India is not only one of the most biodiverse countries in South Asia, but also a fast-growing economy with a rapid-paced urbanization that is resulting in lots of rural and urban boundaries disappearing. To understand the context and relevance of urban foraging across growing urban sprawls, observations from some major cities of Uttar Pradesh were consolidated. These include Lucknow, Allahabad, Kanpur, Varanasi, Gorakhpur, Raebareli, and Mirzapur (Fig. 6.2). All these growing sprawls are represented by different kinds of agroclimatic zones and influx of rural people to growing urban areas. Climate of these cities is predominantly tropical.

6.2 Foraging and Cultural Connect

Apart from diverse nutritional and livelihood benefits, urban foraging has also been seen as an important way to keep cultural identities and traditional rituals alive in the region. The culture of the region dates back to one of the earliest city settlements and has coevolved with the region's ecological diversity. In traditional societies, the usage and sustainability of natural resource management are determined by the beliefs and practices of communities, and local cultures are strengthened by their intimate connections to the natural environment that sustains them (Rist et al. 2003). The inhabitants of North Indian states hold strong religious convictions and observe traditional cultural rituals year-round (Tewari 1982; Sati 2021). As a result of the importance of biodiversity to the ceremonies in this region, foraging for offerings and celebration meals is frequently incorporated into the customs (Garg and Singh 2013; Garg and Singh 2014).

6.3 Foraging Patterns

In Uttar Pradesh, foraging wisdom for underutilized wild plants is observed to be directly influenced by local socio-economics of the communities living near natural ecosystems (Prakash et al. 2016). The practices over the years have provided a buffer to marginalized communities and migrants during times of food shortages and have also reduced their exposure to food insecurity. Many wild and uncultivated plants traditionally have been gathered and used for diverse dietary requirements, nutrition, and food supplements in the state (Maheshwari et al. 1986; Singh and Prakash 1994; Prakash and Singh 2001; Prakash et al. 2016).

The culture of foraging for wild edible plants as food and medicine has been fairly common and exercised by the local inhabitants living in rural, urban, and peri-urban settings. Wild plants have particularly been collected from landscapes, which were easily accessible to locals, and the resources collected were used to diversify the diet and satiate taste buds, simultaneously providing valuable nutrition to locals. With the migration of people for livelihood, education, etc. to the urban spaces, the opportunities for foraging became less because of insufficient access and availability of natural green spaces, but foraging knowledge is still intact with elderly and middle-aged locals. The majority of the marginalized and poor people in the urban areas mostly consume monotonous diets consisting primarily of staple cereals and very limited amounts of vegetables and fruits because of economic and availability reasons and insufficient buying power (Sharma et al. 2020). The need for foraging in the region also supports local livelihood interventions for people who sell the foraged plant parts, a practice that has become widespread in the global south where foraging plays a critical role in the survival, livelihood, nutritional support, and developing resilience for the urban poor (Somesh et al. 2021; Manika 2022). With migration of people from rural to urban and peri-urban settings, traditional practices originating from rural settings have also become part and parcel of the urban culture and are still practiced in various forms for urban foraging.

6.4 Material and Methods

6.4.1 Study Area

Observations from some major cities of Uttar Pradesh were consolidated to understand urban foraging across urban sprawl. These include Lucknow, Allahabad, Kanpur, Varanasi, Gorakhpur, Raebareli, and Mirzapur (Fig. 6.2). Different types of agroclimatic zones characterize these sprawling urban areas, but together, they are identified as major cities of eastern Uttar Pradesh. A tropical climate prevails in these cities.

6.4.2 Interviews

To understand the urban foraging context and motivations in urban sprawls of Uttar Pradesh, we share our observations and experiences of wild edible harvests from urban green spaces (UGSs) and urban forests (UFs) based on our observations and interactions with informal foragers and street vendors selling wild edibles. We conducted semi-structured interviews with 48 foragers in the selected cities. These include the people who were found in active foraging during the survey as well as the vendors who were selling some lesser known wild edibles. 68.75% of the respondents were females, which may not represent the entire pool of urban foragers. As part of the study, observations were made of which species were available for sale in the city along with interview data. Consent was taken to record the interview as audio files, and further it was transcribed in MS Word. The following questions were asked:

Are there plants that you gather in your city aside from in a private garden that can be directly eaten or used for other purposes like medicine or fuelwood?
How did you get into foraging?
Is there anyone else in your family who forages? Did you learn from the person?
Do you harvest primarily for your own use? Do you sell what you forage?
Which species do you collect?
Do any of these species have cultural/religious value?
What are the places where you harvest?
Demographics/background information on (a) age, (b) sex, (c) nativity, and (d) time since they are living in the city.

6.4.3 Taxa Recording, Identification, and Compilation

Notes were recorded on frequently and commonly foraged wild species, parts used, and uses for the harvested species. Botanical, Hindi, and other local vernacular names were added to the information. In addition, specific food items prepared from harvested plant parts were also recorded (e.g., pickle or vegetable preparations). Names of a variety of herbs, shrubs, trees, climbers, and fungi were reported with common or vernacular names. The most common method to identify the species was by common vernacular name and identification from http://www.flowersofindia.net/. The identities of plants were also cross-checked from Common Indian Wild Flowers (Kehimkar 2000), Jungle Trees of Central India (Krishen 2013), and Trees of Uttar Pradesh (Chaudhary et al. 2016). The list of the urban foraged plants includes the genera, species, family, common name, part used, and origin. If there was any uncertainty for identification beyond genus level, the identification was limited to the genus level. Usage category of the species was limited to food (either raw or cooked) to keep the focus of the research paper limited to urban foraging ensuring food security and nutritional requirements of growing urban areas in India.

6.4.4 Salient Quotes

Interviewing each forager in depth created a vivid and complex picture of the situation. To illustrate specific points, pertinent remarks and passages were transcribed.

6.5 Results

6.5.1 Current Foraging Trends

It was observed that most of the urban residents have their second or third generations living in growing urban sprawls, and hence, their food habits are still very traditional. In response to a question about her own food habits, a female vendor selling a lesser known edible fruit "Kaitha" (*Limonia acidissima*) described it this way:

> *Kaitha* was readily available everywhere when we were kids, and we ate it raw. It was often used in side dishes like *chutney* by the elders during seasonal times. Though it is no longer easily available and children cannot luxuriously afford it, making *chutney* out of it is still a practice that persists today.

In urban or peri-urban areas, it is very easy to find wild edibles being sold by vendors (Fig. 6.3) or a person gathering flowers, buds, fruits, leaves from trees, or herbs for personal consumption. Urban foraging is a very valuable process because the nutrition and quality of the food come at no cost to the forager. Additionally, foraging promotes a sense of place and connection to nature.

Fig. 6.3 Pictures showing wild edibles foraged from urban green spaces sold in local urban street markets: (**a**) *Averrhoa carambola* and *Physalis* spp., (**b**) locally collected beans and foraged *Limonia acidissima*, (**c**) man selling foraged *Phyllanthus emblica*. *Picture credit: Authors*

The personal motivation of locals in foraging wild food from different urban green spaces or urban forests has been varied. One of the foragers who was collecting "Sahjan/drumstick" (*Moringa oleifera*) leaves told:

> For my special dishes, I collect Sahjan leaves throughout the year. Despite fruits being consumed by most people, my family is familiar with the fantastic flavour of the leaves, so I collect them regularly.

On the other hand, a teenage boy collecting "Shahtoot" (*Morus alba*) excitedly stated, "This is the long fruit variety and I will get higher rates in the market for this one."

Most of the respondents reported free availability of wild edibles in their neighborhood as one of the important reasons to harvest and include them in their dietary systems. Locals who were earning some alternative economic benefits from wild edibles harvested from urban green spaces were mostly people seated in slums of the urban sprawls belonging to marginalized poor families and women. Women having more traditional knowledge to utilize the wild edibles in various food preparations were also involved in selling them in local vegetable markets for small economic benefits (Fig. 6.3).

6.5.2 Common Urban Foraging Grounds

Urban foraging is reported to take place from a variety of urban green spaces that include managed as well as unmanaged, and public as well as private, and is not limited to parks and urban forests, abandoned areas and wastelands, roadside trees, etc. (Fig. 6.4). Foraging is not necessarily considered a labor-intensive task. The socio-economic importance of home gardens, schoolyards, and backyard foraging has been bolstered by their significant contribution to the region's households. Fast-shrinking UGS, viz. public parks, playground, fallow lands, avenue/roadside plantations, wastelands, urban forests, close to wetlands, and campus forests, are important areas for gathering wild edibles. Nearly 100% of respondents said that they collect edibles along roadsides. A total of 22 respondents said that they look for wild edibles near campus boundaries like universities and colleges. Most of these spaces are used by people from all walks of life to gather wild edibles. More harvest of wild edibles takes place from unmanaged green areas than managed green spaces, as unmanaged public green spaces are more accessible to the marginalized and migrants too. Locations close to traffic and polluted areas such as solid waste dumping site or industrial units are mostly observed to be avoided for foraging. Winter, spring, and autumn were the preferred seasons for urban foraging; however, summer foraging also takes place, especially from fruit-yielding trees.

6.5.3 Preferred Foraged Plant Species

Our study identified 46 wild plants gathered by urban locals belonging to 25 families, many of them angiosperms (Table 6.1). Fabaceae was the most dominant family, represented by more than 17% of the foraged species, followed by Moraceae (11%), Rutaceae (8%), and Sapotaceae (6%) (Fig. 6.5 and Table 6.1).

Fig. 6.4 Urban green spaces commonly used for foraging: (**a**) community parks, (**b**) roadsides and avenues, (**c**) wetlands and waterbodies, (**d**) unconstructed and unmanaged green spaces in abandoned plotted areas. *Picture credit: Authors*

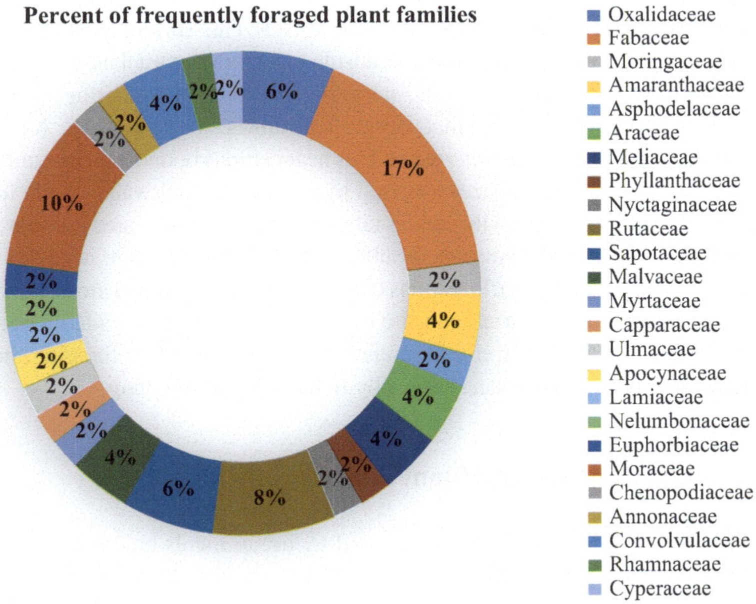

Fig. 6.5 Percentage of important plant families that contribute to urban foraging in UP (species = 46, families = 25)

Table 6.1 Frequently forged plant species in urban sprawls of Uttar Pradesh state of India

Sr. No.	Common Name	Botanical name	Family	Plant part used	Native/ exotic
1	Harita-manjari	*Acalypha indica* L.	Euphorbiaceae	Leaves, stem	Native
2	Bel	*Aegle marmelos* (L.) Corrêa	Rutaceae	Fruit	Native
3	Aloe vera	*Aloe vera* (L.) Burm.f.	Asphodelaceae	Leaves	Exotic
4	Wild chaulai	*Amaranthus viridis* L.	Amaranthaceae	Leaves	Exotic
5	Lal Saag	*Amaranthus tricolor* L.	Amaranthaceae	Leaves	Native
6	Shareefa	*Annona squamosa* L.	Annonaceae	Fruit	Exotic
7	Suran/ Jimikand	*Amorphophallus paeoniifolius* (Dennst.) Nicolson	Araceae	Tubers	Native
8	Kathal	*Artocarpus heterophyllus* Lam.	Moraceae	Fruit	Native
9	Badhal	*Artocarpus lacucha* Buch. Ham.	Moraceae	Fruit, stem bark, leaves	Native
10	Amrakh	*Averrhoa carambola* L.	Oxalidaceae	Fruit	Exotic
11	Neem	*Azadirachta indica* A. Juss.	Meliaceae	Flowers, leaf, bark, twig, gum	Native
12	Kachnar	*Bauhinia variegata* (L.) Benth.	Fabaceae	Flower, buds, stem, roots, bark, seeds, leaves	Native
13	Punarnava	*Boerhavia diffusa* L.	Nyctaginaceae	Leaves	Exotic
14	Palash	*Butea monosperma* (Lam.) Taub.	Fabaceae	Flowers	Native
15	Semal	*Bombax ceiba* L.	Malvaceae	Buds	Native
16	Ker	*Capparis decidua* (Forssk.) Edgew.	Capparaceae	Fruit	Native
17	Karonda	*Carissa carandas* L.	Apocynaceae	Fruit	Native
18	Chakotra	*Citrus maxima* L.	Rutaceae	Fruits	Native
19	Aprajita	*Clitoria ternatea* L.	Fabaceae	Tender pods, flowers	Exotic
20	Bathua	*Chenopodium album* L.	Chenopodiaceae	Leaves	Exotic
21	Arvi/Patta	*Colocasia esculenta* (L.) Schott	Araceae	Leaves, tuber	Exotic
22	Singhara	*Eleocharis dulcis* *Trinius cv. Tuberosa* (Roxb.)	Cyperaceae	Fruits	Native
23	Udumbara	*Ficus glomerata* Roxb.	Moraceae	Flowers and fruits	Native
24	Gular	*Ficus racemosa* L.	Moraceae	Flowers	Native
25	Phalsa	*Grewia asiatica* L.	Malvaceae	Fruit	Native

(continued)

Table 6.1 (continued)

Sr. No.	Common Name	Botanical name	Family	Plant part used	Native/ exotic
26	Chilbil	*Holoptelea integrifolia* (Roxb.) Planch	Ulmaceae	Seeds	Native
27	Jangli shakarkand	*Ipomoea batatas* (L.) Lam.	Convolvulaceae	Tuber	Exotic
28	Naari ka Saag	*Ipomoea aquatica* Forssk.	Convolvulaceae	Leaves	Native
29	Kaitha	*Limonia acidissima* L.	Rutaceae	Fruit	Native
30	Sahjan	*Moringa oleifera* Lam.	Moringaceae	Leaves, roots, seed, bark, fruit, flowers, and immature pods	Native
31	Kadi Patta	*Murraya koenigii*	Rutaceae	Leaves	Native
32	Khirni	*Manilkara hexandra* (Roxb.) Dubard	Sapotaceae	Fruit	Exotic
33	Mahua	*Madhuca longifolia* (J.Konig) J.F.Macbr.	Sapotaceae	Flower, fruit, leaves	Native
34	Shahtoot	*Morus alba* L.	Moraceae	Fruits	Native
35	Lotus	*Nelumbo nucifera* Gaertn.	Nelumbonaceae	Roots, seeds	Exotic
36	Khatti Buti	*Oxalis corniculata* L.	Oxalidaceae	Leaves	Exotic
37	Sorel	*Oxalis hispidula* Zucc.	Oxalidaceae	Leaves	Exotic
38	Wild amla	*Phyllanthus emblica* L.	Phyllanthaceae	Leaves, fruits	Native
39	Jungle Jalebi	*Pithecellobium dulce* (Roxb.) Benth.	Fabaceae	Flower, bark	Exotic
40	Ajwain patta	*Plectranthus amboinicus* (Lour.) Spreng.	Lamiaceae	Leaves	Exotic
41	Agati	*Sesbania grandiflora* (L.) Pers.	Fabaceae	Flowers	Native
42	Sanai	*Senna tora* (L.) Roxb.	Fabaceae	Seeds	Exotic
43	Jamun	*Syzygium cumini* L.	Myrtaceae	Fruits	Native
44	Imli	*Tamarindus indica* L.	Fabaceae	Fruit/leaves	Exotic
45	Clover	*Trifolium* L.	Fabaceae	Leaves, stem, pod, flower	Exotic
46	Ber	*Ziziphus mauritiana* Lam.	Rhamnaceae	Fruits	Native

There are many more plants that could not be identified as they were harvested during other seasons than when the study was conducted. Nonavailability of their pictures and unclear taxonomical description were other major reasons.

Some of the frequently reported foraged plants in urban sprawls of the state were *Moringa oleifera* (Sahjan), *Murraya koenigii* (kadi patta), *Tamarindus indica* (Imli), *Ziziphus mauritiana* (wild ber), *Syzygium cumini* (Jamun), *Carissa carandas* (Karonda), *Artocarpus heterophyllus* (Kathal), *Aegle marmelos* (Bel), *Averrhoa carambola* (Amrakh), etc. These are gathered for personal consumption as well as to be sold in local markets by the urban poor to earn their livelihood. More than 60% of the respondents agreed that they gather these species from urban settings.

It was found that a lot of school-going children used to prefer *Holoptelea integrifolia* (Chilbil) seeds during later spring and early summers. However, loss of the trees from green spaces is also taking away foraging of these seeds that had immense nutritional benefits.

Previously, *Madhuca longifolia* (Mahua), *Bombax ceiba* (Semul), *Bauhinia variegata* (Kachnar), *Sesbania grandiflora* (Agati), *Artocarpus heterophyllus* (Badhal), *Ficus racemosa* (Gular), *Pithecellobium dulce* (Jungle Jalebi), and *Limonia acidissima* (Kaitha) were some of the lesser known wild edibles that were preferred for consumption, but due to loss of green spaces and lesser presence of these plants, foraging of these plant parts has also drastically reduced. However, there are still families who like to collect and make traditional recipes out of them (Fig. 6.6).

Foraged plants are mostly cooked before consumption, but there are a few that are also consumed raw and are mostly fruits, viz. *Moringa oleifera* (Sahjan),

Fig. 6.6 Some wild edibles frequently foraged from urban locations: (**a**) *Grewia asiatica,* (**b**) *Limonia acidissima,* (**c**) *Pithecellobium dulce,* (**d**) *Phyllanthus emblica,* (**e**) *Artocarpus heterophyllus,* (**f**) *Acalypha indica,* (**g**) *Moringa oleifera,* (**h**) *Nelumbo nucifera,* (**i**) *Madhuca longifolia,* (**j**) *Averrhoa carambola,* (**k**) *Oxalis corniculata,* (**l**) *Ipomoea aquatica,* (**m**) *Ziziphus mauritiana,* (**n**) *Bombax ceiba,* (**o**) *Colocasia esculenta,* (**p**) *Butea monosperma,* (**q**) *Amorphophallus paeonifolius,* (**r**) *Chenopodium album,* (**s**) *Bauhinia variegata,* (**t**) *Murraya koenigii. Picture credit: Authors*

Tamarindus indica (Imli), *Ziziphus mauritiana* (wild ber), *Syzygium cumini* (Jamun), *Averrhoa carambola* (Amrakh), *Manilkara hexandra* (Khirni), *Limonia acidissima* (Kaitha), *Artocarpus heterophyllus* (Kathal), *Aegle marmelos* (Bel), *Pithecellobium dulce* (Jungle Jalebi), *Madhuca longifolia* (Mahua), *Grewia asiatica* (Phalsa), *Morus alba* (Shahtoot), etc. Foraged from urban green spaces, most wild plants were observed to be prepared in traditional ways, and knowledge was passed from one generation to forthcoming generation, and an important intergenerational partnership was observed in conserving this knowledge base. Food preparations from foraged plants are not limited to fritters, pickles, and jams. Foraged plant parts are also used in preparing the main course as well as many side dishes. For example, "choka" made from *Amorphophallus paeoniifolius* (Jimikand) is a very popular dish in even urban areas of eastern Uttar Pradesh like Gorakhpur. Likewise, "sherbet" made from fruits of *Aegle marmelos* (Bel) and flowers of *Butea monosperma* (plash) is very popular and is preferred during summer and spring, respectively.

Women in the region are especially adept at foraging wild species and have developed efficient systems for identifying, harvesting, and preparing wild foods (Singh and Singh 2009). A few traditional recipes from foraged plants in urban sprawls of India are presented in Fig. 6.7.

6.5.4 Customary Consumption of Foraged Species

In Uttar Pradesh, celebrations and accompanying rituals are intricately tied to the consumption of indigenous foods. Elephant yam or Suran or oal (*Amorphophallus paeoniifolius*), for instance, is a must-be-prepared vegetable or pickle for "Diwali," one of the biggest Indian festivals. According to numerous religious beliefs, this tuber is considered sacred since it grows in the soil and some of its corms stay in the soil after harvesting that gives rise to the plant in the next season. Since Diwali also represents growth and preservation of wealth and health, this vegetable is seen as auspicious and is believed to bring prosperity. There is a tradition of eating the early

Fig. 6.7 Traditional recipes from foraged species. (**a**) Vegetable made out of *Ipomoea aquatica* leaves. (**b**) Traditional vegetable from *Bauhinia variegata* tender pods. (**c**) "Choka" made out of *Amorphophallus paeoniifolius* tuber. *Picture credit: Authors*

buds of Semul (*Bombax ceiba*) in the "Terai" region including urban sprawls in the area because of local and cultural beliefs. According to a female respondent:

"For a newlywed bride, preparing the season's first 'Semul bud curry' is a must. It symbolizes the arrival of the new season and good news for the coming year."

Also, during and after "Amla Ekadashi," it is mandatory to start consuming Amla (*Phyllanthus emblica*) on a regular basis throughout the winters in many forms as it is a rich source of vitamin C and helps keep away colds.

During spring, mahua (*Madhuca longifolia*) is cherished during its blooming and fruiting season. Juice of mahua flowers is extracted to prepare desserts. Tender fruits are used as a vegetable to accompany flatbread. Locally, it is known as kolaiya. Even though mahua fruits themselves have no flavor, mahua flower dishes are delicious. Mahua fruit is especially popular with the elderly because of its health benefits.[1]

As the summers arrive, sherbet (drink) made from palash (*Butea monosperma*) blossoms is prepared during "Holi," the celebration of colors. This beverage aids digestion and is a coolant; thus, it may have been customary to take it during this summer event. Likewise, Bel (*Aegle marmelos*) and Phalsa (*Grewia asiatica*) are commonly used as summer drinks during the peak of the season. In contrast to the sweeter bel *sherbet*, the phalsa *sherbet* is a tangy concoction made from a variety of spices and sugar. Since both of these fruits are coolants and have heat-resistant properties, they provide respite from harsh hot and dry summers in the region.

Monsoon brings some other customary cuisines with it. During late summers and early monsoons, Karonda (*Carissa carandas*) is eaten raw as well as in cooked form and is also being preserved in the form of pickles to be used throughout the year. Certain fasting rituals, such as "harchat," necessitate the consumption of Naari ka Saag (*Ipomoea aquatica*) with pseudocereals and curd (Fig. 6.6). This is a wonderful approach to vary the diet for micronutrients, and *Ipomoea aquatica* is very much foraged by urbanites who still follow traditional customs in the state. Table 6.2 illustrates the customary consumption of some important foraged species and their time of consumption. Thus, in this way, many species are consumed throughout the year in connection with a variety of rituals. The promotion of locally wild-harvested plants can help to protect habitats. Furthermore, it promotes nature connectedness and place-based attachment among locals, which in turn facilitates the conservation of biodiversity and enhances ecosystem services. If the demand for these products continues, it may result in the conservation of these species because the supply chain will have to be maintained to meet the demand, and this is only possible if the main resource is protected. This will lead to better protection of their habitat, as well as stricter regulations, which will help the population of these species remain stable. In addition, the economic incentives of supplying these products will likely result in investments in sustainable practices. For instance, it was agreed by all foragers that protecting nature is crucial to sustaining supplies of resources, which in turn can assist human health. 19 foragers suggested partnering directly with cafeterias or canteens for some of the seasonal species in order to ensure fair trade and better income opportunities.

[1] https://www.downtoearth.org.in/news/food/intoxicatingly-delicious-57241

Table 6.2 Customary consumption of forged plant species during festivities and religious rituals in Uttar Pradesh followed and practiced in fast-expanding urban areas

Sr. No.	Occasion/ festival	Month/ period	Local name	Scientific name	Part eaten
1	Vasant panchami	February–March	Ber	*Ziziphus mauritiana* Lam.	Fruits
2	Amalki Ekadashi	March	Amla	*Phyllanthus emblica* L.	Leaves, fruits (food is cooked under and/or near amla trees)
3.	Holi	March–April	Kathal	*Artocarpus heterophyllus* Lam.	Fruits
4	Holi	March–April	Palash	*Butea monosperma* (Lam.) Taub.	Flowers
5	Mesh Sankranti	April	Bel	*Aegle marmelos* (L.) Corrêa	Fruits
6	Aashad Amavasya	May–June	Neem	*Azadirachta indica* A. Juss	Ripe fruits
7	Naag panchami	July–August	Neem	*Azadirachta indica* A. Juss	Tender leaves
8	Harchat	August–September	Naari ka Saag	*Ipomoea aquatica* Forssk.	Leaves
9	Harchat	August–September	Kamalgatta/ Kamal kakdi	*Nelumbo nucifera* Gaertn.	Seeds/fruits/ stem
10	Navratri	September–October	Kamal kakdi	*Nelumbo nucifera* Gaertn.	Stem
11	Diwali	October–November	Suran/ Jimikand	*Amorphophallus paeoniifolius* (Dennst.) Nicolson	Tuber
12	Govardhan Pooja	October–November	Arvi/Patta	*Colocasia esculenta* (L.) Schott	Tuber
13	Kartik Poornima	October–November	Shareefa	*Annona squamosal* L.	Fruits
14	Devuthani ekadashi	October–November	Shareefa	*Annona squamosal* L.	Fruits
15	Chath pooja	October–November	Chakotra	*Citrus maxima* L.	Fruits
16	Chath pooja	October–November	Singhara	*Eleocharis dulcis Trinius cv. Tuberosa* (Roxb.)	Fruits
17	Chath pooja	October–November	Shareefa	*Annona squamosal* L.	Fruits

6.5.5 Foraging for Livelihood and Living

Due to increasing urbanization, altering lifestyles, and Internet commerce, India in general is a very desirable market for multinational food merchants. Nonetheless, the majority of urban customers purchase food from local, family-run businesses, fresh markets, and sellers with pushcarts (Mittal 2018), and this is why the foraged foods are sold for a profit that helps sustain the family. In the region, foraging and thereby selling the produce in small and big markets are a common component of a suite of several seasonal livelihood choices that supplement household income. However, as everywhere else, in this region, foraging is not regarded as a primary source of income, but rather a supplement to other activities. Fruits such as Imli (*Tamarindus indica*), Sahjan (*Moringa oleifera*), Amrakh (*Averrhoa carambola*), Jamun (*Syzygium cumini*), Shahtoot (*Morus alba*), and Desi ber (*Ziziphus mauritiana*) are generally foraged within the city limits and sold in the market to supplement the household income. Nevertheless, a few years ago, the vendor basket was considerably richer and included diverse species like Jungle Jalebi (*Pithecellobium dulce*), Kaitha (*Limonia acidissima*), Badhal (*Artocarpus lacucha*), and others that are now hard to find.

As a result of the high purchasing power of the metropolitan populace, the approach of selling the local foraged species enhances the odds for sustaining a living in urban regions. It cannot be disregarded that it provides a unique source of income for women by allowing them to save and earn money. By accumulating financial wealth, urban women from disadvantaged socio-economic origins can achieve self-sufficiency and control over their home economy. It contributes to achieving gender equality to some extent.

Additionally, urban foraging has also helped to improve the physical health of city dwellers. A total of three respondents said that they sometimes collect medical plants and have traditional knowledge about how to treat common ailments. This decentralized approach to medicine may help people who are unable to access healthcare institutions for a variety of reasons. Furthermore, there are therapeutic and enriching advantages to the process of foraging. In the global North, during the past few decades, urban green spaces have been seen and constructed primarily to provide recreation and calm (Brandner and Schunko 2022), which should be adapted in the region as the cities expand in global south specially in cities of India that are going to be highly urbanized by 2050.

6.6 Discussion

6.6.1 Challenges to Foraging in Expanding Cities

The practice of foraging is central to societal well-being as well as the motivation to preserve natural resources. Popularity and sustainability of this practice are challenged by several varied, unpredictable, and frequently conflicting challenges coming from damaged landscapes, deteriorating cultural values, climate change, and regional population expansion. If we talk in context of the cities in Uttar Pradesh, there is a major loss in the urban green spaces in the region. Urban agglomerations in the state have not performed well in the ClimateSmart Cities Assessment Framework (CSCAF) (Abbasi 2022).

Land-use change studies show a drastic increase in built-up areas in the cities of this region. For instance, between 1985 and 2015, the land use in Lucknow saw some dramatic changes where the built-up area increased from 28.1% to 77.4%. This change has led to a loss of biodiversity due to the conversion of vegetation to built-up areas (Verma and Garg 2022). Eastern Uttar Pradesh has also seen rapid population growth in the last decade, and the fate is similar for many other fast-growing tier II and tier III cities (Dhyani et al. 2021; Misra and Mishra 2017).

The rapid expansion of urban centers has led to unsustainable urbanization altering land use, degraded groundwater, eaten-up green spaces (Raju et al. 2022), increased extreme pollution, and exacerbated economic disparities (World Bank 2016). There is a need to rethink conventional ways of planning and designing urban open spaces in the landscape of north Indian cities like Uttar Pradesh, based on the conditions of urban poverty in the North Indian cities and their growing population pressure (Ghose 2013; Mukherjee and Chatterjee 2019). Now with the recent decision that the geographical boundaries of the development authorities in these major cities will be expanded along with proper planning (The Pioneer 2022), the town planners should include the provision of properly planned UGS to save the knowledge and culture of biodiversity-related traditional foraging and conservation practices.

Planning of all kinds still overlooks hunger and food security as important considerations in urban planning of all kinds (Tacoli 2017). By incorporating edible species into urban green areas and by improving access to greenery in the form of UGS, nutrition and well-being can be improved. In order to address sustainability and land efficiency, urban planners should place emphasis on conservation, rejuvenation, and improving blue and green cover. Considering the complex social-ecological context within which foraging food has been embedded, it is evident that to revitalize the traditional practices like foraging, it is imperative to integrate development and nature, as it not only provides food and nutrition security, but also mental and emotional well-being. By combining development and nature, we not only gain tangible benefits but also ensure the continuation of important cultural practices and beliefs.

6.6.2 Way Forward: Future of Urban Foraging

The ClimateSmart Cities Assessment Framework (CSCAF) of the Indian Ministry of Housing and Urban Affairs (MoHUA) has, for instance, a theme of urban planning, green cover, and biodiversity that focuses on nature-based solutions and conservation of green blue ecosystems in cities in order to mitigate and adapt to climate change. In this way, these urban centers have hope for the future.

Green infrastructure programs have proliferated in cities across the nation under which climate-smart cities are aimed at creating a green mindset in cities to incentivize a holistic, climate-responsive development. The initiative of ClimateSmart Cities Assessment Framework (CSCAF) of the Ministry of Housing and Urban Affairs (MoHUA) in India, has, for instance, a theme of urban planning, green cover, and biodiversity that focuses on nature-based solutions and conservation of green blue ecosystems in cities in order to mitigate and adapt to climate change. In this way, these urban centers have hope for the future. A similar program, the India Cooling Action Plan[2] (ICAP), was implemented to provide sustainable cooling and thermal comfort for all while securing environmental and socio-economic benefits to society. As an added contribution, the Nagar Van scheme,[3] to develop 500 urban forests, has been implemented across the country recently. These structured implementations could be instrumental in creating a much-needed blue green infrastructure across the Indian cities and could aid in the edible sustenance (Fig. 6.1).

Promotion of publicly managed gardens in the local community contributes to the conservation of local biodiversity in a manner that provides food security and economic value to households. There is a need to include food garden systems into nationally recognized conservation as well as nutrition plans since they hold the potential for both conserving species that might otherwise be neglected and providing food of high nutritional value. For instance, the government schools in Uttar Pradesh have launched a program in which students will establish their own kitchen gardens and use the produce for their midday meals (Agarwal 2019). With the increasing number of apartment dwellers who have limited space, one can opt for either vertical gardens or terrace gardens (Sofo and Sofo 2020), which may be expensive for some, or they can go foraging in nearby green spaces, which offer a wide variety of plants and does not require monetary input.

In local context of the fast-expanding state and also one of the most populous states, foragers and public parks bring some positive concerns that can help to co-manage urban green spaces as a viable option (Hajzeri and Kwadwo 2019) if these fast-growing cities are able to include edible plants into urban spaces. However, due to the interest of the local marginalized population and the pressure on these trees for alternative and direct livelihood benefits, there are chances that this free resource can also be exploited at large. Though community participation is key to building long-term resilience, co-designing solutions can help fostering and inculcating a sense of ownership. Foraging and greening possibilities can be useful for multiple stakeholder involvement to make focused interventions for addressing sustainable use of wild species in urban settings. Any strategy that empowers and supports community to get involved in sharing their knowledge and experiences for conserving urban green spaces can be considered effective.

By effective decentralization of responsibilities, establishing and understanding local priorities by strengthening the capacity of local authorities to engage urban residents can be helpful in developing foraging safe spaces in cities. Edible city projects around the world have demonstrated significantly greater levels of emancipation and inclusive urban regeneration (Säumel et al. 2019). From this study, it was clear that fast-depleting green patches in urban areas are the main foraging grounds. Urban residents are mostly driven by traditional, cultural, and religious knowledge systems and practices to forage for diverse wild flowers, fruits, leaves, etc. for food, rituals, and other purposes. The practice of foraging

[2] https://pib.gov.in/PressReleaseIframePage.aspx?PRID=1805795#:~:text=The%20India%20Cooling%20Action%20Plan,going%20programmes%2F%20schemes%20of%20the

[3] https://pib.gov.in/PressReleasePage.aspx?PRID=1846954

depends on the availability of specific species in a particular season. Limited access to foraging resources adversely impacts the livelihoods and economic conditions of the urban poor. In some cases, however, the collection may not be sustainable because of factors such as overharvesting or collecting from polluted areas. Therefore, green spaces should be thoughtfully considered when planning urban spaces. The results clearly support that foraging in fast-growing cities of the Global South is driven by traditional rural roots and foundations of urban residents, and they still feel connected to them by finding these resources in their urban settings and foraging for them.

Under conditions of rapid and continuing urbanization, it is increasingly important for the city planners and municipal corporation to reflect on the priorities and needs of urban residents, for promoting a sense of belongingness and connectedness to maintain socio-ecological and economic vibrancy and environmental stability.

6.7 Conclusion

Sustainable edible cities require the convergence of biodiversity, urban environments, and stakeholders to meet the localized SDGs. Continuing to understand the multiple pressures being exerted by rapid urbanization in Uttar Pradesh is crucial for policymakers. However, if the cities of this region are designed in a way that biodiversity is maintained in the novel urban ecosystems, we believe that the fertile land in the region and other biotic and abiotic conditions of the region would facilitate the process. For both urban residents and urban nature, it offers a promising arena that helps reconcile urbanization with biodiverse edible green structures. In this way, sustainably designed cities which preserve green-blue infrastructure (the use of blue elements, like rivers, canals, ponds, wetlands, and floodplains, and green elements, such as trees, forests, fields, and parks) can play an essential role in meeting the Convention on Biological Diversity (CBD) target of stemming biodiversity losses. Citizens' well-being and long-term sustainability are premised on the green spaces in the cities, which offer a suite of regulatory services and nature-based solutions. It has been shown that UGS offers opportunities for residents to interact with nature and spend time in activities such as foraging even if they are not intended for recreational purposes. It is also possible for cities such as Lucknow, Varanasi, Allahabad, and Gorakhpur, which have rich cultures and histories, to lose their charm and local cuisines if these spaces are not properly managed, and therefore it is essential that UGS is properly managed for long-term sustainability of these attributes. The culture of these cities is intertwined with biodiversity, so there would be a loss of cultural aspects as well. In order to maintain urban green spaces for future generations, policymakers and land managers should recognize and encourage foraging. Foraging is an important part of the region, as it provides an opportunity for people to learn about the ecology of their local environment and know more about their customary foods and provides a source of healthy and sustainable diet. Furthermore, encouraging foraging helps to keep green spaces vibrant and healthy, as it encourages people to be active within them.

Acknowledgements First author is thankful to DST-SERB research grant (PDF/2015/000030). Second author is thankful to Knowledge Resource Centre of CSIR-NEERI, Nagpur, for plagiarism check under the number.

Conflict of Interest The authors declare no conflicts of interest.

References

Abbasi A (2022) CSCAF in monitoring urban green spaces. https://niua.in/c-cube/blog/content/cscaf-monitoring-urban-green-spaces. Accessed 13 Feb 2023

Agarwal A (2019) UP Govt schools encourage students to grow kitchen gardens & supplement their mid-day meals. https://www.storypick.com/up-schools-kitchen-garden/. Accessed 13 Feb 2023

Athar S, White R, Goyal H (2022) Financing India's urban infrastructure needs: constraints to commercial financing and prospects for policy action (English). Washington, D.C.: World Bank Group. http://documents.worldbank.org/curated/en/099615110042225105/P17130200d91fc0da0ac610a1e3e1a664d4. Accessed 10 Feb 2023

Athira M (2020) COVID-19: how households in Kerala are foraging for veggies in their own backyard during coronavirus lockdown. The Hindu https://www.thehindu.com/life-and-style/homes-and-gardens/many-families-in-kerala-are-foraging-for-nutritious-vegetables-around-their-houses-during-covid-19-lockdown/article31256597.ece. Accessed 10 Feb 2023

Babu BV, Kusuma YS, Sivakami M, Lal DK, Marimuthu P, Geddam JB, Khanna A, Agarwal M, Sudhakar G, Sengupta P, Borhade A, Khan Z, Kerketta AS, Akoijam BS (2017) Living conditions of internal labour migrants: a nationwide study in 13 Indian cities. Int J Migr Border Stud 3(4):328–351. https://doi.org/10.1504/IJMBS.2017.10006517F

Basu M, Rajarshi D, Pankaj K, Shalini D (2021) Home gardens moderate the relationship between Covid-19-induced stay-at-home orders and mental distress: a case study with urban residents of India. Environ Res Commun 3(10):105002. https://doi.org/10.1088/2515-7620/ac2ab2

Behera SK, Mishra S, Sahu N, Manika N, Singh SN, Anto S, Kumar R, Husain R, Verma AK, Pandey N (2022) Assessment of carbon sequestration potential of tropical tree species for urban forestry in India. Ecol Eng 181:106692. https://doi.org/10.1088/2515-7620/ac2ab2

Brandner A, Schunko C (2022) Urban wild food foraging locations: understanding selection criteria to inform green space planning and management. Urban For Urban Green 73:127596. https://doi.org/10.1016/j.ufug.2022.127596

Chaudhary LB, Kushwaha AK, Bajpai O (2016) Trees of Uttar Pradesh (Part 1). Bennett: Coleman & Co Ltd, Lucknow, India

Das C, Shukla J, Dhyani S (2022) Endorsing city biodiversity index (CBI): assessing ecosystem health in urban sprawls and eco-DRR-inclusive urban planning. In: Dhyani S, Basu M, Santhanam H, Dasgupta R (eds) Blue-green infrastructure across Asian countries. Springer, Singapore, pp 9–20

Dasgupta R, Dhyani S, Basu M, Kadaverugu R, Hashimoto S, Kumar P, Johnson BA, Takahashi Y, Mitra BK, Avtar R, Mitra P (2021) Exploring indigenous and local knowledge and practices (ILKPs) in traditional jhum cultivation for localizing sustainable development goals (SDGs): a case study from Zunheboto District of Nagaland, vol 72. Environ Manage, India, p 147. https://doi.org/10.1007/s00267-021-01514-6

Dhyani S, Dasgupta R, Kumar P, Pujari P, Gupta AK (2019) Exploring the current and future potential of urban agriculture in growing urban sprawls of India: strengths and challenges. Clim Change Environ Sustain 7(1):97–101. https://doi.org/10.5958/2320-642X.2019.00012.7

Dhyani S, Kadaverugu R (2020) Food security and cultural benefits from urban green spaces: exploring urban foraging as a silently growing global movement. Clim Change Environ Sustain 8(2):233–241. https://doi.org/10.5958/2320-642X.2020.00022.8

Dhyani S, Lahoti S, Khare S, Pujari P, Verma P (2018) Ecosystem based disaster risk reduction approaches (EbDRR) as a prerequisite for inclusive urban transformation of Nagpur City, India. Int J Disaster Risk Reduc 32:95–105. https://doi.org/10.1016/j.ijdrr.2018.01.018

Dhyani S, Majumdar R, Santhanam H (2021) Scaling-up nature-based solutions for mainstreaming resilience in Indian cities. In: Mukherjee M, Shaw R (eds) Ecosystem-based disaster and climate resilience. Disaster and risk research: GADRI book series. Springer, Singapore, pp 279–306

Garekae H, Shackleton CM (2020a) Urban foraging of wild plants in two medium sized south African towns: people, perceptions and practices. Urban Forestry Urban Greening 49:126581. https://doi.org/10.1016/j.ufug.2020.126581

Garekae H, Shackleton CM (2020b) Foraging wild food in urban spaces: the contribution of wild foods to urban dietary diversity in South Africa. Sustainability 12(2):678. https://doi.org/10.3390/su12020678

Garg A, Singh V (2013) Mandu sacred grove in upper ganga Ramsar site, Uttar Pradesh. Curr Sci 104:409–410

Garg A, Singh V (2014) Siddhwari sacred grove in Upper Ganga Ramsar site of Uttar Pradesh 105(8):1039–1040

Ghose DS (2013) Urban poverty in Bihar in 21st century: a case study of Patna municipal corporation area. In: Urban poverty: issues and challenges. Rajesh Publication, New Delhi, pp 176–195

Hajzeri A, Kwadwo VO (2019) Investigating integration of edible plants in urban open spaces: evaluation of policy challenges and successes of implementation. Land Use Policy 84:43–48. https://doi.org/10.1016/j.landusepol.2019.02.029

Indian National Commission on Population (2020) Population projections for India and states 2011–2036. Indian Ministry of Health & Family & Welfare, New Delhi, India

Kehimkar ID (2000) Common Indian wild flowers. Bombay natural history society. Oxford University Press

Krishen P (2013) Jungle trees of Central India: a field guide for tree spotters. Penguin Books India

Landor-Yamagata JL, Kowarik I, Fischer LK (2018) Urban foraging in Berlin: people, plants and practices within the metropolitan green infrastructure. Sustainability 10(6):1873. https://doi.org/10.3390/su10061873

Maheshwari JK, Singh KK, Saha S (1986) Ethnobotany of Tribals of Mirzapur District, Uttar Pradesh. India. Economic Botany Information Service, National Botanical Research Institute, Lucknow, India

Manika N (2022) Creating foraging-friendly cities can strengthen people-biodiversity connections. https://www.downtoearth.org.in/blog/wildlife-biodiversity/creating-foraging-friendly-cities-can-strengthen-people-biodiversity-connections-83649. Accessed 11 Feb 2023

McLain RJ, Hurley PT, Emery MR, Poe MR (2014) Gathering "wild" food in the city: rethinking the role of foraging in urban ecosystem planning and management. Local Environ 19(2):220–240. https://doi.org/10.1080/13549839.2013.841659

Misra HN, Mishra A (2017) Urbanization and regional development in eastern Uttar Pradesh: some policy imperatives. Int J Environ Sci Nat Resour 6(3):60–72. https://doi.org/10.19080/IJESNR.2017.06.555687

Misra S, Maikhuri RK, Kala CP, Rao KS, Saxena KG (2008) Wild leafy vegetables: a study of their subsistence dietetic support to the inhabitants of Nanda Devi biosphere reserve. Ind J Ethnobiol Ethnomed 4:15. https://doi.org/10.1186/1746-4269-4-15

Mittal S (2018) The changing face of food retail in India. When food becomes immaterial, confronting the digital age: focus on the Global South. pp 32-39. https://focusweb.org/publications/the-changing-face-of-food-retail-in-india/

Mukherjee N, Chatterjee B (2019) Poverty and inequality in urban India with special reference to West Bengal: an empirical study. In: Jayanthakumaran K, Verma R, Wan G, Wilson E (eds) Internal migration, urbanization and poverty in Asia: dynamics and interrelationships. Springer, Singapore. https://doi.org/10.1007/978-981-13-1537-4_7

Paumgarten F, Locatelli B, Witkowski ETF (2018) Wild foods: safety net or poverty trap? A south African case study. Hum Ecol 46(2):183–195. https://doi.org/10.1007/s10745-018-9984-z

Prakash A, Manika N, Yadav RK (2016). Observations on useful underutilized wild food plants among the Tharu tribal of Uttar Pradesh for sustaining local people, their livelihood and maintaining biodiversity: Uttar Pradesh State Biodiversity Board, pp 31-43

Prakash A, Singh KK (2001) Observation on some threatened plants and their conservation in Rajaji National Park Dehradun, Uttaranchal. J Econ Tax Bot 25(2):363–366

Raju R, Nanda R, Singh A, Malik K (2022) Multi-temporal analysis of groundwater depletion-induced land subsidence in central ganga alluvial plain. Northern India Geocarto Int 37(26):11732–11755. https://doi.org/10.1080/10106049.2022.2060322

Rao N, Narain N, Chakraborty S, Bhanjdeo A, Pattnaik A (2020) Destinations matter: social policy and migrant workers in the times of covid. Eur J Dev Res 32(5):1639–1661. https://doi.org/10.1057/s41287-020-00326-4

Rist S, Delgado F, Wiesmann U (2003) The role of social learning processes in the emergence and development of Aymara land use systems. Mt Res Dev 23(3):263–270. https://doi.org/10.1659/0276-4741(2003)023[0263:TROSLP]2.0.CO;2

Ritchie H, Roser M (2018) Urbanization. OurWorldInData.org. https://ourworldindata.org/urbanization

Roberts S, Shackleton C (2018) Temporal dynamics and motivations for urban community food gardens in medium-sized Towns of the Eastern Cape, South Africa. Land 7(4):146. https://doi.org/10.3390/land7040146

Säumel I, Reddy SE, Wachtel T (2019) Edible City solutions—one step further to foster social resilience through enhanced socio-cultural ecosystem Services in Cities. Sustainability 11:972. https://doi.org/10.3390/su11040972

Sati VP (2021) Major fairs and festivals. In: The Ganges. Springer geography. Springer, Cham, pp 119–127

Shackleton CM, Hurley PT, Dahlberg AC, Emery MR, Nagendra H (2017) Urban foraging: a ubiquitous human practice overlooked by urban planners, policy, and research. Sustainability 9(10):1884. https://doi.org/10.3390/su9101884

Sharma M, Kishore A, Roy D, Joshi K (2020) A comparison of the Indian diet with the EAT-lancet reference diet. BMC Public Health 20(1):812. https://doi.org/10.1186/s12889-020-08951-8

Singh KK, Prakash A (1994) Indigenous phytotherapy among the Gond tribe of Uttar Pradesh, India. Ethnobotany 6:37–41

Singh RK, Singh A (2009) Women's wisdom and indigenous human healthcare practices. Indian J Trad Knowl 8(2):262–269

Sofo A, Sofo A (2020) Converting home spaces into food gardens at the time of Covid-19 quarantine: all the benefits of plants in this difficult and unprecedented period. Hum Ecol Interdiscip J 48(2):131–139. https://doi.org/10.1007/s10745-020-00147-3

Somesh D, Rao R, Murali R, Nagendra H (2021) Patterns of urban foraging in Bengaluru city. Urban For Urban Green 57:126940. https://doi.org/10.1016/j.ufug.2020.126940

Tacoli C (2017) Food (in) security in rapidly urbanising, low-income contexts. Int J Environ Res Public Health 14(12):1554. https://doi.org/10.3390/ijerph14121554

Tewari LG (1982) Women's fasts and festivals from Uttar Pradesh: their art, rituals, and stories. S Asia J S Asian Stud 5(1):42–50. https://doi.org/10.1080/00856408208723028

The Pioneer (2022) UP Govt gears up to expand boundaries of development authorities. https://www.dailypioneer.com/2022/state-editions/up-govt-gears-up-to%2D%2D%2D%2Dexpand-boundaries-of%2D%2D%2D%2Ddevelopment-authorities.html

UN Habitat (2022) World cities report 2022: envisaging the future of cities. UN-Habitat, Kenya, Nairobi. https://unhabitat.org/world-cities-report-2022-envisaging-the-future-of-cities

UN Permanent Forum on Indigenous Issues (UNPFII) (2015) State of the World's indigenous peoples. 2nd Volume. https://www.refworld.org/docid/55c89dac4.html

UN, United Nations Department of Economic and Social Affairs Population Division (2019) World urbanization prospects: The 2018 revision (ST/ESA/SER.A/420): New York: United Nations

Verma R, Garg PK (2022) Multi-temporal urban growth analysis with expansion indicators in Lucknow constituency by open-source data, India. J Urban Manag 11(4):412–423. https://doi.org/10.1016/j.jum.2022.07.001

WHO, World Health Organization (2013a) Comprehensive mental health action plan 2013–2020. World Health Organization, Geneva. https://www.who.int/publications/i/item/9789241506021
WHO, World Health Organization (2013b) Comprehensive mental health action plan 2013–2020. World Health Organization, Geneva. https://www.who.int/publications/i/item/9789240031029
World Bank Group (2016) Uttar Pradesh–poverty, growth and inequality (English). India state briefs. http://documents.worldbank.org/curated/en/187721467995647501/Uttar-pradesh-Poverty-growth-and-inequality
Zurba M, Stucker D, Mwaura G, Burlando C, Rastogi A, Dhyani S, Koss R (2020) Intergenerational dialogue, collaboration, learning, and decision-making in global environmental governance: the case of the IUCN intergenerational partnership for sustainability. Sustainability 12(2):498. https://doi.org/10.3390/su12020498

N Manika (PhD) is an Eco-consultant and Scientific Program Associate at Flora Fauna Science Foundation, India. She is a member of IUCN Commission on Ecosystems Management and SER. Her research interests revolve around biological and bio-cultural conservation.

Shalini Dhyani (PhD) is a Principal Scientist at the CSIR-National Environment Engineering Research Institute, India. She is Asia Vice Chair and Steering Committee member in the Commission on Ecosystems Management of IUCN and Visiting Scientist at CIFOR-ICRAF.

Contribution of Underutilized Wild Edibles Extracted from Urban Landscapes for Food Security and Socio-Ecological Resilience in Kashmir

7

Ishtiyak Ahmad Peerzada, M. A. Islam, T. H. Masoodi, S. A. Gangoo, P. A. Sofi, A. R. Malik, Tahir Mushtaq, N. A. Pala, Mir Muskan Un Nisa, Tuybia Bilal, and Humara Gulzar

Abstract

The significance of wild edibles in supporting the food and nutritional security of underprivileged communities has been acknowledged. However, there is a lack of adequate information on the extraction, consumption, and trade patterns of wild edibles in urban landscapes, and their actual contribution has not been estimated in Kashmir. Therefore, this study was conducted to explore the diversity, extraction and consumption patterns, and trade of wild edibles in urban landscapes of the Srinagar district of Kashmir Himalayas. A multistage random sampling technique was employed to select a sample of (106) households for the survey. Data were collected through structured interviews, non-participant observation, and focus group discussions. Simple descriptive statistics including frequency, percentages, averages, and others were used for data analysis to identify the key socio-economic characteristics of sample respondents, quantify the wild edibles extracted, estimate its economic value, and determine the potential of wild edibles in securing the food and nutritional security. Results revealed that wild edibles constituted 28 species of the flora and fauna belonging to 28 genera and 20 families. Wild edibles were consumed mostly as vegetables (14) followed by wild fruits (7), wild nuts (3), fish (2), mushroom (1), and honey (1). The total economic value of household annual consumption of foods was ₹ 6,121,460.50

I. A. Peerzada (✉) · M. A. Islam · T. H. Masoodi · S. A. Gangoo · P. A. Sofi · A. R. Malik · T. Mushtaq · N. A. Pala · M. M. U. Nisa · T. Bilal · H. Gulzar
Faculty of Forestry, Sher-e-Kashmir University of Agricultural Sciences & Technology, Benhama, Ganderbal, Jammu & Kashmir, India
e-mail: drishtiyak@skuastkashmir.ac.in

S. Dhyani, M. Sardeshpande (eds.), *Urban Foraging in the Changing World*,
https://doi.org/10.1007/978-981-97-0345-6_7

@ ₹57,749.63 per household per year, of which wild edibles contributed the lowest share (2.26%) and ranked 9th. It is therefore concluded that the potential of wild edibles needs due prioritization and exploitation in urban landscapes for providing food and nutritional security to the local inhabitants.

Keywords

Wild edibles · Vegetables · Fruits · Mushrooms · Economic value · Urban landscape Food security · Kashmir

7.1 Introduction

Throughout human history, food has played a critical role in both biological and sociocultural aspects, providing essential nutrients and energy. For millions of years, humans have had a close relationship with wild edible sourcing (Gosden and Hather 2004). Humans were able to acquire a deep understanding of the environment and access a wide variety of animal and plant-based foods through innovative techniques, due to the practice of wild edible sourcing (Harris and Hillman 2014; Ray et al. 2020). Since ancient times and throughout various stages of human adaptation, wild edibles have played a significant role in human societies. Humans have demonstrated their capacity to adjust to natural habitats and engage with nature and societal conditions, with a particular emphasis on consuming wild edibles. Wild edibles can be obtained not only from forests but also from managed landscapes such as fallow lands, agroforestry systems, and urban green spaces. In these landscapes, wild foods complement and diversify food production and income and offer variety in diets while also improving ecosystem services and climate resilience (Mbow et al. 2014; Shackleton 2014). Besides food and nutrition, utilization of wild edibles is prevalent as coping strategies during scarcity, particularly in developing countries where food insecurity is acute. Many wild edibles have the potential to enhance the nutritional value, sustainability, and resilience of our food systems. Therefore, it is important to explore and utilize the underutilized wild edibles (Hunter and Fanzo 2013; Powell et al. 2015).

Wild edibles are plants that grow naturally in self-sustaining groups within natural or semi-natural environments, without the need for direct human intervention (Heywood 1999). However, wild edibles in an urban context refer to edible plants that grow spontaneously in urban environments, such as parks, vacant lots, or even along sidewalks. These plants are often considered weeds but can provide a source of fresh food for those interested in foraging and exploring the natural resources available in urban areas. However, it is important to note that the availability and diversity of wild edibles may vary depending on the specific region and climate. Approximately one billion people around the world are reported to consume wild foods to fill health, nutritional, and ethno-medicinal needs, primarily derived from plants, on a daily basis (Lulekal et al. 2011). Wild edibles play an important role in dietary divergence, healthcare, livelihoods, poverty eradication, food security, crop diversification, income generation, and alleviation of malnutrition also (Godfray et al. 2010). It is estimated that around two billion people, in all countries across the globe, suffer from a lack of micronutrients, which not only increases their vulnerability to diseases but also impedes economic growth (FAO 2012). Wild edibles are a rich source of antioxidants and typically offer additional sources of essential nutrients, such as proteins, fibre, minerals, vitamins, and micronutrients, which enhance the quality of diets (Ogle 2001; Bvenura and Sivakumar 2017). People living in extreme poverty collect wild edibles in considerable quantities for their subsistence, cash income, safety net, and prevention of famine (Islam and Quli 2016; Peerzada et al. 2022).

Wild edibles, which include bushmeat, insects, honey, fungi, wild vegetables, and wild edible fruits, are a type of non-timber forest products (NTFPs) and provide essential nutrition for approximately one-sixth of the global population (Vira et al. 2015). They are especially important during times of crop shortages (Broegaard et al. 2017) and are consumed in both rural and urban settings (Erskine et al. 2015). Certain wild edibles possess medicinal properties and are utilized for their therapeutic benefits in treating various ailments (Cunningham and Shackleton 2004; Hickey et al. 2016). In numerous nations, medicinal plants are frequently consumed as dietary supplements, daily foods, and functional foods in order to enhance overall health. Wild plants contain a broad range of plant secondary metabolites, including polyphenols, terpenoids, and polysaccharides, which make them suitable as nutraceuticals or functional foods that have the potential to promote good health (García-Herrera et al. 2014). In addition to their medicinal and nutritional value, wild edibles can also generate income and employment opportunities through the sale or exchange of their fruits, leaves, juice, and local beverages. Additionally, they may possess higher nutritional value compared to cultivated plants (Kang et al. 2013; Islam and Quli 2016).

7.1.1 Wild Edibles in Kashmir, India

Wild edibles have by tradition occupied an important position in the cultural, religious, and health sector of Indians. The presence of diverse climatic zones and ecological diversity creates a basis for rich phyto-diversity, and this fact is strongly supported by various studies carried out on wild edibles by various researchers throughout India. Kashmir valley situated in the lap of Indian Himalayas is home to more than 2000 flowering plant species and is considered one of India's 28 hotspots for endemic plant diversity. The region has a variety of wild edibles that provide essential nutrients such as proteins, carbohydrates, fats, vitamins, and minerals, which are necessary for the well-being of the poor communities living in the area. The consumption of wild edibles plays a crucial role in supplementing the diets of the local population. Additionally, these resources serve as a source of income, as well as medicinal purposes, making them important for the overall well-being of the communities in the region (Koul et al. 1982).

7.1.2 Significance of This Study

So far, the lack of attention given to wild edibles in agricultural research, development, and policy may have negatively impacted the nutrition of rural households, as the production, consumption, and variety of wild edibles have decreased (Dansi et al. 2008). The need to protect local indigenous knowledge systems related to the collection and utilization of wild edibles is urgent and time-sensitive (Tabuti et al. 2004). The knowledge of traditional wild edibles is not only connected to the local biodiversity and their availability, but it is also strongly ingrained in daily food habits that are influenced by various sociocultural factors, including the prevalence of industrialized food, socio-economic status and food security, and significance of cultural identity (Pawera et al. 2020; Aziz et al. 2021). Moreover, climate change poses a threat to food security, and wild edibles play a crucial role in the diets of millions of people, particularly in low-income and rural communities. However, the vulnerability of wild edibles to climate change remains poorly understood (Wessels et al. 2021).

The contribution of wild edibles in food systems of rural communities is well established. There is relatively limited information available on the contribution of wild edibles from urban landscapes. They provide a cost-effective source of nutrition for both rural and semi-urban populations across different cultures and continents (Rowland et al. 2016; Jones 2017). It is, therefore, important to emphasize on the significance of traditional plant foraging in comprehending its contribution towards maintaining sustainable food systems (Khan et al. 2021) in fragile regions like Kashmir. Conservation efforts have focused on the utilization of wild edibles by households, as there are concerns about the sustainability of their exploitation. However, there is limited empirical evidence regarding the economic value and significant contribution of wild edibles to household food security in the temperate urban landscape of Kashmir Himalaya. Therefore, for the purpose of this study, all the edibles/foods that are collected, rather than intentionally grown, are classified as wild edibles, regardless of their location. This includes wild edibles of both plant and animal origin that are harvested from agricultural land, uncultivated areas, wetlands, urban green spaces, parks, or forested regions, following Termote et al. (2011).

7.2 Materials and Methods

7.2.1 Study Area

This study was conducted in the Harwan area of the Srinagar district of J&K UT. The area is situated on the northern fringe of the Srinagar district between 34.1017° N and 74.8067° E geo-coordinates at an altitude of 1600 m asl (Fig. 7.1). Harwan is a historical and scenic area located in the outskirts of Srinagar city in the Indian union territory of Jammu and Kashmir. It is situated at a distance of about 18 km from the city centre on the banks of Dal Lake. The area is known for its beautiful natural surroundings, which include forests, mountains, and streams. The region experiences temperate, sub-alpine, and alpine weather patterns and is

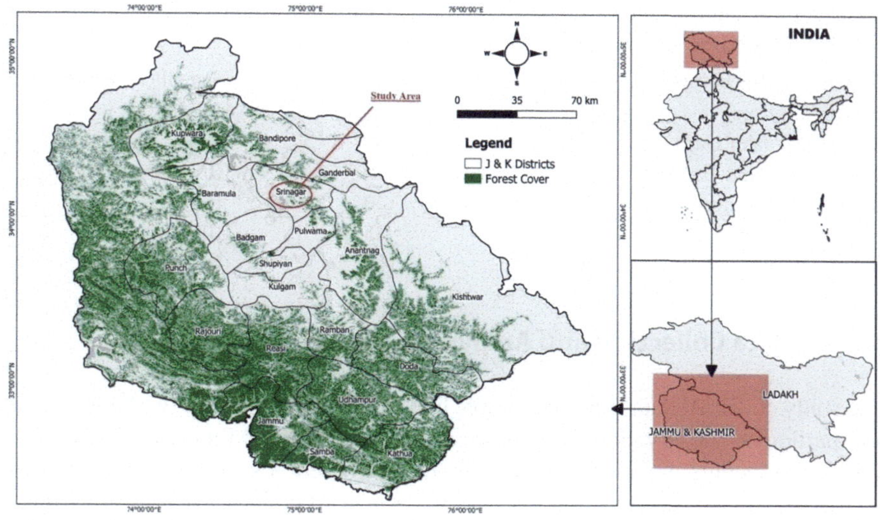

Fig. 7.1 Location map of the study area

characterized by the presence of pine forests intermixed with a variety of valuable wild edibles. People of Harwan have a rich tradition of using these wild edibles for food, nutrition, and medicinal purposes.

7.2.2 Methodology

The aim of the household survey was to create a map of the presence of wild edible plants, assess the extent to which people rely on them for their livelihoods, evaluate their economic value and contribution, investigate current patterns of extraction and consumption, prioritize their use, and analyse their usage patterns. This study was designed to investigate the interdependence between the local community's reliance on these resources and their ability to adapt to various pressures.

7.2.3 Sampling Technique

A reconnaissance survey of the entire study area was carried out to identify the sampling units prior to the experimental layout, which is based on the people's dependence on wild edibles. The study employed a multistage sampling approach (Ray and Mondol 2004) to identify the block, panchayats, and households in Srinagar District, J&K. In the first stage, Harwan block was chosen. In the second stage, five panchayats were selected based on the households' reliance and dependence on wild edibles for their livelihoods. In the third stage, 100 households engaged in the extraction, consumption, and sale of wild edibles were chosen randomly. The household head or the eldest member was interviewed as the respondent (Table 7.1).

Table 7.1 Sampling framework and sample size distribution in the study area (no. of households = 106)

S. No.	Panchayats	Sampling frame	Sample size	Sampling intensity
1.	Dara	170	17	10%
2.	Fakir Gujri	276	28	10%
3.	Gand Taal	184	18	10%
4.	Khimber	204	20	10%
5.	Theed	226	23	10%
Total	**05**	**1060**	**106**	**10%**

7.2.4 Data Collection and Analysis

The study collected data from the sample households through structured interviews, non-participant observation, and focus group discussions (Mukherjee 1993). The interview schedule included questions about the households' socio-economic characteristics; collection, consumption, and trade of wild edibles; sources of income; and economic contribution of wild edibles. The socio-economic variables included age, education level, social membership, household size, household labour, farm size, livestock ownership, main occupation, wealth status, gross annual income, proximity to different landscapes, and visits. The variables were measured using a modified socio-economic status scale (Peerzada et al. 2022). Non-participant observation helped to have first-hand on-the-scene contact with the respondents, examine the behaviour in natural situation, and study the situation-based features of conduct. The focus group discussions involved 10–12 participants, including village elders with knowledge of wild edible identification and use. The observations from focus group discussions were used to validate and interpret the data collected through household surveys. The data on the nutritional value and edibility ratings of wild edible plants were gathered from a secondary source called the "Plants for a Future" database (https://pfaf.org/user/Plant.aspx?LatinName).

The study used descriptive statistics including frequency, percentages, means, standard deviations, and ranges to summarize the socio-economic characteristics of households; estimate the economic value of wild edible collection, consumption, and trade and income generation; and evaluate contribution to household food security (Snedecor and Cochran 1967). The data were collected in local units and then converted into International System Units before being analysed using the statistical software package SPSS Ver. 21.0. The findings are presented in tables and graphs.

7.3 Results

7.3.1 Socio-Economic Profile of the Sample Households

Results of the socio-economic characteristics of the households for the five panchayats are presented in Table 7.2. The findings indicate that the respondents were between the age group of 28 and 74 years, with mean age of 47.64 years. The

Table 7.2 Socio-economic profile of the sample households

Variables	Measurement	Type	Mean	St. Dev.	Maximum	Minimum
Age	Age in the year of household head or respondent	Continuous	47.64	14.57	74	28
Education	0 = illiterate, 1 = <primary, 2 = primary, 3 = middle, 4 = high school, 5 = intermediate, 6 = graduate and over	Categorical	1.56	1.75	6	0
Social participation	0 = no membership, 1 = membership of 1 organization, 2 = membership of >1 organization, 3 = office bearer, 4 = public leader	Categorical	1.35	0.76	4	0
Household size	1 = ≤ 5 members, 2= > 5 members	Categorical	1.67	0.48	2	1
Household labour	1 = 1 worker, 2 = 2 workers, 3 = 3 workers, 4= >3 workers	Categorical	3.15	0.91	4	1
Farm size	0 = landless, 1 = marginal (up to 1.0 ha), 2 = small (1.1–2.0 ha), 3 = medium (2.1–4.0 ha),4 = large (>4.0 ha)	Categorical	1.67	0.48	2	1
Livestock possession	0 = no livestock, 1 = ≤5 livestock, 2 = 6 to 10 livestock, 3= >10 livestock	Categorical	2.28	1.16	3	0
Main occupation	1 = wage labour, 2 = caste occupation, 3 = cultivation, 4 = business, 5 = service, 6 = any other	Categorical	3.28	1.32	6	1
Wealth status	0 = poor, 1 = medium, 2 = rich	Categorical	1.44	0.76	2	0
Annual income	₹/household/annum	Continuous	68,622.44	58,672.46	389,523.00	6400.00

(continued)

Table 7.2 (continued)

Variables	Measurement	Type	Mean	St. Dev.	Maximum	Minimum
Proximity to forest	Distance of home to forests (km)	Continuous	0.78	0.41	1.96	0.38
Forest visits	0 = never, 1 = occasionally, 2 = frequently, 3 = very frequently	Categorical	1.56	0.88	3	0

average level of education the sample respondents had was primary (score, 1.56), which indicates that the literacy levels in terms of formal education are quite low. The mean social participation among the interviewees was low (score, 1.35) because the majority of the respondents had no social memberships. The average household size of the sampled respondents was about 5.0, whereas the average number of workers per household was 3. Majority of the sampled households were marginal and small landholders, with a mean landholding size of 0.73 ha only. The mean value of livestock possession (2.29) shows that the sampled respondents had livestock ranging from 5 to 10 per household. Agriculture is the prevalent main occupation among most of the sampled households (mean score, 3). The medium wealth status (mean score, 1.44) of sampled household indicates that the quality and quantity of domestic items and farm implements possessed by them were indigenous and low cost. The annual income of the sample respondents ranged from ₹ 5400.00 to ₹ 309,523.00 with a mean of ₹ 68,622.44. The average distance between the sample household's home and the nearby forest was 0.78 km, which indicates that the respondents lived very close to the forests and visit forests frequently (mean score, 1.56) (Table 7.2).

7.3.2 Diversity and Use Pattern of Wild Edibles

The utilization pattern of the wild edibles in the study area is influenced by a variety of factors such as access to resources, availability of markets, diversity of species present in the local landscapes, and traditional knowledge associated with the species. However, the use of wild edibles varies significantly between different areas and even between households within the same community. Consequently, the wild edibles collected by local households have been categorized into different groups (Table 7.3) based on the ways in which each species is utilized (e.g. as vegetables, wild fruits, wild nuts, mushrooms, wild honey, fishes) and which parts of the plant are harvested (e.g. leaves, stems, rhizomes, tubers, fruits, nuts, kernel, fruiting body, nectar, fillets). The scientific name, local name/common name, family, habit, parts used, edibility rating, source of nutrition, and landscapes of wild edibles extracted from urban landscapes for household food security are summarized. The investigation further revealed that the wild edibles constituted 28 species of the flora and fauna belonging to 28 genera and 20 families.

Table 7.3 Diversity and use pattern of wild edibles exploited from urban landscapes in the sample households ($N = 106$)

Species (family)	Local name	Habit	Part used	Edibility rating	Source of nutrition	Landscapes
Vegetables (14)						
Taraxacum officinale H.H. Weber ex Wiggers (Asteraceae)	Hand/ dandelion	Herb	Leaves/roots	4/5	Vitamins A, C, D, E, iron, phosphorus, calcium, sodium, phosphorus, and carotenoids	Parks and gardens/agrarian
Silene vulgaris (Moench.) Garcke (Caryophyllaceae)	Whatkram	Herb	Leaves	2/5	Vitamin C, carbohydrates, protein, carotenes	Parks and gardens
Polygonum aviculare Linn. (Polygonaceae)	Drubhaak	Herb	Leaves	2/5	Vitamins A and C, phosphorus, zinc, potassium, and manganese	Parks and gardens
Capsella bursa-pastoris (L.) Medik (Brassicaceae)	Kralmond	Herb	Leaves	3/5	Vitamins A, C, B1, and B2, protein, calcium, phosphorus, potassium	Parks and gardens
Plantago lanceolata Linn. (Plantaginaceae)	Gule	Herb	Leaves	2/5	Phenylethanoids iridoid, glycosides aucubin and catalpol	Parks and gardens/agrarian
Nasturtium officinale R. Br. (Brassicaceae)	Nag-Babber	Herb	Leaves	4/5	Vitamins A, C, B1, and B2, calcium, phosphorus, potassium, sodium	Parks and gardens
Rumex nepalensis Spreng. (Polygonaceae)	Ubaj	Herb	Leaves	1/5	Oxalic acid, tannins, carbohydrates Fats and dietary fibre	Parks and gardens/agrarian
Portulaca oleracea Linn. (Portulacaceae)	Nunar	Herb	Leaves/stems	4/5	Vitamins A, B1, and B2, carbohydrates, protein, calcium, iron, alpha-linolenic acid, omega-3 fatty acid	Agrarian
Diplazium esculentum (Retz.) SW. (Athyriaceae)	Dade	Herb	Leafy stems	4/5	Folic acid, calcium, iron, and phosphorus	Woodland/orchards

(continued)

Table 7.3 (continued)

Species (family)	Local name	Habit	Part used	Edibility rating	Source of nutrition	Landscapes
Vegetables (14)						
Nelumbo nucifera Gaertn. (Nelumbonaceae)	Nadur	Herb	Root stem	4/5	Protein, carbohydrates, starch	Marshy/wetland
Nymphaea alba Linn. (Nymphaeaceae)	Brimposh/ Buem	Herb	Leaves/rhizomes	3/5	Nitrogen, potassium, calcium, sodium, protein, fibre	Marshy/wetland
Helianthus tuberosus Linn. (Asteraceae)	Farm gogaej/ Jerusalem artichoke	Herb	Tubers	4/5	Vitamins C and B6, potassium, magnesium, carbohydrates, fibre, protein, iron	Parks and gardens/agrarian
Dipsacus inermis Wall. (Caprifoliaceae)	Wopal-Hakh	Herb	Tender leaves	1/5	Crude protein, fibre, carbohydrates, sodium, calcium, iron, phosphorus	Hillsides
Cichorium intybus Linn. (Asteraceae)	Kasini/chicory	Herb	Tender plant	4/5	Crude protein, carbohydrates, fibre, sodium, calcium, iron, potassium, magnesium	Parks and gardens/agrarian
***Wild fruits* (7)**						
Ficus auriculata Lour. (Moraceae)	Trinbal/ Roxburghaii fig	Tree	Fruits	4/5	Protein, carbohydrates, calcium, magnesium, potassium, and phosphorus	Parks and gardens/agrarian
Rubus ellipticus Smith. (Rosaceae)	Chhansh/ yellow Himalayan raspberry	Shrub	Fruits	2/5	Vitamins A, C, and B6, thiamine, riboflavin, calcium, zinc, protein, lipids, carbohydrates, fibre	Parks and gardens/agrarian
Morus alba Linn. (Moraceae)	Tul/mulberry	Tree	Fruits	4/5	Vitamins K, C, A, and E, riboflavin, crude protein, carbohydrates, fibre, calcium, magnesium, potassium, zinc, and phosphorus	Parks and gardens/agrarian

Diospyros lotus Linn. (Ebenaceae)	Amlok/date palm	Tree	Fruits	5/5	Vitamin C, protein, carbohydrates, fat, calcium, potassium, manganese	Parks and gardens/sacred groves
Punica granatum Linn. (Lythraceae)	Anardana/ pomegranate	Tree	Fruits	4/5	Vitamins A, C, E, K, and B12, thiamin, riboflavin, protein, carbohydrates, fat, fibre, calcium, potassium, manganese	Parks and gardens/agrarian/ sacred groves
Ziziphus jujuba Mill. (Rhamnaceae)	Brey/Indian jujube	Tree	Fruits	4/5	Vitamin C, carbohydrates, proteins, dietary fibre, calcium, manganese, sodium, phosphorus	Parks and gardens/agrarian/ sacred groves
Crataegus songarica K. Koch (Rosaceae)	Ringa/ hawthorn	Tree	Fruits	3/5	Vitamin C, carotene, protein, lipids, carbohydrates, fibre	Parks and gardens/agrarian/ sacred groves
Wild nuts **(3)**						
Trapa natans Linn. (Lythraceae)	Singhara/ Ghaarie/water chestnut	Aquatic floating herb	Nuts	4/5	Vitamin B, starch, protein, carbohydrates, calcium, magnesium, zinc, and potassium	Wetlands
Castanea sativa Mill. (Fagaceae)	Gour/chestnut	Tree	Nuts	5/5	Vitamins A, B, C, and E macro- (K, P, Mg, Ca, Na), and micronutrients (Mn, Fe, Zn, and Cu), calcium, protein, carbohydrates, starch, sucrose	Parks and gardens/sacred groves
Juglans regia Linn. (Juglandaceae)	Doon/walnut	Tree	Kernel	4/5	Vitamins A and B, protein, carbohydrates, omega-3 fatty acids, ascorbic acids	Parks and gardens/agrarian/ sacred groves
Mushrooms **(1)**						
Morchella esculenta (Linn.) Pers. (Morchellaceae)	Guchi/Morrels	Fungus	Fruiting body	4/5	Vitamins A, B, C, and D, protein, lipids, carbohydrates, magnesium, calcium, potassium, sodium, iron, zinc	Woodland/orchards

(continued)

Table 7.3 (continued)

Vegetables (14)

Species (family)	Local name	Habit	Part used	Edibility rating	Source of nutrition	Landscapes
Wild honey (1)						
Apis cerana Fabricius (Apidae)	Maanch	Beneficial insect	Regurgitated enzymatically processed nectar	5/5	Vitamins B3, B2, B6, and C, iron, manganese, selenium, copper, zinc, potassium, phosphorus, magnesium, sodium, carbohydrate, fructose, glucose, protein	Woodland/orchards/sacred groves/pastures/parks rich in flora: *Plectranthus rugosus, Robinia pseudoacacia, Malus domestica, Rosa webbiana*
Fish **(2)**						
Schizothorax niger Heckel, 1838 (Cyprinidae)	Kashirgard/ Snowtrout	Herbivorous	Fillets	5/5	Vitamins A, D, and E, calcium, iron, omega 3	Wetlands/lakes
Cyprinus carpio Linnaeus, 1758 (Cyprinidae)	Panjaebgard / common carp	Omnivorous	Fillets	4/5	Vitamins A, D, and E, calcium, iron, omega 3	Wetlands/lakes/rivers

#Edibility rating as per "Plants for a Future" database: https://pfaf.org/user/Plant.aspx?LatinName

The wild edibles comprised vegetables (14), fruits (7), nuts (3), mushrooms (1), honey (1), and fish (2). The parts consumed as food were leaves (10), fruits (8), nuts (2), fillets (2), fruiting body (1), kernel (1), leafy stems (1), regurgitated enzymatically processed nectar (1), rhizome (1), root stem (1), root (1), whole plant (1), and tuber (1). The life forms of the wild edibles included herb (14), tree (8), shrub (1), aquatic floating herb (1), beneficial insect (1), fungus (1), herbivorous fish (1), and omnivorous fish (1). The family-wise representation of the wild edibles was as follows: Asteraceae (3), Brassicaceae (2), Cyprinidae (2), Lythraceae (2), Moraceae (2), Polygonaceae (2), Rosaceae (2), Apidae (1), Athyriaceae (1), Caprifoliaceae (1), Caryophyllaceae (1), Ebenaceae (1), Fagaceae (1), Juglandaceae (1), Morchellaceae (1), Nelumbonaceae (1), Nymphaeaceae (1), Plantaginaceae (1), Portulacaceae (1), and Rhamnaceae (1). The wild edibles are rich sources of alpha-linolenic acid, ascorbic acids, calcium, carbohydrate, carotene, copper, protein, fibre, fat, folic acid, fructose, glucose, iridoid, glycosides aucubin and catalpol, iron, omega 3, lipids, magnesium, manganese, nitrogen, fatty acid, oxalic acid, phenylethanoids, phosphorus, carotenoids, potassium, protein, riboflavin, selenium, sodium, starch, sucrose, tannins, thiamin, vitamins, zinc, etc. Out of 28 species, around 4 species have 5/5, 15 species have 4/5, 3 species each have 3/5 and 2/5, and 2 species have 1/5 edibility rating, respectively. The major landscapes exploited for wild edible extraction were parks and gardens, agrarian, woodland, orchards, marshy wetland, hillsides, sacred groves, pastures, lakes, and rivers.

7.3.3 Quantification of Wild Edibles Collected and Consumed in the Sample Households

The data pertaining to the household collection and consumption of wild edibles in sample households are summarized in Table 7.4. Vegetables, fruits, and nuts are the main wild edibles collected by a considerable number of households (7.55–71.69%), whereas the involvement of households in the collection of mushrooms, honey, and fish is low (2.70–25.47%).

7.3.4 Valuation of Wild Edibles Collected and Consumed in the Sample Households

Current study revealed that the total monetary value of wild edibles extracted from urban landscapes was ₹ 137,760.50/year including both subsistence (₹ 64,908.00/year) and cash income (₹ 72,852.50/year) @ ₹ 1299.62/household/year (Table 7.5). Of the total monetary value of wild edibles, mushrooms contributed the highest share (35.66%), followed by fish (33.00%), nuts (15.57%), vegetables (8.54%), honey (4.50%), and fruits (2.73%). Involvement of households in the sale of wild edibles was found to be for nuts (33.01%), fish (16.03%), vegetables (13.21%), mushrooms (9.43%), fruits (4.71%), and honey (1.88%).

Table 7.4 Quantification of wild edibles collected and consumed in the sample households ($N = 106$)

S. No.	Wild edibles	HHs involved in collection (%)	Quantity collected (kg/year)	Quantity consumed (kg/year)	Mode of consumption
1.	Vegetables	76 (71.69)	235.50	179.65	Cooked
2.	Fruits	56 (52.83)	37.55	21.45	Raw
3.	Nuts	58 (54.72)	214.48	117.38	Raw
4.	Mushrooms	18 (16.98)	24.56	13.42	Cooked
5.	Honey	08 (7.55)	7.75	2.55	Raw
6.	Fish	27 (25.47)	181.85	52.65	Cooked

Table 7.5 Valuation of wild edibles collected and consumed in the sample households ($N = 106$)

Wild edibles	HHs involved in marketing (%)	Rate (₹/kg)	Quantity marketed (kg)	Subsistence value (₹)	Cash income (₹)	Total monetary value (₹)	Relative contribution (%)
Vegetables	14 (13.21)	50.00	155.85	8982.50	2792.50	11,775.00	8.54
Fruits	5 (4.71)	100.00	26.10	2145.00	1610.00	3755.00	2.73
Nuts	35 (33.01)	100.00	197.10	11738.00	9710.00	21,448.00	15.57
Mushrooms	10 (9.43)	2000.00	21.14	26,840.00	22,280.00	49,120.00	35.66
Honey	2 (1.88)	800.00	6.20	2040.00	4160.00	6200.00	4.50
Fish	17 (16.03)	250.00	149.20	13,162.50	32,300.00	45,462.50	33.00
Total	–	–	–	64,908.00	72,852.50	137,760.50	100.00
Average	–	–	–	612.33	687.28	1299.62	–

7.3.5 Contribution of Wild Edibles to Food Security in the Sample Households

The findings (Table 7.6) revealed that the total monetary value of household annual consumption of foods from all the off-farm and on-farm sources was ₹ 6,121,460.50 among the sample households, @ ₹57,749.63 per household per year. Among the food categories, rice contributed the largest share (20.78%, rank 1st) to the household food security followed by eggs (18.18%, rank 2nd), meat (16.88%, rank 3rd), fruits (13.85%, rank 4th), oils (13.33%, rank 5th), vegetables (7.79%, rank 6th), fish (4.33%, rank 7th), grains (2.60%, rank 8th), and wild edibles (2.26%, rank 9th). Thus, wild edibles are the least important constituent of food items for household food security.

Table 7.6 Contribution of wild edibles to food security in the sample households ($N = 106$)

Edibles	Quantity consumed (kg)	Rate (₹/kg)	Total absolute contribution (₹)	Average absolute contribution (₹)	Relative contribution (%)	Rank
Rice	31,800.00	40.00	1,272,000.00	12,000.00	20.78	1st
Grains	1060.00	150.00	159,000.00	1500.00	2.60	8th
Oils	3710.00	220.00	816,200.00	7700.00	13.33	5th
Meat	1590.00	650.00	1,033,500.00	9750.00	16.88	3rd
Fish	1060.00	250.00	265,000.00	2500.00	4.33	7th
Eggs[§] and milk[#]	7420.00	150.00[§] and 35[#]	1,113,000.00	10,500.00	18.18	2nd
Vegetables	15,900.00	30.00	477,000.00	4500.00	7.79	6th
Fruits	8480.00	100.00	848,000.00	8000.00	13.85	4th
Wild edibles	701.69	–	137,760.50	1299.62	2.26	9th
Total	71,721.69	–	6,121,460.50	57,749.63	–	–
Average	676.62	–	57749.63	–	–	–

§ = dozen, # = litre

7.4 Discussion

Despite the abundant diversity of wild edibles in the urban landscapes of the Kashmir Himalaya, the research findings suggest that these plants do not play a significant role in the dietary habits of households in the study area. The utilization of wild edibles is infrequent and limited due to the availability of sufficient household food and their higher utilization in regular diets. Nonetheless, during the harsh winter months when agricultural produce is scarce and households face the risk of food shortages, processed and preserved wild edibles sourced from urban landscapes assume significance in people's diets. In fact, wild edibles such as vegetables, fruits, nuts, mushrooms, and honey are highly influenced by seasonality, with their harvesting season aligning with agricultural seasons, except for fish, which remains available throughout the year. The studies conducted by Garekae and Shackleton (2020) and Palliwoda et al. (2020) have also underscored the potential role of wild edibles in addressing food security requirements, particularly during the lean season in urban landscapes.

Moreover, when it comes to wild edibles like vegetables, fruits, nuts, and mushrooms, the majority of their consumption is for subsistence purposes, with a smaller portion being sold for cash income. On the other hand, in the case of honey and fish, a significant quantity is sold for cash income, while a smaller portion is retained for subsistence consumption. The economic valuation of wild edibles suggests that they hold greater significance as a source of income rather than as a primary source of food. This observation particularly applies to wild edibles such as honey and fish. Hence, the importance of wild edibles as a source of cash income becomes more prominent during the lean season. This finding is supported by the studies conducted by Poe et al. (2013), Somesh et al. (2021), and Ghosh-Jerath et al. (2021).

These studies also highlight the substantial economic value associated with the utilization of wild edibles.

This study further reveals that the value of wild edibles for both consumption and market sales is greater in the poorer households than in the wealthier families in the study area. This trend can be attributed to the fact that food security is a pressing concern for many poor families and marginalized sections of society in the urban landscape. As a result, households with lower socioeconomic backgrounds actively engage in collecting, consuming, and trading a wide range of wild edibles as a regular practice to address their food security needs. The availability of wild edibles serves as a crucial safety net or essential coping strategy for the urban poor to address food shortages, in addition to earning money. Previous research (Kang et al. 2013; Kunwara et al. 2020) has identified a similar trend, highlighting that wild edibles play a more significant role in ensuring food security for the poorest households compared to wealthier ones. Wild edibles in urban landscapes are confronted with various risks arising from human activities. Individuals have noted a reduction in the accessibility of wild edibles due to heightened market demands, conversion of urban spaces for alternative purposes, swift expansion of the population, and existence of numerous opportunities for livelihoods based on wild edibles. Therefore, it is essential to implement alternative interventions to conserve wild edibles and adapt to the ongoing changes and future challenges. The promotion of domestication, value addition, and commercialization of wild edibles, as well as the planting of climate-resilient alternative edible plants, can help create sustainable food systems, conserve wild edibles, reinforce indigenous traditional knowledge (ITK), and diversify sources of income. These findings further support previous studies (Gupta et al. 2013; Synk et al. 2017) that emphasize the unsustainability of the current utilization of wild edibles.

7.5 Conclusion

It is important to note that while wild edibles can be a great source of nutrition and flavour, it is essential to properly identify them before consuming them. Some wild plants can be poisonous or have harmful side effects, so it is important to exercise caution and seek expert advice before consuming them. Wild edibles extracted from urban landscapes are plants that grow in urban environments, such as parks, abandoned lots, and roadside areas, that are edible and can be harvested for consumption. These wild edibles are often highly nutritious and can provide an important source of food for urban communities, especially in areas where access to fresh, healthy food is limited. They also provide environmental benefits, such as improving soil quality and supporting biodiversity, and can contribute to a sense of community and connection to nature. However, it is important to note that not all plants found in urban landscapes are safe to consume and some may be contaminated with pollutants. It is important to properly identify and harvest wild edibles and to avoid collecting plants from areas that may be contaminated. The collection of wild edibles in urban landscapes has gained popularity in recent years due to a growing interest in local and sustainable food systems, as well as a desire to reconnect with nature in urban environments. Wild edibles can provide a low-cost and nutritious source of food, as well as offer other benefits such as increased physical activity and exposure to green spaces. However, it is important to exercise caution when collecting wild edibles from urban landscapes, as they may be contaminated with pollutants or other hazardous substances. It is also important to respect private property and obtain permission before harvesting from someone else's land. Moreover, urban sprawling and its impact on green spaces are a common challenge faced by many cities including Srinagar. If left unchecked, it can lead to environmental degradation, loss of biodiversity vis-à-vis wild edibles, and decreased quality of life for residents.

Acknowledgements Authors are thankful to the people of the area for cooperation and hospitality, as well as spending valuable time for interviews during the study.

Conflict of Interest The authors declare no conflict of interest.

References

Aziz MA, Ullah Z, Al-Fatimi M, De Chiara M, Sõukand R, Pieroni A (2021) On the trail of an ancient middle eastern ethnobotany: traditional wild food plants gathered by Ormuri speakers in Kaniguram, NW Pakistan. Biology 10:302

Broegaard RB, Rasmussen LV, Dawson N, Mertz O, Vongvisouk T, Grogan K (2017) Wild food collection and nutrition under commercial agriculture expansion in agriculture-forest landscapes. Forest Policy Econ 84:92–101

Bvenura C, Sivakumar D (2017) The role of wild fruits and vegetables in delivering a balanced and healthy diet. Food Res Int 99:15–30

Cunningham AB, Shackleton CM (2004) Use of fruits and seeds from indigenous and naturalised plant species. In: Lawes M, Eeley H, Shackleton CM, Geach B (eds) Indigenous forests and woodlands in South Africa: policy, people and practice. University of KwaZulu-Natal Press, Pietermaritzburg, South Africa, pp 603–626

Dansi A, Adjatin A, Adoukonou-Sagbadja H, Faladé V, Yedomonhan H, Odou D, Dossou B (2008) Traditional leafy vegetables and their use in the Benin Republic. Genet Resour Crop Evol 55(8):1239–1256. https://doi.org/10.1007/s10722-008-9324-z

Erskine W, Ximenes A, Glazebrook D, da Costa M, Lopes M, Spyckerelle L, Nesbitt H (2015) The role of wild foods in food security: the example of Timor-Leste. Food Secur. 7:55–65

FAO [Food and Agriculture Organisation] (2012) The state of food insecurity in the world: economic growth is necessary but not sufficient to accelerate reduction of hunger and malnutrition. FAO, Rome, Italy. http://www.fao.org/publications/sofi/2012/en/

García-Herrera P, Sánchez-Mata MC, Cámara M et al (2014) Nutrient composition of six wild edible Mediterranean Asteraceae plants of dietary interest. J Food Compos Anal 34(2):163–170

Garekae H, Shackleton CM (2020) Urban foraging of wild plants in two medium-sized south African towns: people, perceptions and practices. Urban For Urban Green 49:126581

Ghosh-Jerath S, Kapoor R, Barman S, Singh G, Singh A, Downs S, Fanzo J (2021) Traditional food environment and factors affecting indigenous food consumption in Munda tribal Community of Jharkhand, India. Front Nutr 7:600470. https://doi.org/10.3389/fnut.2020.600470

Godfray HCJ, Beddington JR, Crute IR, Haddad L, Lawrence D, Muir JF, Pretty J, Robinson S, Thomas SM, Toulmin C (2010) Food security: the challenge of feeding 9 billion people. Science 327(5967):812–818. https://doi.org/10.1126/science.1185383

Gosden C, Hather JG (2004) The prehistory of food: appetites for change. Routledge, London. https://doi.org/10.4324/9780203203385

Gupta SK, Sharma OP, Raina NS, Sehgal S (2013) Ethno-botanical study of medicinal plants of Paddar valley of Jammu and Kashmir, India. Afr J Tradit Complement Altern Med 10(4):59–65. https://doi.org/10.4314/ajtcam.v10i4.10

Harris DR, Hillman GC (2014) Foraging and farming: the evolution of plant exploitation. Routledge, New York, NY. https://doi.org/10.4324/9781315746425

Heywood V (1999) Use and potential of wild plants in farm households. FAO Farm Systems Management Series. Food and Agriculture Organisation, Rome, Italy. http://www.fao.org/docrep/003/w8801e/w8801e00.htm

Hickey GM, Pouliot M, Smith-Hall C, Wunder S, Nielsen MR (2016) Quantifying the economic contribution of wild food harvests to rural livelihoods: a global-comparative analysis. Food Policy 62:122–132

Hunter D, Fanzo J (2013) Introduction: agricultural biodiversity, diverse diets and improving nutrition. In: Fanzo J, Hunter D, Borelli T, Mattei F (eds) Diversifying Food and Diets. Routledge, New York, NY, pp 33–46

Islam MA, Quli SMS (2016) Non-timber forest products (NTFPs) supporting food security in tribal Jharkhand. Jharkhand J Dev Manag Stud 14(1):6855–6864

Jones AD (2017) Critical review of the emerging research evidence on agricultural biodiversity, diet diversity, and nutritional status in low- and middle-income countries. Nutr Rev 75:769–782. https://doi.org/10.1093/nutrit/nux040

Kang Y, Łuczaj Ł, Kang J, Zhang S (2013) Wild food plants and wild edible fungi intwo valleys of the Qinling Mountains (Shaanxi, Central China). J Ethnobiol Ethnomed 9:26. http://www.ethnobiomed.com/content/9/1/26

Khan S, Hussain W, Sulaiman S, Hussain H, Altyar AE, Ashour ML, Pieroni A (2021) Overcoming tribal boundaries: the biocultural heritage of foraging and cooking wild vegetables among four Pathan groups in the Gadoon Valley, NW Pakistan. Biology 2021(10):537

Koul AK, Karihaloo JL, Hamal IA (1982) Wild edible plants of Kashmir–some less known vegetable substitutes and beverages. Bull Bot Surv India 24(1–4):67–69

Kunwara RM, Fadimana M, Thapab S, Acharyac RP, Camerond M, Bussmanne RW (2020) Plant use values and phytosociological indicators: implications for conservation in the Kailash sacred landscape, Nepal. Ecol Indic 108:105679

Lulekal E, Asfaw Z, Kelbessa E, Van Damme P (2011) Wild edible plants in Ethiopia: a review on their potential to combat food insecurity. Africa Focus 24:2

Mbow C, Smith P, Skole D, Duguma L, Bustamante M (2014) Achieving mitigation and adaptation to climate change through sustainable agroforestry practices in Africa. Curr Opin Environ Sustain 6:8–14

Mukherjee N (1993) Participatory rural appraisal. Methodology and applications. Concept Publishing Company, Delhi, India

Ogle BM (2001) Wild vegetables and micronutrient nutrition: studies on the significance of wild vegetables in Women's diets in Vietnam. (Doctoral dissertation,. Acta Universitatis Upsaliensis

Palliwoda J, Banzhaf E, Priess JA (2020) How do the green components of urban green infrastructure influence the use of ecosystem services? Examples from Leipzig, Germany. Landscape Ecol 35:1127–1142. https://doi.org/10.1007/s10980-020-01004-w

Pawera L, Khomsan A, Zuhud EA, Hunter D, Ickowitz A, Polesny Z (2020) Wild food plants and trends in their use: from knowledge and perceptions to drivers of change in West Sumatra, Indonesia. Food 9:1240

Peerzada IA, Islam MA, Chamberlain J, Dhyani S, Reddy M, Saha S (2022) Potential of NTFP based bioeconomy in livelihood security and income inequality mitigation in Kashmir Himalayas. Sustainability 14:2281. https://doi.org/10.3390/su14042281

Poe MR, McLain RJ, Emery M, Hurley PT (2013) Urban forest justice and the rights to wild foods, medicines, and materials in the city. Hum Ecol 41:409–422. https://doi.org/10.1007/s10745-013-9572-1

Powell B, Thilsted S, Ickowitz A, Termote C, Sunderland T, Herforth A (2015) Improving diets with wild and cultivated biodiversity from across the landscape. Food Secur 7:535–554. https://doi.org/10.1007/s12571-015-0466-5

Ray A, Ray R, Sreevidya EA (2020) How many wild edible plants do we eat—their diversity, use, and implications for sustainable food system: an exploratory analysis in India. Front Sustain Food Syst 4:56. https://doi.org/10.3389/fsufs.2020.00056

Ray GL, Mondol S (2004) Research methods in social sciences and extension education. Kalyani Publishers, New Delhi, India, pp 66–76

Rowland D, Ickowitz A, Powell B, Nasi R, Sunderland TCH (2016) Forest foods and healthy diets: quantifying the contributions. Environ Conserv 44:102–114. https://doi.org/10.1017/S0376892916000151

Shackleton S (2014) Impacts of climate change on food availability: non-timber forest products. In: Freedman B (ed) Global environmental change. Handbook of global environmental pollution, vol 1. Springer, Dordrecht, The Netherlands, pp 695–700

Snedecor G, Cochran WG (1967) Statistical methods. Iowa State Univ. Press, Ames, Iowa, USA, pp 17–36

Somesh D, Rao R, Murali R, Nagendra H (2021) Patterns of urban foraging in Bengaluru city. Urban For Urban Green 57:126940

Synk CM, Kim BF, Davis CA, Harding J, Rogers V, Hurley PT, Emery MR, Nachman KE (2017) Gathering Baltimore's bounty: characterizing behaviors, motivations, and barriers of foragers in an urban ecosystem. Urban For Urban Green 28:97. https://doi.org/10.1016/j.ufug.2017.10.007

Tabuti JRS, Dhillion SS, Lye KA (2004) The status of wild food plants in Bulamogi County, Uganda. Int J Food Sci Nutr 55(6):485–498. https://doi.org/10.1080/09637480400015745

Termote C, Van Damme P, Djailo BD (2011) Eating from the wild: Turumbu, Mbole and Bali traditional knowledge on non-cultivated edible plants, district Tshopo, DR Congo. Genet Resour Crop Evol 58(4):585–618. https://doi.org/10.1007/s10722-010-9602-4

Vira B, Wildburger C, Mansourian S (2015). Forests, trees and landscapes for food security and nutrition. A global assessment report; International Union of Forest Research Organizations (IUFRO): Vienna, Austria, 33

Wessels C, Merow C, Trisos CH (2021) Climate change risk to southern African wild food plants. Reg Environ Chang 21:29

Ishtiyak Ahmad Peerzada (PhD) is an Assistant Professor cum Scientist, Division of Forest Products and Utilization, Faculty of Forestry, Sher-e-Kashmir University of Agricultural Sciences and Technology of Kashmir, India.

M.A. Islam (PhD) is a Professor at Faculty of Forestry, Sher-e-Kashmir University of Agricultural Sciences and Technology of Kashmir, India.

T.H. Masoodi is Registrar at Sher-e-Kashmir University of Agricultural Sciences and Technology of Kashmir, India.

S.A. Gangoo (PhD) is Dean at Faculty of Forestry, Sher-e-Kashmir University of Agricultural Sciences and Technology of Kashmir, India

P.A. Sofi (PhD) is Head Division of Forest Products and Utilization, Faculty of Forestry, Sher-e-Kashmir University of Agricultural Sciences and Technology of Kashmir, India.

A.R. Malik (PhD) is Assistant Professor cum Scientist at Faculty of Forestry, Sher-e-Kashmir University of Agricultural Sciences and Technology of Kashmir, India.

Tahir Mushtaq (PhD) is an Assistant Professor cum Scientist at the Faculty of Forestry, Sher-e-Kashmir University of Agricultural Sciences and Technology of Kashmir, India.

N.A. Pala (PhD) is an Assistant Professor cum Scientist at the Faculty of Forestry, Sher-e-Kashmir University of Agricultural Sciences and Technology of Kashmir, India.

Mir Muskan Un Nisa is a PhD Scholar at the Faculty of Forestry, Sher-e-Kashmir University of Agricultural Sciences and Technology of Kashmir, India.

Tuybia Bilal is a PhD Scholar at the Faculty of Forestry, Sher-e-Kashmir University of Agricultural Sciences and Technology of Kashmir, India.

Humara Gulzar is a PhD Scholar at Faculty of Forestry, Sher-e-Kashmir University of Agricultural Sciences and Technology of Kashmir, India.

Diversity and Distribution of Forageable Tree Species Across Green Space Types in Fast-Growing City in India: Lessons from Nagpur, India

8

Shruti Lahoti, Ashish Lahoti, Shalini Dhyani, and Osamu Saito

Abstract

Foraging is collecting wild food in natural areas and is common across urban areas worldwide. However, research on urban foraging has mainly focused on developed countries in the northern hemisphere, and there is insufficient information available from rapidly urbanizing cities of India. The study potentially supports urban foraging in fast-growing urban areas, by providing insights about the relevance of urban green spaces (UGSs) in supporting and strengthening human-nature interactions. The study uses a case study approach following field surveys to investigate the urban foraging potential of UGSs in Nagpur, India, by documenting plant species diversity, richness, their functional attributes, and usage value of trees under different UGS classes. Our findings reveal that the UGS in the study city is rich in plant diversity but varies among different UGS classes. A high diversity of foraging species was recorded from the avenue trees, while there is a need to improve it along water courses, public parks, as well as gardens, playgrounds, and urban forests. Our findings indicate towards dominance of native foraging species such as *Azadirachta indica*, *Ficus religiosa*, *Ficus benghalensis*, *Dalbergia sissoo*, *Alstonia scholaris*, *Bauhinia variegata*, and *Mitragyna parvifolia*, which have potential for foraging for various purposes. The UGS managers should consider growth rate, resistance to drought and pollution, and high usage value especially edible and foraging followed by functional attributes when selecting species for plantation. The study attempts to investigate urban trees in different UGS typologies in Nagpur systematically to

S. Lahoti (✉) · A. Lahoti · O. Saito
Institute for Global Environmental Strategies (IGES), Tokyo, Japan

S. Dhyani
CSIR-National Environmental Engineering Research Institute, Nagpur, Maharashtra, India

S. Dhyani, M. Sardeshpande (eds.), *Urban Foraging in the Changing World*,
https://doi.org/10.1007/978-981-97-0345-6_8

explore their foraging potential. Further research is recommended to explore who engages in foraging activities, what they forage for, the ecology of UGSs that support foraging activities, and the characteristics of foraged plants.

Keywords

Urban foraging · Urban green space (UGS) · Traditional knowledge · Local communities · UGS planning

8.1 Introduction

Urban green areas are important assets worldwide, as more than half of the global population live in urban regions, especially in developing nations (UNDESA 2018). While UGSs are crucial, they are also a disputed element of the urban environment due to ongoing densification and development (Haaland and Konijnendijk 2015; Kabisch et al. 2015; Lin and Fuller 2013). With limited UGSs, only a minority of urban dwellers have direct experiences with nature (Cox et al. 2017). Daily interaction with urban biodiversity is identified as a challenge (Schuttler et al. 2018), despite recognizing that engaging with nature is crucial for mental, physical, psychological, and social well-being (MacKerron & Mourato 2013; Shanahan et al. 2015). In general, daily nature experiences are decreasing in fast-growing cities especially in the Global South (Soga and Gaston 2016; Schuttler et al. 2018). Hence, UGS planning emphasizes reintroducing nature into cities and promoting sustainable urban ecosystems by facilitating enhanced and improved urbanites' connection with nature (Erickson 2010; Beatley 2011).

As per Lin et al. (2018), urban gardening helps people connect with nature through an improved understanding of providing natural resources. The collection of edible plants outside of managed gardens (known as "urban foraging," as described by Shackleton et al. 2017) can complement these benefits and provide opportunities for interacting with urban environments (McLain et al. 2014; Shackleton et al. 2017). Foraging has been an integral part of human-nature interaction since ancient times and is not limited to only urban settings (Ladio and Lozada 2004; Sujarwo et al. 2016) especially in developing and underdeveloped countries. Recent research clearly highlighted and indicates that urban foraging is a valuable but underappreciated way to foster such interaction (Shackleton et al. 2017). Urban foraging can be motivated by various factors, including nutritional and health benefits, access to culturally significant foods, increased dietary variety, psychological benefits, nature connectedness, and self-provisioning (McLain et al. 2014; Mollee et al. 2017; Palliwoda et al. 2017; Poe et al. 2013). Furthermore, Ceuterick and Vandebroek (2017), Poe et al. (2013), and Unnikrishnan and Nagendra (2015) have shown that urban foraging also supports cultural identity and provides a connection to nature. Overall, urban foraging can be an important way to encourage human-nature interaction in urban environments, and it offers a range of benefits to those who participate (Dhyani and Kadaverugu 2020).

The traditional practice of human-nature interactions through foraging has experienced a resurgence in urban areas (urban foraging), where urban dwellers are engaging with urban landscapes, vegetation, and discovering edible, medicinal, and craft-related plants from either formally or informally managed UGSs (Severson 2009; Craft 2010). Studies have recorded species harvested from urban environments across the world, including from big cities like Berlin, Bangalore, Kampala, and New York (McLain et al. 2014; Mollee et al. 2017). Further studies have also indicated the land use types where foraging occurs, like parks, planting strips, alleyways, linear spaces, vacant lots, street trees, and along railroad tracks as potential foraging places (Jahnige 2004; Gabriel 2006; Grabbatin et al. 2011; Hurley et al. 2013; Poe et al. 2013). Wehi and Wehi (2010) further explained that urban foraging happens in public and private lands, as well as common lands, with land use types varying from formally managed parks (Palliwoda et al. 2017) to vacant lots and interstitial spaces (Hurley et al. 2015; Poe et al. 2013). Although urban foraging is a common practice, it is not well-addressed or comprehended in the areas of urban policy and planning, as well as urban ecology and sustainability studies, with some exceptions (McLain et al. 2012; McLain et al. 2014; Shackleton et al. 2017). Some studies have identified that urban foraging is banned or discouraged in public open areas (McLain et al. 2012; Shackleton et al. 2017; Synk et al. 2017; Hurley and Emery 2018). Whereas in the Indian context there is no such ban imposed, a regulated approach is always suggested to ensure sustainable harvesting. Many marginalized migrants to cities as well as slum dwellers are also dependent on UGS and practice foraging as alternative livelihood or primary livelihood.

UGSs in Indian context are primarily designed to fulfill the aesthetic, economic, and social requirements of urban residents. Conventionally, UGSs have been perceived as places for leisure and enjoyment rather than recognized as providers of essential goods and human health advantages (Gobster 2007; Hurley et al. 2015; Unnikrishnan and Nagendra 2015; Clark and Nicholas 2013; McLain et al. 2012). However, there is change observed in this thinking, and there is huge interest in greening the cities to address various climate and disaster risks in fast-growing urban sprawls. The importance of UGS as a source of natural resources is widely acknowledged, but there needs to be more focus to understand the foraging context in Indian urban areas, as it is mostly overlooked or not sufficiently valued. Numerous studies, including those by Poe et al. (2013), Hurley et al. (2015), McLain et al. (2012), Mollee et al. (2017), and Shackleton et al. (2017), have highlighted this issue. Additionally, while urban foraging is a common practice worldwide, the literature on the topic is predominantly from the Global North, as noted by Poe et al. (2013), McLain et al. (2014), Short-Gianotti and Hurley (2016), and Shackleton et al. (2017). Policymakers in Global South cities, where there is significant socioeconomic inequality, tend to overlook the potential of urban foraging (Nagendra et al. 2012). With increasing urbanization, it is crucial to recognize the significance of foraging in cities of the Global South, as highlighted by some recent studies in Africa (Sardeshpande and Shackleton 2023; Adeyemi and Shackleton 2023). It could be a vital means of survival and resilience for the urban poor, especially migrants and slum dwellers in cities of the Global South (Somesh et al. 2021).

Nevertheless, there is growing need to have more field-based data concerning urban foraging practices from a variety of trees preferred and planted in these urban spaces, the socioeconomic characteristics of foragers, and the effects of urbanization on foraging (ibid.). This information will be relevant to improve socio-ecological connections in urban planning efforts that can help to understand and improve food insecurity concerns by promoting and endorsing the role of UGS as a place for strengthening sustainable food systems.

In response, in light of the relevance of urban foraging as a promising pathway to enhance biodiversity experience and benefits, urban dwellers can be benefitted in many ways, such as gaining a deeper connection with nature, accessing fresh and locally sourced food, and contributing to sustainable practices within their communities. To understand how urban areas can support urban foraging, we attempted to study the vegetation in different classes of the UGSs. The study follows a case study approach (Crowe et al. 2011) with field data collection. By focusing on a fast-growing tier II Indian city that was considered one of the greenest cities that is not rapidly losing its green spaces (Dhyani et al. 2018), the study prepares an inventory of tree species intending to understand the potential of UGSs to support urban foraging with changing purposes of city landscapes. According to our knowledge, this is the first research paper systematically investigating urban trees in different UGS typologies with reference to understanding and exploring its foraging potential. Moreover, there is a scarcity of information regarding tier II cities in India that are experiencing rapid development. Only a few studies are available that focus on specific locations within these urban ecosystems, such as the NEERI Campus in Nagpur (Gupta et al. 2008; Dhyani et al. 2020) and the Indian Institute of Science Campus in Bangalore (Mhatre 2008). With this context, the study aims to answer the below questions:

1. What kind of species diversity and richness are prevalent across different classes of UGSs?
2. Which tree species dominate the urban landscape, and what are the city garden department of the municipal corporation's selection criteria?
3. How can UGSs foster urban dweller interactions with urban biodiversity by improving provisioning benefits supported by urban foraging?

8.2 Methods

8.2.1 Study Area

Nagpur is a fast-growing city in Central India, surrounded by dense forests known for their diverse plants and animals (Fig. 8.1). It is also famous as the "Tiger Capital of India." Nagpur is considered the greenest city, with various natural areas such as water bodies, farmland, parks, forests, and wastelands. However, as population and development increase, urbanization affects the environment inside and outside the city. A study on land use and land cover (LULC) shows that the city's development is

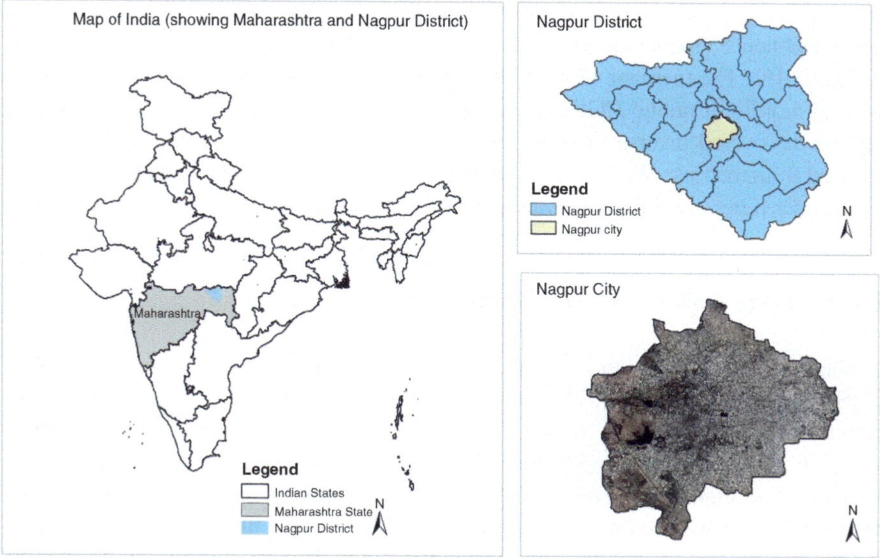

Fig. 8.1 The study city Nagpur, Maharashtra, India

spreading outward, affecting the city's biodiversity and ecosystem services (Dhyani et al. 2018). The per capita recreational public urban green space is only 3.65 m^2 on average, with significant provisions varying across the city and underserved areas (Lahoti et al. 2019b). For Nagpur, the available vegetation data have been recorded as a plant inventory list (Chaturvedi et al. 2013), as well as the vegetation structure, composition, and carbon sink potential of different UGS typologies (Lahoti et al. 2020). A few specific studies include the NEERI campus's floristic diversity (Gupta et al. 2008; Dhyani et al. 2020) and the tree carbon stock of Seminary Hills urban forest (Dhyani et al. 2021). The floristic survey has revealed the presence of 124 tree species belonging to 38 plant families, indicating diversity in tree flora, however, with varying tree cover across different zones of the city (Chaturvedi et al. 2013).

8.2.2 Study Zones

UGSs consist of both public and private areas that vary in terms of vegetation and landscape depending on factors such as ownership, management, and available resources (Threlfall et al. 2016). Private greens also contribute to the urban environment, but for the purposes of this study, only public UGSs were considered due to accessibility issues and also because for urban foraging mostly these public green spaces are available and accessible to local people. To capture the overall vegetation conditions of the city indicates towards tree species within different UGS classes that promote foraging in urban landscapes. These classes of UGSs used are parks and garden, playground, lakeside, forest, riverside, institutional premises, roadside, and vacant lands that are also potential foraging grounds for diversity of urban dwellers. The UGS classes acted as pre-stratification, allowing increased efficiency as variation among the stratum was lower and hence smaller sample data can represent larger parcel of the entire stratum (i-Tree 2018). The thematic map of UGSs was used to adequately represent the finely grained mosaic of public UGSs for pre-stratification (Lahoti et al. 2019a). In order to capture the overall diverse tree vegetation of the city, three representative zones were selected from the East, West, and Center (Lahoti et al. 2019b; Chaturvedi et al. 2013), as these zones represent different levels of UGS availability.

8.2.3 Data Collection and Sampling

A team of botanists and landscape planners collected data of tree species in different UGS classes through fieldwork. During the data collection (December 10th, 2018–January 20th, 2019), 246 plots were surveyed. The plots were of different shapes (circular or linear) and sizes (300 m^2 or 314 m^2) depending on the UGS class. For road networks, rectangular plots measuring 100 m by 1.5 m were selected on both sides of the road. In contrast, rectangular plots measuring 100 m by 3 m were used for lakeside and riverside edges. For other typologies, circular plots with a radius of 10 m were used. Plot selection was random in the identified zones and the UGS classes, and marked using GPS, as shown in Fig. 8.2. The number of sample plots of the UGS class varied based on the vegetation structure and composition (i-Tree 2018), considering observations from previous field visits (February–April 2018). The i-Tree manual was consulted to determine a rough estimate of the number of plots required for the inventory, which recommends a minimum of 20 plots per UGS class to achieve a standard error of 10% and represent the entire city. The log sheet used for the data collection is in *Appendix A*. In addition to noting down the tree species and various other aspects of trees, including their potential role in providing provisioning benefits, this chapter incorporates information on the types of tree species and their value in terms of supporting urban foraging potential. The potential role in providing provisioning benefits was derived from desk-based analysis conducted at a later stage.

8.2.4 Data Analysis

The Shannon diversity index (H′) and Simpson's diversity index (1/D) were computed using Eqs. (1) and (2), respectively, as described by Borah et al. (2013). Menhinick's index was employed to determine species richness, which involves dividing the number of species found in a sample by the square root of the total number of species in the sample. The inventory of the tree species identified in the different UGS typologies was prepared. We used the importance value index (IVI) of tree species to identify the dominant tree species. IVI is a widely accepted measure for evaluating the ecological significance of tree species in forest ecosystems (Curtis and McIntosh 1951). IVI was calculated by summation of a total of relative abundance, relative density, and relative frequency (Curtis and McIntosh 1951). IVI was used to understand the share of an individual in UGS classes. Based on IVI, the

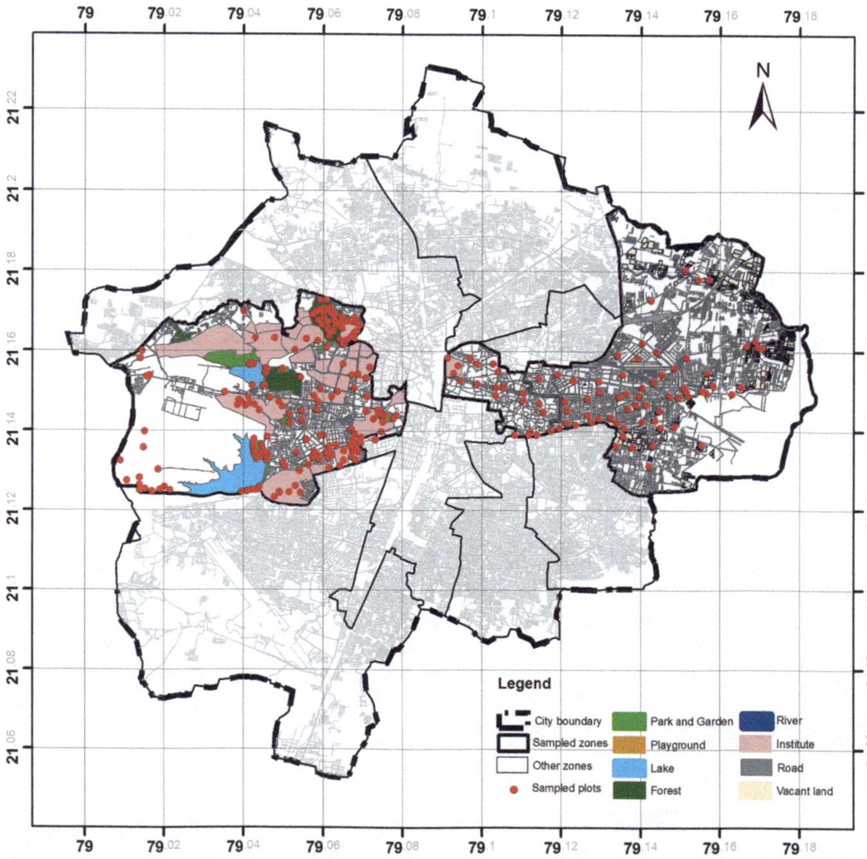

Fig. 8.2 UGS map showing different classes and sampled plots

dominant tree species were identified (IVI above 20). After identifying the dominant tree species in the city, a literature review was conducted to gain more insights into the functional attributes and usage of these species that also included its foraging potential and also preference by local people. A comprehensive search of the identified dominant tree species allowed us to understand their role in urban ecosystems, the general practice of species selection in UGS, and their potential to support urban foraging.

$$H' = \sum_{i=1}^{s} p_i lnln\, p_i \ldots.. \tag{8.1}$$

$$D = \frac{1}{\sum_{i=1}^{s} pi^2} \ldots.. \tag{8.2}$$

In Eq. (8.1), H′ is Shannon-Wiener diversity index, (8.2) D is Simpson's index, p is the proportion (n/N) of individuals of one particular species found (n) divided by the total number of individuals found (N), ln is the natural log, Σ is the sum of the calculations, and s is the number of species.

8.3 Results

During the study, a total of 2362 individuals were observed, representing 86 different tree species across eight UGS typologies. Of the 86 species, 73 were identified and classified into 58 genera and 22 families. Some of the most abundant species identified during the fieldwork are illustrated in Fig. 8.3. However, there were 13 species that could not be identified and 9 plots that had no trees, mostly in the riverside and vacant land typologies.

Fig. 8.3 Photo plate showing some of the identified dominant tree species of the *city (1: Ficus benghalensis, 2: Ceiba pentandra, 3: Ziziphus mauritiana, 4: Hardwickia binata, 5: Azadirachta indica, 6: Polyalthia longifolia, 7: Pongamia pinnata, 8: Ficus religiosa, 9: Eucalyptus globulus)*

Table 8.1 Identified species, phytosociology, and diversity attributes of eight UGS classes

UGS classes	No. of species	No. of families	No. of sample plots	No. of species	No. of families	S (species richness)	Shannon (H)	Simpson (1-D)
Parks and garden	31	29	30	31	29	1.89	3.06	0.94
Playground	24	19	18	24	19	2.08	2.77	0.92
Lakeside	22	20	20	22	20	1.91	2.70	0.92
Forest	33	25	50	33	25	1.20	1.84	0.64
Riverside	30	25	25	30	25	2.36	2.98	0.94
Institutional premises	46	35	32	46	35	2.72	3.14	0.93
Roadside	49	39	53	49	39	2.11	3.30	0.95
Vacant land	23	20	18	23	20	2.52	2.70	0.92

8.3.1 Species Richness and Diversity

Table 8.1 shows that the highest level of diversity was observed on roads, with 46 species present. Institutions and parks and garden were closely behind in terms of diversity. As measured by the Simpson's index, the dominance index also indicated high levels of diversity in roads, followed by parks and garden and the riverside. The Shannon index, which considers even rare species, showed that roads were the most diverse, potential locations for having tree diversity followed by institutions, parks, and gardens. The data demonstrate that roads are significantly more diverse than other UGSs, as indicated by the Shannon and Simpson indices. Institutional premises also showed relatively high levels of diversity. Forests, however, were found to be comparatively low in diversity, with lower levels of evenness and high variation in the abundance of species. This was because forests were predominantly plantations of trees or were encroached by invasives like Subabool (*Leucaena leucocephala*). When the sample sizes of both forests and roads were compared, it was clear that roads were less homogeneous. Vacant land showed a high degree of evenness, with all species in roughly equal proportions and very low variation in abundance. Riverside and playgrounds also showed relatively high levels of evenness. As measured by the number of species present, species richness varied significantly among the different types of urban green spaces. The highest species richness was found in institutional premises, with a value of 2.72, while the lowest was found in forests, with a value of 1.2.

8.3.2 General Characteristics of Dominant Tree Species

The general characteristics of dominant tree species identified in the study based on their IVI are recorded in Table 8.2, including information about their origin, phenology, canopy size, and growth rate. Among the dominant tree species, the majority (73%) were found to be native species, while the introduced species were naturalized to India. The finding highlights the use of native species in UGSs, as they are

Table 8.2 General characteristics of dominant tree species

	Species	Common name	Height (m)	Crown	Origin	Phenology	Canopy size	Growth
					Native, introduced	Evergreen/ deciduous	Narrow/small/ medium/large	Fast/ intermediate/ slow
1	*Acacia catechu*	Khair[2]	Up to 15[3]	Light feathery crown[13]	N[4]	D[1]	S-M[9]	F[9]
2	*Albizia lebbek*	Siris[5]	18–30[1]	Open wide spreading[8]	I[1]	D[1]	L[8]	F[8]
3	*Albizzia odoratissima*	Black siris [1]	15–25[1]	Spreading[17]	N[1]	D[1]	M[10]	F[1]
4	*Alstonia scholaris*	Shaitan Ka Jhar[2]	Up to 40[1]	Pagoda shaped, tiered, dense and round[12]	N[1]	E[1]	M-L[8]	I[8]
5	*Azadirachta indica*	Neem[1]	15–20[1]	Wide spreading, dense, ovoid[8]	N[1]	E[1]	M[2]	F[1]
6	*Boswellia serrata*	Salai [1]	Up to 20[2]	Light, spreading crown[8]	N[1]	D[2]	M[8]	S[8]
7	*Ceiba pentandra*	Kapok[8]	Up to 30[8]	Thin pagoda shaped[8]	I[1]	D[8]	L[8]	F[8]
8	*Cassia siamea*	Kassod[1]	Up to 18[1]	Irregular and spreading[8]	N[1]	E[1]	M[1]	I[8]
9	*Dalbergia sissoo*	Shisham[1]	Up to 25[1]	Oval with irregular outline[11]	N[1]	D[1]	M-L[1]	F[1]
10	*Delonix regia*	Gulmohar[6]	10–18[8]	Umbrella shaped and spreading[8]	I[1]	E/D[8]	M[8]	F[8]
11	*Eucalyptus globulus*	Nilgiri[1]	Up to 45[1]	Narrow[8]	I[1]	E[1]	L[8]	F[8]
12	*Ficus benghalensis*	Banyan tree[5]	20–30[8]	Dense wide spreading[8]	N[1]	E[8]	L[8]	F[8]
13	*Ficus glomerata*	Cluster fig [1]	20–30[8]	Irregular crown[8]	N[1]	SD[8]	L[8]	F[2]
14	*Ficus religiosa*	Peepal[1]	15–30[8]	Wide spreading[8]	N[1]	SE[1]	L[2]	F[2]

15	*Hardwickia binata*	Anjan[1]	25–30[1]	Columnar to widening[8]	N[1]	D[8]	M[1]	S[14]
16	*Leucaena leucocephala*	Subabool[1]	5–10[1]	Fairly open and rounded[8]	I[1]	E[1]	M[1]	F[1]
17	*Moringa oleifera*	Drumstick tree[1]	10–12[1]	Wide, umbrella shaped, open crown[8]	N[1]	D[1]	S-M[8]	F[1]
18	*Peltophorum pterocarpum*	Copper pod[8]	15–25[1]	Dense, spreading, umbrella shaped[8]	N[1]	E[8]	L[8]	F[1]
19	*Pithecellobium dulce*	Jungle jalebi[2]	10–15[1]	Broad and spreading or rounded[8]	I[1]	SD[15]	M[15]	F[8]
20	*Polyalthia longifolia*	False ashoka[1]	Up to 30[1]	Narrow and columnar[16]	N[8]	E[1]	N[1]	F[16]
21	*Pongamia pinnata*	Karanj[2]	15–25[1]	Broad crown[8]	N[5]	SE[1]	M[8]	F[8]
22	*Tectona grandis*	Teak[1]	Upto 40[1]	Open crown[8]	N[1]	D[1]	L[1]	F[8]
23	*Terminalia catappa*	Jangli badam[2]	Up to 35[1]	Spreading and vase shape[1]	N[1]	D[1]	L[1]	F[16]
24	*Vachellia farnesiana*	Sweet acacia[1]	5–10[1]	Wide and low[8]	I[1]	D[8]	S[8]	F[8]
25	*Vachellia nilotica*	Babul[7]	5–20[1]	Broad and rounded (rarely umbrella shaped)[8]	N[1]	E[8]	S[8]	F[8]
26	*Ziziphus mauritiana*	Ber[2]	Up to 15[8]	Spreading[8]	N[8]	E[8]	M[8]	I[8]

Sources: 1: Wikipedia.org; 2: Indian Biodiversity Portal; 3: www.fao.org; 4: International Legume Database & Information Service (ILDIS); 5: United States Department of Agriculture (USDA); 6: Cowen, D. V. (1984). Flowering Trees and Shrubs in India (Sixth ed.). Bombay: THACKER and Co. Ltd.; 7: www.merriam-webster.com; 8: http://tropical.theferns.info; 9: Indian Council of Forestry Research and Education, Dehradun; 10: https://www.flowersofindia.net/; 11: http://hort.ufl.edu/trees/DALSISA.pdf; 12: National parks flora and fauna (https://www.nparks.gov.sg/florafaunaweb); 13: Indian Council of Forestry Research and Education, Dehradun. Khair; 14: ENVS newsletter by Institute of Forest Genetics and Tree Breeding; 15: https://toptropicals.com/; 16: https://www.wikilawn.com; 16: Species Profiles for Pacific Island Agroforestry (www.traditionaltree.org); 17: https://www.flowersofindia.net

better adapted to local environmental conditions and provide various ecological benefits. The phenology revealed a balanced plantation of evergreen (42%) and deciduous (38%) trees, wherein evergreen trees provide year-round shade and other ecological benefits. In contrast, deciduous trees shed their leaves during the dry season, helping to conserve water and reducing the risk of wildfire. The fast-growing tree (81%) domination indicates growth rate as a key criterion considered by UGS managers to establish UGSs quickly.

Regarding the vertical structure of vegetation, the trees were either medium or large sized, wherein the larger trees provide a range of ecological benefits, such as carbon sequestration, habitat for wildlife that also supports their foraging requirements, and stormwater management, and medium-sized trees benefit with shade and aesthetic value. Medium-sized trees are more appropriate for planting in residential areas or along sidewalks.

8.3.3 Functional Attributes and Usage of Dominant Tree Species

Further, the functional attributes of tree species were studied through literature as shown in Table 8.3. Functional attributes refer to the specific roles and benefits that trees provide in urban ecosystems. These attributes can include but are not limited to carbon sequestration, air quality improvement, water regulation, and aesthetic value. Among the various functional attributes of tree species, ornamental characteristics were identified as one of the most important considerations during the selection and planting of trees in urban areas with less and almost no consideration to foraging requirements. Species like *Moringa, Ziziphus, Pithecellobium,* and *Azadirachta* can be considered important trees that are preferred for foraging for edible purposes. Ornamental trees provide visual appeal and enhance the aesthetic value of UGSs, which in turn can have positive impacts on mental health and well-being. Shade and shelter characteristics were also found to be important functional attributes, as they provide protection from the sun and wind, respectively, and create more comfortable microclimates for people and wildlife. Other functional attributes that were studied included the ability of trees to improve soil quality, regulate temperature and humidity, reduce noise pollution, and support biodiversity. The findings of this study highlight the importance of considering a wide range of functional attributes when selecting and managing trees in urban ecosystems, in order to maximize the benefits that they provide to both the environment and society; however, foraging was not the main consideration for plantations.

The usage as listed in Table 8.3 indicates that most of the dominant tree species identified in the city provide a variety of provisioning benefits such as timber, food, medicine, fuelwood, fiber, and other products. This highlights the importance of urban vegetation not only for its ecological significance but also for its economic and cultural value. Timber provisioning refers to the use of trees for the production of wood, which can be used for construction, furniture, and other purposes. Other uses of trees may include the production of fruits, nuts, and other non-timber forest products, which can be used for food, medicine, and other purposes. Medicinal

Table 8.3 Functional attributes of dominant tree species

S. no.	Species	Functional attributes				Usage						
		Shade/ shelter	Erosion/soil management	Soil improver	Ornamental	Timber	Fuelwood/ fodder	Fiber	Agroforestry	Edible	Medicinal	Other
1	*Acacia catechu*	–	–	–	–	●[3]	●[3]	–	●[2]	●[2]	●[2]	●[2]
2	*Albizia lebbek*	●[3]	●[3]	●[3]	●[3]	●[3]	●[3]	–	●[2]	–	●[2]	●[2]
3	*Albizzia odoratissima*	●[3]	●[3]	●[3]	–	●[3]	●[3]	–	●[2]	●[2]	–	●[2]
4	*Alstonia scholaris*	–	–	–	●[3]	●[3]	●[3]	●[3]	–	●[2]	●[2]	●[2]
5	*Azadirachta indica*	●[3]	●[3]	●[3]	–	●[3]	●[3]	–	●[2]	●[2]	●[2]	●[2]
6	*Boswellia serrata*	–	–	–	●[3]	●[3]	●[3]	●[3]	●[2]	●[2]	●[2]	●[2]
7	*Ceiba pentandra*	–	–	–	●[3]	●[3]	–	●[3]	●[2]	●[2]	●[2]	●[2]
8	*Cassia siamea*	●[3]	●[3]	●[3]	●[3]	●[3]	●[3]	–	●[2]	●[2]	●[2]	●[2]
9	*Dalbergia sissoo*	●[3]	–	●[3]	●[3]	●[3]	–	●[3]	●[2]	–	●[2]	●[2]
10	*Delonix regia*	●[3]	–	–	●[3]	●[3]	–	–	●[2]	●[2]	●[2]	●[2]
11	*Eucalyptus globulus*	●[3]	●[3]	–	●[3]	●[3]	–	●[3]	●[2]	●[2]	●[2]	●[2]
12	*Ficus benghalensis*	–	–	–	●[5]	●[5]	–	●[5]	●[2]	●[5]	●[2]	●[2]
13	*Ficus glomerata*	●[4]	●[4]	–	–	●[4]	●[4]	●[4]	●[4]	●[4]	●[4]	●[4]
14	*Ficus religiosa*	–	–	–	●[3]	●[3]	●[3]	–	–	●[2]	●[2]	●[2]
15	*Hardwickia binata*	–	–	●[3]	–	●[3]	●[3]	●[3]	●[2]	–	●[2]	●[2]
16	*Leucaena leucocephala*	●[3]	●[3]	●[3]	●[3]	●[3]	●[3]	●[3]	●[2]	●[2]	●[2]	●[2]
17	*Moringa oleifera*	–	●[3]	●[3]	–	●[3]	●[3]	●[3]	●[2]	●[2]	●[2]	●[2]
18	*Peltophorum pterocarpum*	●[3]	–	●[3]	●[3]	●[3]	●[3]	–	–	–	–	●[2]
19	*Pithecellobium dulce*	●[3]	–	–	●[3]	●[3]	●[3]	–	●[2]	●[2]	●[2]	●[2]

(continued)

Table 8.3 (continued)

S. no.	Species	Functional attributes				Usage						
		Shade/ shelter	Erosion/soil management	Soil improver	Ornamental	Timber	Fuelwood/ fodder	Fiber	Agroforestry	Edible	Medicinal	Other
20	*Polyalthia longifolia*	–	–	–	●[1]	–	–	–	–	–	–	●[2]
21	*Pongamia pinnata*	●[3]	●[3]	●[3]	●[3]	●[3]	–	●[3]	●[2]	–	●[2]	●[2]
22	*Tectona grandis*	–	–	–	●[3]	●[3]	–	–	●[2]	●[2]	●[2]	●[2]
23	*Terminalia catappa*	●[3]	●[3]	●[3]	●[3]	●[3]	●[3]	–	●[2]	●[2]	●[2]	●[2]
24	*Vachellia farnesiana*	–	–	–	–	–	–	–	●[2]	●[2]	●[2]	●[2]
25	*Vachellia nilotica*	–	–	●[3]	–	●[3]	–	●[3]	●[2]	●[2]	●[2]	●[2]
26	*Ziziphus mauritiana*	●[3]	●[3]	–	●[3]	●[3]	●[3]	–	●[2]	●[2]	●[2]	●[2]

Sources: 1: Wikipedia (https://en.wikipedia.org); 2: Useful Tropical Plants (http://tropical.theferns.info); 3: World Agroforestry (http://old.worldagroforestry.org/); 4: Plants for a Future (https://pfaf.org); 5: Plants of the World Online (http://powo.science.kew.org)

values of trees refer to their traditional use in herbal medicine and other forms of traditional medicine. Many tree species contain natural compounds that have been used for centuries by indigenous communities for their medicinal properties. These medicinal values are often associated with the high levels of bioactive compounds found in the bark, leaves, and other parts of the tree. Overall, the findings of the study highlight the diverse range of provisioning benefits by UGSs and the importance of considering these benefits in promoting urban foraging. Overall, recognizing the multiple values of trees in UGS can help to promote a more sustainable and resilient urban environment while also providing opportunities for people to connect with nature and benefit from its many values.

8.4 Discussion

8.4.1 Tree Species Diversity and Richness in UGSs

The city is rich in floristic diversity, as compared to Oakland and Athens, where the density is high, but urban tree assemblage is dominated by few species (Nowak 1994; Profous et al. 2012). However, in other medium-sized European cities like Ferrara, Cremona, and Bologna, a high species diversity is recorded in urban landscape (Salinitro et al. 2018; Pellizzari et al. 2015). The tree diversity in Nagpur varied among the classes studied and was found to be lower than the floral diversity recorded by Chaturvedi et al. (2013). Mishra et al. (2015), Nagendra and Gopal (2010), and Mishra et al. (2015) reported that Nagpur's tree diversity was similar to Delhi and Bangalore's managed UGS tree diversity. Nagpur had higher diversity than other cities like Allahabad (Pandey and Kumar 2018). Among the UGS classes, the highest diversity was recorded in the avenue plantation of Nagpur, similar to reports from other cities such as Vantaa, Finland, by Ranta et al. (2015); New York City by Hurley and Emery (2018); and Ballarat in Australia by Champness et al. (2019). However, unlike Lutyens' Delhi and Bangalore roadside plantations, Nagpur's avenues showed lower diversity (Bhalla and Bhattacharya 2017; Nagendra and Gopal 2010). The institutional premises of Nagpur have high diversity and species richness, similar to the campuses of Pune and Bangalore (Waran and Patwardhan 2006; Sudha and Ravindranath 2000). The reserve forest had the lowest species richness and diversity, even lower than the local urban forest of NEERI (Gupta et al. 2008). This suggests that the forest department as well as garden department of Nagpur Municipal Corporation (NMC) have adopted a mass plantation strategy for certain species that are important for forestry uses. It has not been the NMC's consideration to provide foraging benefits to urban dwellers and enhance nutritional and food security from these socio-ecological landscapes. As NMC is the managing and planting authority for these spaces, there is a need to initiate discussions to improve their planning process and include foraging from trees as an important criterion in tree plantation in various classes of UGS. It is also important to highlight that a lot of urban foraging experiences are not limited to trees and are from seasonal herbs, shrubs, grasses, climbers, etc. that were not covered by this study and can be focused in future studies in UGS of India. However, the study provides sufficient local data about tree diversity and highlights areas and classes where more species can be introduced to improve tree diversity, particularly along water bodies, parks, playgrounds, and the reserve forest areas that support foraging requirements of local people.

8.4.2 Tree Species Selection in Urban Environment

Trees in urban areas face various challenges, including exposure to pollutants, elevated temperatures due to the heat island effect, limited space for root growth, and reduced water availability caused by compacted soil (Mansell 2003; Watson and Kelsey 2006). Understanding these challenges helps in selecting suitable tree species based on their attributes, like selection of trees with dense and broad canopies to enhance the aesthetic value as well as for higher aboveground biomass. Additionally, trees with thick foliage effectively remove air pollutants and provide shade, thus aiding in temperature regulation (Jim 2012; McPherson et al. 1997; Pauleit and Duhme 2000). Thus, careful species selection plays a crucial role in supporting urban foraging from public UGS. By choosing appropriate species, the urban environment can be significantly improved, leading to increased benefits, particularly in terms of food and nutritional security, derived from UGSs in India.

In Nagpur, the vegetation comprises both native and introduced species, including a mix of evergreen and deciduous trees. Among the most common native species in all classes is *Azadirachta indica*, which has a high tolerance to drought, fast growth rate, and dense canopy. It is also a potential foraging species. Its flowers during spring are used in food preparations, and leaves and seeds render medicinal benefits for all social classes, including the use of its twigs as tooth cleaners. The most prevalent introduced species is *Polyalthia longifolia*, primarily planted in parks and gardens because of its thick foliage, which creates a screening effect and has a fast growth rate. Ornamental trees such as *Delonix regia*, *Cassia siamea*, *Peltophorum pterocarpum*, and *Bauhinia variegata* are also popular choices due to their flowering characteristics, fast growth rate, and shady canopy. *Bauhinia* is a fast-regenerating and preferred foraged tree for its buds during winters to prepare vegetable preparations or pickles. However, this traditional knowledge is also slowly depleting because of fast urbanization, loss of these trees, and loss of linkages with culture and traditions due to loss of intergenerational connections to learn about foraging. These species are dominant in managed UGSs, as identified in other studies of Indian cities (Sudha and Ravindranath 2000; Anamika and Pradeep 2016; Dhanapal and Chaudhry 2012). In forested areas, deciduous tree species with high drought tolerance are the dominant characteristics. Therefore, *Tectona grandis*, *Hardwickia binata*, and *Boswellia serrata* are the preferred species for mass plantation but have no food value. However, they have value for wild foraging by birds and insects. Overall, growth rate, resistance to drought and pollution, high usage value, and functional attributes are identified to be prominent in tree species selection by UGS managers. Other studies in Bangalore have identified growth rate as an important criterion for species selection (Sudha and Ravindranath 2000), while some researchers have emphasized the importance of productive trees with longer life spans to mitigate carbon concentrations in cities (Nowak and Crane 2002).

8.4.3 Urban Foraging Potential in UGSs of Nagpur and Other Urbanized Cities

Barthel and Isendahl (2013) and Campbell (2016) highlight the limited space for food production in urban areas and the potential benefits of urban foraging. The concept of the "edible cities" further underscores the significance of increasing the availability of edible species in urban areas for localized food production (Sartison and Artmann 2020). Regarding Nagpur's UGSs, the findings reveal prevalence of native tree species like *Azadirachta indica, Ficus religiosa, Ficus benghalensis*, and *Bauhinia variegata*, indicating high potential of urban foraging activities. The dominance of edible plants and fruit trees in Nagpur's UGSs creates an opportunity for urban foraging, thereby contributing to food security. A study conducted in Bangalore, India, observed a positive association between the presence of fruit trees in urban parks and their use for food-related activities, such as picnicking and foraging (Nagendra et al. 2012). This aligns with the preferences of people in Nagpur who visit UGSs for similar purposes, especially during holidays, and for housewives and elderly individuals who frequent these spaces in the mornings or afternoons.

A study conducted in Delhi found that the presence of these native tree species positively influences bird diversity and abundance (Kaur et al. 2019), indicating that Nagpur's UGSs, with their dominant native trees, can provide food and habitat for local wildlife. Additionally, it is important to recognize the significance of introducing wild plants and spontaneous vegetation beyond a sole focus on ornamental tree species. Diversifying the plant composition in UGSs can support ecological balance, enhance biodiversity, and provide additional resources for foraging.

While incorporating foraging as an acceptable activity in urban areas, there are expected challenges such as high concentration of people, restricted access, establishing sustainable harvesting systems, identifying appropriate species, determining appropriate levels of harvest, and addressing contamination issues (McLain et al. 2014). Russo et al. (2017) also highlight challenges in implementing, managing, and promoting edible green infrastructure in cities, emphasizing the need for context-specific knowledge, expertise, and information from local governments and nongovernmental organizations. Despite these challenges, embracing foraging in UGSs offers multiple benefits, including enhanced ecological understanding; intertwining environmental, social, and economic co-benefits; proximity of food sources to urban dwellers; preservation of cultural identities; and potential contributions to food security and community well-being (Barthel et al. 2010; Russo et al. 2017).

Urban foraging can improve the quality of life for marginalized urban residents, particularly women and socially disadvantaged groups, by providing nutritional supplements and honoring traditional cultural needs while incorporating valuable knowledge into urban planning (Somesh et al. 2021). Incorporating foraging as an acceptable activity in UGSs can promote sustainable urban ecosystem management and provide numerous benefits to urban communities, despite the challenges involved.

8.5 Conclusion

UGSs of Nagpur have the potential to foster interactions between urban dwellers and urban biodiversity due to high species diversity and richness. The dominant species can provide food and habitat for local wildlife and edible fruits and leaves and have medicinal properties. Hence, the trees of UGSs in several ways can foster human-nature interactions, by (1) providing opportunities to observe and learn about the different species by installation of signage, nature trails, and educational programs; (2) encouraging citizen science, where urban dwellers can participate in data collection and contribute to scientific research on urban biodiversity; and (3) building awareness about foraging, even among those who may not be dependent on it for food, to help better design and protect urban spaces and make them more accessible and cleaner. To improve our understanding of urban foraging, it is crucial that future research focuses on better understanding of who is engaging in foraging activities and what they are foraging for. Also, investigate the ecology of UGSs that support foraging activities and characteristics of foraged plants.

Acknowledgments We are thankful to Dhiraj Buradkar, Khushbu Singh, Deepika Sarda, Ashwini Rangari, Harsha Nandurkar, and Pooja Fulwadhani for their support in conducting the fieldwork.

Conflicts of Interest The authors declare no conflict of interest with respect to the research, authorship, and/or publication of this chapter.

References

Adeyemi O, Shackleton C (2023) Understanding foraging practices in Lagos metropolis to redesign urban greenspaces in support of human-nature interactions. Urban For Urban Green 79:127805. https://doi.org/10.1016/j.ufug.2022.127805

Anamika A, Pradeep C (2016) Urban vegetation and air pollution mitigation: some issues from India. Chin J Urban Environ Stud 04(01):1650001. https://doi.org/10.1142/S2345748116500019

Barthel S, Isendahl C (2013) Urban gardens, agriculture, and water management: sources of resilience for long-term food security in cities. Ecol Econ 86:224–234. https://doi.org/10.1016/j.ecolecon.2012.06.018

Barthel S, Folke C, Colding J (2010) Social–ecological memory in urban gardens—retaining the capacity for management of ecosystem services. Glob Environ Change 20(2):255–265. https://doi.org/10.1016/j.gloenvcha.2010.01.001

Beatley T (2011) Biophilic cities: integrating nature into urban design and planning. Island Press, Washington, DC. https://doi.org/10.5822/978-1-59726-986-5

Bhalla P, Bhattacharya P (2017) Urban biodiversity and green spaces in Delhi: a case study of new settlement and Lutyens' Delhi. J Hum Ecol 52:83–96

Borah N, Nath AJ, Das AK (2013) Aboveground biomass and carbon stocks of tree species in tropical forests of cachar district, assam, northeast india. Int J Ecol Environ Sci 39:97–106

Campbell LK (2016) Getting farming on the agenda: planning, policymaking, and governance practices of urban agriculture in New York City. Urban For Urban Green 19:295–305. https://doi.org/10.1016/j.ufug.2016.03.011

Ceuterick M, Vandebroek I (2017) Identity in a medicine cabinet: discursive positions of Andean migrants towards their use of herbal remedies in the United Kingdom. Social Sci Med 177:43–51. https://doi.org/10.1016/j.socscimed.2017.01.026

Champness BS, Palmer GC, Fitzsimons JA (2019) Bringing the city to the country: relationships between streetscape vegetation type and bird assemblages in a major regional centre. J Urban Ecol 5(1). https://doi.org/10.1093/jue/juz018

Chaturvedi A, Kamble R, Patil NG, Chaturvedi A (2013) City–forest relationship in Nagpur: one of the greenest cities of India. Urban For Urban Green 12:79–87. https://doi.org/10.1016/j.ufug.2012.09.003

Clark KH, Nicholas KA (2013) Introducing urban food forestry: a multifunctional approach to increase food security and provide ecosystem services. Landsc Ecol 28(9):1649–1669. https://doi.org/10.1007/s10980-013-9903-z

Cox DTC, Hudson HL, Shanahan DF et al (2017) The rarity of direct experiences of nature in an urban population. Landsc Urban Plan 160:79–84. https://doi.org/10.1016/j.landurbplan.2016.12.006

Craft D (2010) Urban foraging: finding and eating wild plants in the city. Serviceberry Press, Memphis, TN

Crowe S, Cresswell K, Robertson A et al (2011) The case study approach. BMC Med Res Methodol 11(1):1–9. https://doi.org/10.1186/1471-2288-11-100

Curtis J, McIntosh RP (1951) An upland forest continuum in the prairie forest border region of Wisconsin. Ecology 32(3):476–496. https://doi.org/10.2307/1931725

Dhanapal G, Chaudhry P (2012) Open spaces for urban sustainability. Sustainability Outlook, Retrieved from: http://sustainabilityoutlook.in/cont

Dhyani S, Fulwadhani P, Gujre N, Wate S (2020) A Score-based Method for Assessing the Importance of Urban Forest and its Ecosystem Services. Int J Plant Environ 6(1):35–36. https://doi.org/10.18811/ijpen.v6i01.04

Dhyani S, Kadaverugu R (2020) Food security and cultural benefits from urban green spaces: exploring urban foraging as a silently growing global movement. Clim Chang Environ Sustain 8(2):233–241. https://doi.org/10.5958/2320-642X.2020.00022.8

Dhyani S, Lahoti S, Khare S, Pujari P, Verma P (2018) Ecosystem based disaster risk reduction approaches (EcoDRR) as a prerequisite for inclusive urban transformation of Nagpur City, India. Int J Disaster Risk Reduct 32:95–105. https://doi.org/10.1016/j.ijdrr.2018.01.018

Dhyani S, Singh A, Gujre N, Joshi RK (2021) Quantifying tree carbon stock in historically conserved Seminary Hills urban forest of Nagpur, India. Acta Ecologica Sinica 41-3:193–203. https://doi.org/10.1016/j.chnaes.2021.01.006

Erickson D (2010) MetroGreen: connecting open space in North American cities. J Agric Environ Ethics 23:491–494. https://doi.org/10.1007/s10806-009-9220-3

Gabriel N (2006) Urban non-timber forest products in Philadelphia. Master's thesis. Temple University, Philadelphia, PA

Gobster PH (2007) Urban park restoration and the museumification of nature. Nat Cult 2(2):95–114. https://doi.org/10.3167/nc.2007.020201

Grabbatin B, Hurley PT, Halfacre AC (2011) I still have the old tradition: the co-production of sweetgrass basketry and coastal development. Geoforum 42(6):638–649

Gupta R, Chaudhari P, Wate S (2008) Floristic diversity in urban forest area of NEERI campus, Nagpur, Maharashtra (India). J Environ Sci Eng 50(1):55–62

Haaland C, Konijnendijk C (2015) Challenges and strategies for urban green-space planning in cities undergoing densification: a review. Urban For Urban Green 14:760–771. https://doi.org/10.1016/j.ufug.2015.07.009

Hurley PT, Grabbatin B, Goetcheus C, Halfacre A (2013) Gathering, buying, and growing sweetgrass (Muhlenbergia sericea): urbanization and social networking in the sweet-

grass basket-making industry of low country South Carolina. In: Vocks R, Rashford J (eds) African ethnobotany in the Americas. Springer, New York, NY, pp 153–173. https://doi.org/10.1007/978-1-4614-0836-9_6
Hurley PT, Emery MR, McLain R, Poe M, Grabbatin B, Goetcheus C (2015) Whose urban forest? The political ecology of foraging urban nontimber forest products. In: Isenhour C, McDonogh G, Checker M (eds) Sustainability in the Global City. Cambridge University Press, Cambridge, UK. https://doi.org/10.1017/CBO9781139923316.014
Hurley PT, Emery MR (2018) Locating provisioning ecosystem services in urban forests: forageable woody species in new York City, USA. Landsc Urban Plan 170:266–275. https://doi.org/10.1016/j.landurbplan.2017.09.025
i-Tree (2018) I-Tree eco user's manual v6.0. Retrieved from https://www.itreetools.org/resources/manuals/Ecov6_ManualsGuides/Ecov6_UsersManual.pdf
Jahnige P (2004) The hidden bounty of the urban forest. In: Elevitch C (ed) The Overstory book: cultivating connections with trees. Permanent Agriculture Resources, Holualoa, USA, pp 291–295
Jim C (2012) Sustainable urban greening strategies for compact cities in developing & developed economies. Urban Ecosyst 16:741–761. https://doi.org/10.1007/s11252-012-0268-x
Kabisch N, Qureshi S, Haase D (2015) Human-environment interactions in urban green spaces–a systematic review of contemporary issues and prospects for future research. Environ Impact Assess 50:25–34. https://doi.org/10.1016/j.eiar.2014.08.007
Kaur A, Kaur S et al (2019) Phenotypic variations alter the ecological impact of invasive alien species: lessons from Parthenium hysterophorus. J Environ Manag 241:187–197. https://doi.org/10.1016/j.jenvman.2019.03.129
Ladio AH, Lozada M (2004) Patterns of use and knowledge of wild edible plants in distinct ecological environments: a case study of a Mapuche community from North-Western Patagonia. Biodivers Conserv 13:1153–1173. https://doi.org/10.1023/B:BIOC.0000018150.79156.50
Lahoti S, Kefi M, Lahoti A, Saito O (2019a) Mapping methodology of public urban green spaces using GIS: an example of Nagpur City, India. Sustainability 11(7):2166. https://doi.org/10.3390/su11072166
Lahoti S, Lahoti A, Saito O (2019b) Benchmark assessment of recreational public urban green space provisions: a case of typical urbanizing Indian city, Nagpur. Urban For Urb Green 44:126424. https://doi.org/10.1016/j.ufug.2019.126424
Lahoti S, Lahoti A, Joshi RK, Saito O (2020) Vegetation structure, Species composition, and Carbon sink potential of urban green spaces in Nagpur city, India. Land 9(4):107. https://doi.org/10.3390/land9040107
Lin B, Fuller AR (2013) Sharing or sparing? How should we grow the world's cities? J Appl Ecol 50:1161–1168. https://doi.org/10.1111/1365-2664.12118
Lin BB, Egerer MH, Ossola A (2018) Urban gardens as a space to engender biophilia: evidence and ways forward. Front Built Environ 4:79
MacKerron G, Mourato S (2013) Happiness is greater in natural environments. Glob Environ Chang 23:992–1000
Mansell MG (2003) Rural and urban hydrology. Thomas Telford Ltd, London
McLain R, Poe MR, Hurley PT, Lecompte-Mastenbrook J, Emery MR (2012) Producing edible landscapes in Seattle's urban forest. Urban For Urban Green 11(2):187–194. https://doi.org/10.1016/j.ufug.2011.12.002
McLain RJ, Hurley PT, Emery MR et al (2014) Gathering wild food in the city: rethinking the role of foraging in urban ecosystem planning and management. Local Environ 19:220–240. https://doi.org/10.1080/13549839.2013.841659
McPherson EG, Nowak D, Heisler G, Grimmond S et al (1997) Quantifying urban forest structure, function and value: the Chicago urban Forest climate project. Urban Ecosyst 1:49–61. https://doi.org/10.1023/A:1014350822458
Mhatre N (2008) Secret lives: biodiversity of the Indian Institute of Science campus. Indian Institute of Science Press, Bangalore, p 229
Mishra AK, Sharma MP, Singh HB (2015) Addition to the Flora of Delhi. Ind J Plant Sci 4(2):1–6

Mollee E, Pouliot M, McDonald MA (2017) Into the urban wild: collection of wild urban plants for food and medicine in Kampala, Uganda. Land Use Policy 63:67–77. https://doi.org/10.1016/j.landusepol.2017.01.020

Nagendra H, Gopal D (2010) Street trees in Bangalore: density, diversity, composition and distribution. Urban For Urban Green 9:129–137. https://doi.org/10.1016/j.ufug.2009.12.005

Nagendra H, Nagendran S, Paul S, Pareeth S (2012) Graying, greening and fragmentation in the rapidly expanding Indian city of Bangalore. Landsc Urban Plan 105(4):400–406. https://doi.org/10.1016/j.landurbplan.2012.01.014

Nowak DJ (1994) Air pollution removal by Chicago's urban Forest. In: McPherson EG et al (eds) Chicago's urban forest ecosystem: results of the Chicago urban Forest ecosystem: results of the Chicago climate project, pp 63–83

Nowak DJ, Crane DE (2002) Carbon storage and sequestration by urban trees in the USA. Environ Pollut 116:381–389. https://doi.org/10.1016/S0269-7491(01)00214-7

Palliwoda J, Kowarik I, von der Lippe M (2017) Human-biodiversity interactions in urban parks: the species level matters. Landsc Urban Plan 157:394–406. https://doi.org/10.1016/j.landurbplan.2016.09.003

Pandey R, Kumar H (2018) Tree species diversity and composition in urban green spaces of Allahabad city (U.P). Plant Archives 18(2):2687–2692

Pauleit S, Duhme F (2000) Assessing the environmental performance of landcover types for urban planning. Landsc Urban Plan 52:1–20. https://doi.org/10.1016/S0169-2046(00)00109-2

Pellizzari R, Piccoli F, La Alessandrini A (2015) flora vascolare urbana di Ferrara. Quaderni del Museo Civico di Storia Naturale di Ferrara 3:55–90

Poe MR, McLain RJ, Emery M et al (2013) Urban forest justice and the rights to wild foods, medicines, and materials in the city. Human Ecol 41(3):409–422. https://doi.org/10.1007/s10745-013-9572-1

Profous GV, Rowntree RA, Loeb RE (2012) The urban forest land- scape of Athens, Greece: aspects of structure, planning and management. Arboric J 12:83–108. https://doi.org/10.1080/03071375.1988.9756380

Ranta P, Kesulahti J, Tanskanen A et al (2015) (2015) roadside and riverside green–urban corridors in the city of Vantaa, Finland. Urban Ecosyst 18:341–354. https://doi.org/10.1007/s11252-014-0402-z

Russo A, Escobedo FJ, Cirella GT et al (2017) Edible green infrastructure: an approach and review of provisioning ecosystem services and disservices in urban environments. Agric Ecosyst Environ 242:53–66. https://doi.org/10.1016/j.agee.2017.03.026

Salinitro M, Alessandrini A, Zappi A et al (2018) Floristic diversity in different urban ecological niches of a southern European city. Sci Rep 8:15110. https://doi.org/10.1038/s41598-018-33346-6

Sardeshpande M, Shackleton C (2023) Urban foraging: land management policy, perspectives, and potential. PLoS One 15(4):e0230693. https://doi.org/10.1371/journal.pone.0230693

Sartison K, Artmann M (2020) Edible cities–an innovative nature-based solution for urban sustainability transformation? An explorative study of urban food production in German cities. Urban For Urban Green 49:126604. https://doi.org/10.1016/j.ufug.2020.126604

Schuttler SG, Sorensen AE, Jordan RC et al (2018) Bridging the nature gap: citizen science reverse the extinction of experience? Front Ecol Environ 16:405–411. https://doi.org/10.1002/fee.1826

Severson K (2009) Neighbour, can you spare a plum? The New York Times [online]. Available from: http://www.nytimes.com/2009/06/10/dining/10Fruit.html. Accessed 11 Feb 2023

Shackleton C, Hurley P, Dahlberg A et al (2017) Urban foraging: a ubiquitous human practice overlooked by urban planners, policy, and research. Sustainability 9:1884. https://doi.org/10.3390/su9101884

Shanahan DF, Lin BB, Bush R et al (2015) Toward improved public health outcomes from urban nature. Am J Public Health 105:470–478. Print ISSN: 0090-0036

Short-Gianotti AG, Hurley PT (2016) Gathering plants and fungi along the urban-rural gradient: uncovering differences in the attitudes and practices among urban, suburban, and rural landowners. Land Use Policy 57:555–563. https://doi.org/10.1016/j.landusepol.2016.06.019

Soga M, Gaston KJ (2016) Extinction of experience: the loss of human–nature interactions. Front Ecol Environ 14:94–101. https://doi.org/10.1002/fee.1225

Somesh D, Rao R, Murali R, Nagendra H (2021) Patterns of urban foraging in Bengaluru city. Urban For Urban Green 57:126940. https://doi.org/10.1016/j.ufug.2020.126940

Sudha P, Ravindranath N (2000) A study of Bangalore urban forest. Landsc Urban Plan 47(1–2):47–63. https://doi.org/10.1016/S0169-2046(99)00067-5

Sujarwo W, Arinasa IBK, Caneva G et al (2016) Traditional knowledge of wild and semi-wild edible plants used in Bali (Indonesia) to maintain biological and cultural diversity. Plant Biosystems 150:971–976. https://doi.org/10.1080/11263504.2014.994577

Synk CM, Kim BF, Davis CA et al (2017) Gathering Baltimore's bounty: characterizing behaviors, motivations, and barriers of foragers in an urban ecosystem. Urban For Urban Green 28:97–102. https://doi.org/10.1016/j.ufug.2017.10.007

Threlfall C, Ossola A, Hahs A, Williams N, Wilson L, Livesley S (2016) Variation in vegetation structure and composition across urban green space types. Front Ecol Evol 4(June):1–12. https://doi.org/10.3389/fevo.2016.00066

Unnikrishnan H, Nagendra H (2015) Privatizing the commons: impact on ecosystem services in Bangalore's lakes. Urban Ecosyst 18(2):613–632. https://doi.org/10.1007/s11252-014-0401-0

UNDESA (2018) United nations, Department of Economic and Social Affairs, population division. World urbanization prospects: the 2018 revision, online edition. Accessed 03 Jun 2018

Waran A, Patwardhan A (2006) Urban carbon burden of Pune City: a case study from India. (Masters Thesis). Department of Environmental Sciences, University of Pune

Watson GW, Kelsey P (2006) The impact of soil compaction on soil aera- tion and fine root density of Quercus palustris. Urban Forestry and Urban Greening 4(2):69–74. https://doi.org/10.1016/j.ufug.2005.08.001

Wehi PM, Wehi WL (2010) Traditional plant harvesting in contemporary fragmented and urban landscapes. Conserv Biol 24(2):594–604. https://doi.org/10.1111/j.1523-1739.2009.01376.x

Shruti Lahoti is policy researcher at the Institute for Global Environmental Strategies (IGES). Earlier she worked with IGES as a research fellow, and before joining IGES, she was at United Nations University for 8 years. She works as editorial associate for the Sustainability Science journal. She is a landscape planner, and holds a Ph.D. in Sustainability Science from Institute for Advanced Study of Sustainability (UNU-IAS), Tokyo, Japan. Her current research interest is in Urban landscapes, Nature-based solutions, Green infrastructure and other ways to enhance urban green spaces for the well-being of the urban dwellers.

Ashish Lahoti is policy researcher at the Institute for Global Environmental Strategies (IGES). Earlier she worked with IGES as a research fellow, and before joining IGES, she was at United Nations University for 8 years. She works as editorial associate for the Sustainability Science journal. She is a landscape planner, and holds a Ph.D. in Sustainability Science from Institute for Advanced Study of Sustainability (UNU-IAS), Tokyo, Japan. Her current research interest is in Urban landscapes, Nature-based solutions, Green infrastructure and other ways to enhance urban green spaces for the well-being of the urban dwellers.

Shalini Dhyani (PhD) is a Principal Scientist at the CSIR-National Environment Engineering Research Institute, India. She is Asia Vice Chair and Steering Committee member in the Commission on Ecosystems Management of IUCN and Visiting Scientist at CIFOR-ICRAF.

Osamu Saito (PhD) is Principal Policy researcher at the Institute for Global Environmental Strategies (IGES) and visiting associate professor at the University of Tokyo (UoT), Japan. He has earlier worked as Academic Director for the Advanced Study of Sustainability (UNU-IAS), Japan. He has been a managing editor of the Sustainability Science journal published by Springer, and his research interests include socio-ecological studies on the ecosystem services provided by traditional rural production landscapes (Satoyama) in both Japan and other Asian countries.

Part III

Where Is It Foraged?

9 Accessing Culturally Significant Species in New York City, USA's Urban Forest: The Case of *Ginkgo biloba* and *Morus* spp. Harvesting from Street Trees by Chinese-American Residents and Their Descendants

Sarah Becker, Patrick T. Hurley, Marla R. Emery, and Joana Chan

Abstract

Emerging scholarship on urban forests indicates that diverse green spaces provide immigrant populations opportunities to maintain their cultural identities. Modes of engagement in these spaces vary, but urban foraging represents one culturally significant mode for a number of sociocultural groups. While foraging has been recognized as a ubiquitous practice, studies of biocultural aspects of foraging remain underdeveloped. For example, it remains unclear whether or not culturally significant species are accessible to the particular sociocultural groups that harvest them. To examine this research gap, we draw on interviews with New York City foragers who self-identify as Chinese or Chinese-American and who indicate that fruits from ginkgo (*Ginkgo biloba*) and mulberry (*Morus* spp.) are important foraged biocultural items. To assess accessibility, we generated GIS hot spots of each species from the city's street tree census and assessed the overlap of these hot spots with neighborhood tabulation areas (NTAs) with high percentages of residents who self-identify as Chinese or Chinese-American. Our results suggest that *Ginkgo* street trees in Manhattan are highly accessible, while

S. Becker · P. T. Hurley (✉)
Ursinus College, Collegeville, PA, USA
e-mail: phurley@ursinus.edu

M. R. Emery
Norwegian Institute for Nature Research & Rubenstein School of Environment and Natural Resources, University of Vermont USA, Burlington, VT, USA

J. Chan
University of Vermont USA, Burlington, VT, USA

S. Dhyani, M. Sardeshpande (eds.), *Urban Foraging in the Changing World*,
https://doi.org/10.1007/978-981-97-0345-6_9

a subset in Brooklyn are moderately accessible and a smaller subset in Queens are relatively inaccessible. These findings suggest that street trees have the potential to support biocultural foraging practices, even in some of the most developed portions of our cities.

Keywords

Foraging · Neighborhood · Motivation · Biocultural · Hot spots

9.1 Introduction

Scholars of urban foraging are increasingly examining both demand and supply perspectives of this ubiquitous practice (Shackleton et al. 2017; Hurley and Emery 2018, 201; Fischer and Kowarik 2020; Hurley et al. 2022), although much of the existing research centers on the practices of foragers and their harvests (Shackleton et al. 2017; Fischer and Kowarik 2020; Schunko and Brandner 2022). Demand-side perspectives focus largely on identifying which foraging practices exist among city residents, the range of species associated with these practices, the motives behind their continued practice, and how foraging benefits city residents and community members. This literature focuses predominantly on the material benefits foragers obtain, such as for food (see, e.g., contributions to residents' diets (Synk et al. 2017; Stark et al. 2019; Garekae and Shackleton 2020)), but also nonmaterial benefits such as potential support of food and medicinal sovereignty (Poe et al. 2013) and the ability to maintain cultural and family identities by engaging in harvesting (Hurley et al. 2015; Landor-Yamagata et al. 2018; Marquina et al. 2022). Likewise, there is also a growing recognition of the psychological benefits foragers gain from this practice through greater connections to nature (Poe et al. 2014; Fischer and Kowarik 2020; Marquina et al. 2022). Together, the practice has the potential to foster social-ecological resilience among urban residents (Sardeshpande et al. 2021) and support biocultural diversity (Dhyani and Kadaverugu 2020).

Yet within the literature on forager practices and the species they harvest (i.e., forager demand), there has been little attention to the biocultural aspects of the practice (see Marquina et al. 2022 for an exception). Biocultural approaches are most often associated with research on indigenous peoples and the complex set of knowledge, associated values, and beliefs that shape interactions with plants, animals, and places and with studies in rural places (Baumflek et al. 2021; Turner et al. 2022). However, biocultural approaches have also been deployed in cities to consider the practices of non-indigenous communities, such as culturally distinctive immigrant groups (Nesbitt et al. 2020). Where aspects of cultural maintenance have been documented in previous urban foraging research, usually in terms of motivations for foraging, we know that distinct cultural groups do use certain potentially distinctive forageable species in their daily lives and celebrations (Poe et al. 2014; Marquina et al. 2022). For instance, purslane plants feature frequently in Mexican diets (Emery and Hurley, unpublished data), while ginkgo nuts are served during special occasions—like weddings—in Chinese culture (Hurley et al. Forthcoming) and mulberries may be harvested to make a tea (Emery and Hurley, unpublished data). This can be a powerful tool for immigrants in unfamiliar countries, since it provides them with one link among others to integrate their two realities (Jay and Schraml 2009; Marquina et al. 2022).

Research on supply-side dynamics, or material availability, generally focuses on the species composition of urban forests themselves. This research approach seeks to document and understand how existing urban forests function as sites of plant material harvests that provide foods, medicines, and other natural resources (Kaoma and Shackleton 2015; Garekae et al. 2022; Hurley et al. 2022; Adeyemi and Shackleton 2023). This research has assessed the distribution of desirable species from a diversity of green spaces within the urban forest, ranging from parks (Fischer and Kowarik 2020; Stark et al. 2019) and green infrastructure (Russo et al. 2017) to street trees (Hurley and Emery 2018), fruit trees throughout the city (Larondelle and Strohbach 2016), and the entire urban forest (Hurley et al. 2022), as well as differences among green space types (Garekae et al. 2022). Still other research on supply has examined the productivity of desirable species for foragers in urban settings (Bunge et al. 2019). Meanwhile, a separate stream of research advocates for and assesses how managers can enhance the accessibility of food and other resources from urban forests for foraging through both design-oriented interventions that improve species selection for harvesting (Clark and Nicholas 2013) and through changes to policies to improve availability to existing species and spaces (Kowalski and Conway 2019; Shortly and Kepe 2021; Sardeshpande and Shackleton 2023). Here, too, biocultural examinations of urban forests in relation to urban foraging practices are an understudied dimension of this practice (Dhyani and Kadaverugu 2020).

At the intersection of understanding foraging, whether in terms of forager practices or the availability of forageable species, are questions of access. First, there is a rich literature on access to and use of natural resources (Peluso and Ribot 2020), particularly in rural areas. Collectively, this work points to the importance of formal and informal rules associated with places where valued resources occur, social norms that condition harvesting practices, and uneven enforcement of rules and regulations. Specific studies on foraging point further to the importance of the ability to locate and harvest species of significance (Baumflek et al. 2015), while there is an understanding that foraging may be illegal in some countries (Hurley et al. 2015; Shortly and Kepe 2021) or viewed as a problematic practice by some urban green space managers (Schunko et al. 2021). In urbanizing areas, fringe ecologies, or the presence of species in areas of unclear ownership or management jurisdiction, provide opportunities for harvests of otherwise inaccessible species (Hurley et al. 2008, 2013). These informal green spaces have been demonstrated to support foraging practices in South Africa, with foragers sometimes walking long distances to access desired plant materials (Sardeshpande and Shackleton 2023).

Second, there is also a rapidly evolving literature on questions of access to urban green space, which more generally considers equitable, or even, levels of green space for city residents in terms of proximity. Studies consistently find that, when compared to other residents, immigrants and other minoritized groups frequently lack access to these urban green spaces. For example, in Berlin, Germany—a city that strives to provide its residents with 6 square meters of urban green space per person—access is highly uneven among different cultural groups (Kabisch and Haase 2014). Limited access to urban green spaces with culturally important resources in this way, and along inequitable lines, suggests an inability for individuals who self-identify with particular cultures to maintain cultural identities and practices. Without these opportunities for harvest of culturally important species, it can be increasingly difficult to preserve the socio-ecological resilience of urban residents (Calderón-Contreras and White 2020; Hurley et al. 2013).

This chapter examines the extent to which culturally significant species are accessible to urban foragers. The research uses a case study approach, focusing on urban foragers living in New York City, USA, who identify as either Chinese or Chinese-American and who forage for materials from the tree species, *Ginkgo biloba* (nuts) and mulberry (*Morus* spp., berries). We specifically examine the distribution of these two street tree species and consider their accessibility to foragers seeking their fruits. The choice to examine street trees is largely limited by data availability, but this focus offers an important opportunity to consider access through a biocultural lens. Specifically, street trees are a widely distributed and easily accessible resource—in terms of approaching a preferred species by means of generally well-maintained walkways and without the presence of gates or other physical obstacles. We note that while street trees are under the jurisdiction of the New York Parks Department and are subject to citywide provisions against their harvesting (Hurley and Emery 2018), street trees exist in areas that are not regularly monitored by park personnel and exist as de facto commons for many urban foragers (Hurley et al. 2013). In short, street trees exhibit the types of characteristics associated with fringe ecologies described above.

Our research uses a grounded visualization approach to explore the questions of accessibility, drawing on qualitative data about harvesting by the case study community as well as quantitative datasets on the distribution of street trees, patterns of residency among our case study community, and New York City's public transportation system. In the process, we demonstrate a geographic information system (GIS) methodology that can be used to provide a coarse-level assessment of accessibility to culturally significant species for particular cultural groups. We suggest that this method might be adopted by managers in other cities to begin exploring how, in this case, street trees support biocultural dimensions of urban foraging.

9.2 Data and Methods

Our analysis examines the case of access to culturally significant woody species by Chinese immigrants and their descendants in New York City's urban forest. To do so, we employ a four-step process (Fig. 9.1). We note that we report the results of interviews with these foragers in the methods section, as part of step one, since the subsequent steps derive from this critical information. While our methodological process is generally linear in nature, we do employ aspects of analysis common in

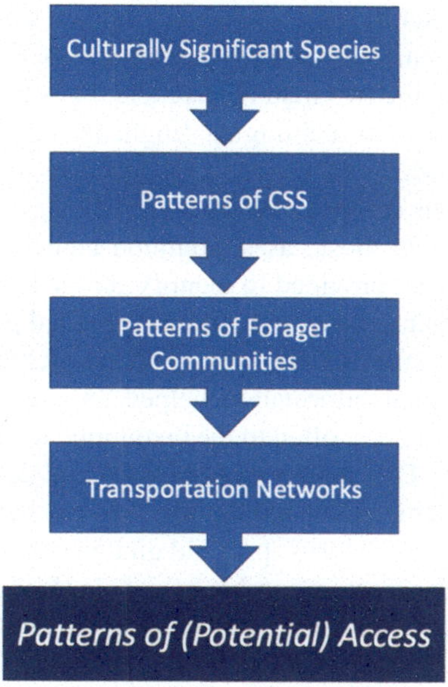

Fig. 9.1 Methodological and analytical approach

grounded visualization approaches to understanding urban social-ecological dynamics (Knigge and Cope 2006; Hurley et al. 2008), following similar analyses undertaken on questions relating to the ways specific urban forests relate to foraging practices (Hurley et al. 2008; Hurley and Emery 2018).

We first draw on interviews with New York City foragers who self-identify as Chinese or Chinese-American to identify culturally significant species for this sociocultural group. Second, we use the city's publicly available 2015 Street Tree Census database to identify patterns for these culturally significant woody species. Third, we use neighborhood tabulation areas (NTAs), a type of census data, to provide a proxy for this cultural group, allowing us to approximate how accessible these significant species to residents are by way of walking. In this step, we consider areas of direct overlap, where accessibility might be possible on foot. Fourth, we expand our considerations of accessibility to include the connectivity that public transportation routes provide. Together, these steps allow us to consider correspondence among these patterns of overlap and suggest areas where different types of access to these plant materials are enhanced (or not) by different modes of transportation. In the section that follows, we discuss each of these data sources and their associated analytical steps.

9.2.1 Identifying Culturally Significant Species Harvested by Chinese Immigrants and Their Descendants

We begin by examining interview data to identify culturally significant species for further analysis. In 2014, semi-structured interviews were conducted with 21 individuals living in New York City who self-identified as Chinese or Chinese-American, with interviewees living in Queens or Brooklyn. Interviews were conducted in interviewees' choice of English, Cantonese, or Mandarin. Interviews conducted in Cantonese and/or Mandarin were simultaneously translated and transcribed by one of the two field researchers: (1) a native Cantonese speaker also fluent in Mandarin and (2) a bilingual English/Cantonese speaker. All transcribed interviews were coded in NVivo using a primarily a priori coding system based on theory and previous empirical findings.

In most cases, common names—in either English, Cantonese, or Mandarin—were provided to identify foraged species. While we did not collect voucher specimens, some respondents provided pictures of harvested materials and indicated the locations from which they frequently harvested specific plant materials. This additional information helped an expert in New York City flora to confirm species matches, often to the taxonomic level of genus, despite the lack of voucher samples.

Based on common names and descriptions of plants, mushrooms, and their uses, participants in the study reported foraging approximately 52 species. Of these, mulberry (*Morus* spp.) and ginkgo (*Ginkgo biloba*) were mentioned as being foraged by the greatest number of people (8 and 6, respectively), more so than other woody or herbaceous species and, thus, signaling these two species as culturally significant to this group of foragers. Since studies surveying urban foragers in the USA and Europe suggest that herbaceous species harvests predominate (Synk et al. 2017; Fischer and Kowarik 2020), the frequency with which members of this sociocultural group reported harvesting ginkgo and mulberry is noteworthy. We note, however, that research from South Africa drawing on reports from urban forest managers suggests that foraging from fruit trees is common (Sardeshpande et al. 2021). Woody tree species are often better documented and mapped than herbaceous species by management agencies in US cities, thus making it easier to assess the accessibility of culturally significant tree species in comparison with other plant types. Because these are common and distinctive species not easily confused with others, we are confident that they are correctly identified.

9.2.2 2015 Street Tree Census

New York City's publicly available 2015 Street Tree Census database and geographic information system (GIS) shapefile were compiled by the NYC Parks Department (New York City Open Data 2018a). This database—the third compiled with volunteer help in the past 30 years—documents the location, size, species, and condition of all street trees on public curbsides in the city (New York City Parks 2017). Over the 17-month process, NYC parks achieved 96% accuracy in their volunteer data collection. These improved results, as compared to the 1995 and 2005 street tree surveys, came from improved volunteer training and a partnership with TreeKIT to mitigate issues in spatial accuracy when mapping trees.

The street tree database maps 652,129 living trees comprised of 130 different species across all five boroughs (New York City Parks 2017). To qualify as a street tree, trees in this database must be planted in the city right-of-way, as opposed to being planted on private property. Trees in official green spaces (e.g., parks) do not qualify, despite the fact that both tree classifications are part of the NYC park system. The database only documents tree species—excluding shrubs and nonwoody herbaceous species from their census. It is worth noting that this study is narrowly focused on one facet of foraging practices, since foragers frequently harvest plant materials from both shrub and herbaceous species.

Of these 130 species, the common names from interview respondents correspond to two entries in the 2015 tree census data: *Ginkgo biloba* and *Morus* spp. *Ginkgo biloba* is the only surviving species of ginkgo, and the database does not identify mulberry trees beyond their genus. Ginkgo is the more widely abundant street tree species of the two, with 21,024 trees (3.22% of the street tree population) across the city. In comparison, mulberry street trees only number 1157 (0.18% of the street tree population). The difference in relative citywide abundance between these species provides insight into questions about the overall availability of culturally important species and, importantly, the implications that geographic abundance plays in influencing this accessibility.

Our analysis of these trees carries two important cautions about these species: First, any results are complicated by the fact that ginkgo trees are dioecious, with only the female trees capable of producing the harvestable fruit and inner ginkgo nut. Second, and importantly, managers are predisposed to planting male trees to avoid the "nuisance" of the smell and the mess from fallen fruits on the sidewalk created by female trees. Mulberry trees present a similar, though somewhat different, challenge; this taxon can be either dioecious or monoecious. In the section below, we describe how these cautions shape key assumptions about interpreting access.

9.2.3 Optimized Hot Spot Analysis

Using the ArcMap 10.7.1 GIS software to visualize dense clusters of specific street tree species in the city, we employed the optimized hot spot analysis to create a map of statistically significant hot and cold spots from a set of incident points (e.g., individual street trees). The resulting optimized hot spot analysis uses the Getis-Ord Gi* statistic to aggregate the data in a grid pattern according to a spatial scale that the tool determines is appropriate for the data. By doing so, the tool should ideally account for any spatial dependence in the data. This tool generated a [scale] grid from the ginkgo street tree data and a [scale] grid from the mulberry data, generating 0.217 km^2 and 0.213 km^2 grids for ginkgo and mulberry, respectively.

We conducted optimized hot spot analyses for both mulberry and ginkgo trees, and for all street trees as another check on spatial dependence. The extent of hot spots and cold spots generated by the optimized hot spot analysis is represented by confidence intervals of 99% (Gi bins of +3 or −3), 95% (+/−2 bins), and 90% (+/−1 bins). Gi bins of 0 indicate that there is neither a high nor a low likelihood of encountering a given street tree species in that grid space. Portions of New York that are not gridded or associated with a Gi bin do not possess street trees. These are usually regions with trees that do not meet the criteria for street trees (e.g., public parks), as explained above.

Given the two cautions described above, we assume that hot spots increase the likelihood of encountering fruiting female ginkgo trees within the identified geographic perimeter. Thus, we draw no conclusions about the accessibility of ginkgo trees elsewhere in the city that are not contained in one of the resulting hot spots. We also assume that hot spots identify areas where there is a higher likelihood of encountering fruiting trees and refrain from drawing conclusions about mulberry accessibility when the trees are not contained within a hot spot.

9.2.4 Neighborhood Tabulation Areas

To approximate the walkable accessibility of street tree hot spots to city residents, we overlaid the map of hot spots with neighborhood tabulations areas (NTAs) and assessed the overlap. The publicly available GIS shapefile of NTAs from the NYC Department of City Planning (at the time the study was conducted) used data from the 2010 census to draw neighborhood boundaries according to a combination of historical delineations and minimum population requirements (New York City Open Data 2020). Most studies assessing green space accessibility rely on various buffers and distances to mimic walkability, though they tend to fall around 300 m (Jarvis et al. 2020; Kabisch and Haase 2014). Instead of setting a predetermined walking distance, we use the boundaries set by NTAs as a proxy for walkability. While using imposed administrative units to model human behavior may overlook some of the complexities of that behavior, the NYC Parks Department uses NTAs for projects like their "Community Parks Initiative," which indicates that these boundaries likely inform street tree management (New York City Open Data 2018c).

We obtained NTA-specific demographic data from the 2008–2012 American Community Survey (ACS) to assess the accessibility of culturally significant species to residents who self-identify as Chinese. All data on race from the ACS refers to the categories they use to differentiate between people who self-identify with a given social group, usually based on some consideration of national origin and sociocultural identities. For the purpose of this study, Chinese residents refer to ACS respondents who self-identified as Chinese or Chinese-American (United States Census Bureau 2012).

ArcMap automatically generated breaks in the percentage of NTA residents who self-identify as Chinese with the Jenks natural breaks method. This optimization method clusters the data into five classes that minimize the deviation within each class while maximizing the deviation between the average means of each class. We focus on NTAs with high proportions of residents who self-identify as Chinese. Overlap between street tree hots spots and NTAs acts as a proxy for walkable accessibility, with the understanding that proximity to natural resources is not necessarily correlated with accessibility and usage of those resources. Further extrapolation on this point is provided in the discussion.

While walkability is a vital component of accessibility, it does not account for foragers who may be willing to leave their neighborhoods to harvest these culturally significant species elsewhere in the city. As a metropolitan hub, NYC possesses an interconnected public transportation network that extends through the five boroughs. We overlaid hot spot and NTA results with GIS shapefiles of subway entrances and bus shelters to assess the connectivity of NTAs to specific hot spots via public transportation routes (New York City Open Data 2018d; New York City Open Data 2018b). While the city of New York does not provide geographically referenced models of entire train lines or bus routes, these sets of point data act as proxies for those systems.

9.3 Results

9.3.1 Hot Spots of *Ginkgo biloba* and *Morus* spp. Trees

By examining the hot spot results for all of New York City's 652,129 street trees, we gain a basic understanding of the underlying patterns that govern individual tree species' distribution throughout the city (Fig. 9.2a). If we begin by looking at all 130 street tree species, we see a combination of hot spots and cold spots in each borough (Fig. 9.2b.). Manhattan, Brooklyn, and Queens have more hot spots than cold spots, while the Bronx has a more equal balance between the two, and Staten Island has more cold spots than hot spots. All five boroughs also have some street treeless areas, as indicated by the lack of gridded output. Staten Island has the greatest street treeless area, followed by Queens, the Bronx, Brooklyn, and Manhattan. Most of these areas are either private property (e.g., LaGuardia and JFK Airports) or public parks that do not meet street tree criteria (e.g., Central Park).

Hot spot results for ginkgo and mulberry trees vary widely in geographic location, extent, and statistical significance. There are three main hot spots for ginkgo street trees in New York: (1) Manhattan, (2) Brooklyn, and (3) Queens (Fig. 9.2c). The first and largest hot spot (39.40 km^2 at 99%; 40.71 km^2 at 95%; 41.36 km^2 at 90%) covers the majority of Manhattan, excluding the northernmost portion of the borough, a small portion of its southern tip, and Central Park at its interior. The second largest hot spot (12.15 km^2 at 99%; 14.11 km^2 at 95%; 14.76 km^2 at 90%) is located in the northwestern portion of Brooklyn. The third hot spot (4.99 km^2 at 99%; 5.21 km^2 at 95%) is in the northeastern corner of Queens.

All three *Ginkgo* street tree hot spots share some similarities with the underlying distribution of street trees across the city (Fig. 9.2b). These similarities mostly extend to the hot spots' geographic locations, not to their geographic extent or size. In Manhattan, the ginkgo hot spot expands well beyond the edge of the hot spot in that borough for all city street trees. The Brooklyn and Queens hot spots, in contrast, are both smaller than their respective counterparts. The Queens hot spot is closest in size to the underlying street tree distribution than the other two.

By contrast, there are no extensive hot spots of mulberry street trees, just isolated grid cells with slightly elevated probability of encounter (Fig. 9.2d). These grids are found in a small area in the southern portion of the Bronx of 0.43 km^2 with a confidence interval of 95%. Interestingly, this area does not fall within any of the hot spots for the wider street tree population. The lack of

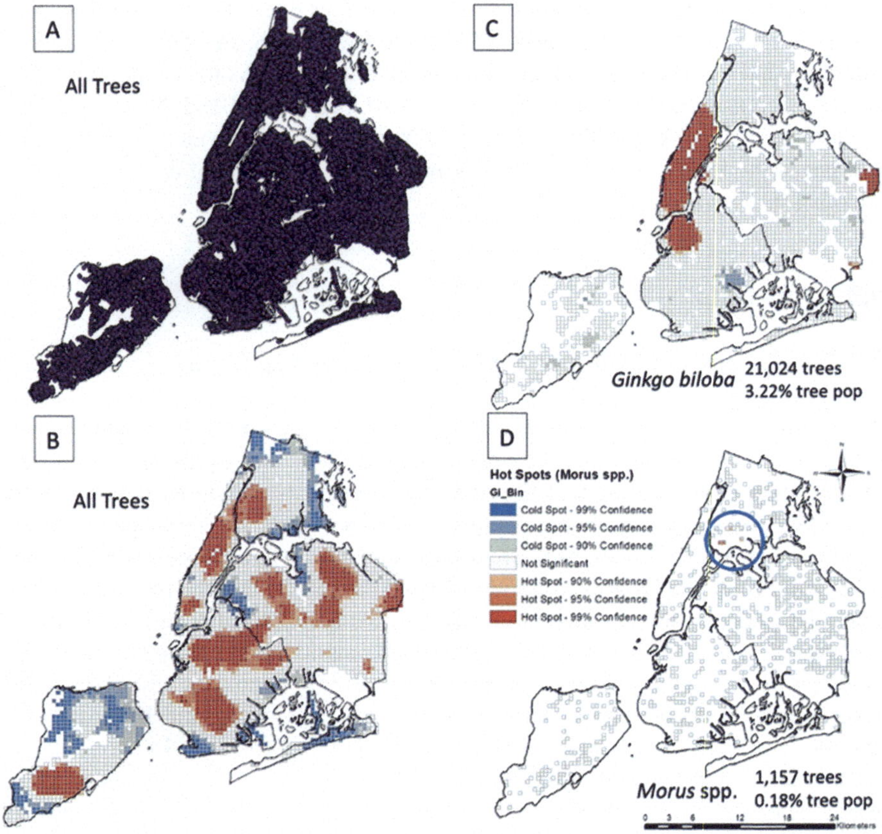

Fig. 9.2 (**a**) All street trees, (**b**) street tree hot spots, (**c**) hot spots for *Ginkgo biloba* trees, (**d**) hot spots for *Morus* spp

mulberry hot spots indicates that, despite the species' presence as a street tree elsewhere in the city, there are not enough mulberry trees planted close enough together to significantly increase a forager's likelihood of encountering multiple mulberry trees.

9.3.2 Neighborhood Tabulation Areas

There are three contiguous, or nearly contiguous, clusters of NTAs with high proportions of residents who self-identify as Chinese or Chinese-Americans (Fig. 9.3). The cluster in Manhattan is comprised of 4 NTAs (Fig. 9.4), the cluster in Brooklyn is comprised of 12 NTAs (Fig. 9.4), and the cluster in Queens is comprised of 23 NTAs (Fig. 9.4). Each of these clusters has a similar configuration. The NTAs at their center have the highest proportion of residents self-identifying as Chinese

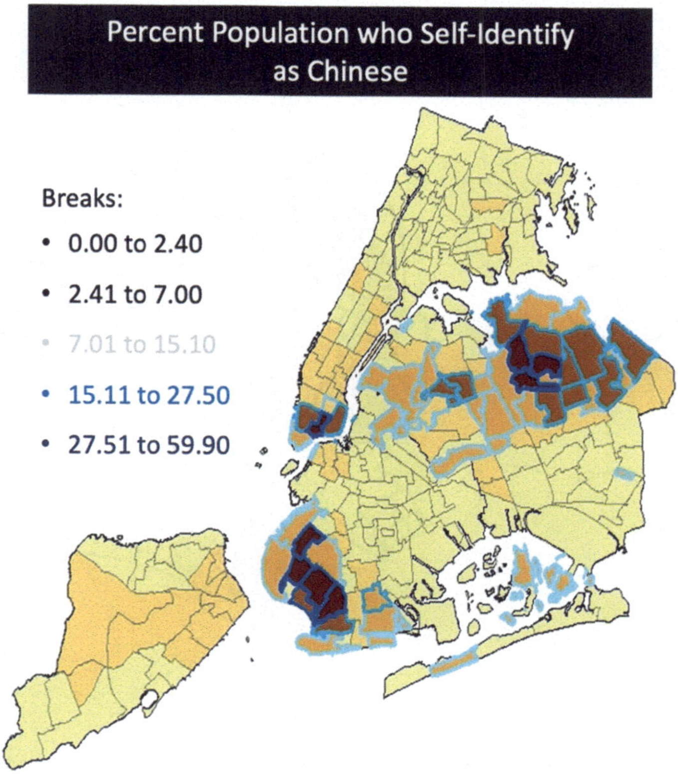

Fig. 9.3 Neighborhoods with relatively high, high, and very high percentages of residents self-identifying as Chinese or Chinese-American

(27.51–59.90%), with the surrounding NTAs declining in concentration as they radiate out from those centers.

Only one area in the entire city features correspondence between these two variables: the ginkgo hot spot and NTA cluster in Manhattan overlap with one another (Fig. 9.4). In all other areas of the city, either there are no hot spots in communities where there are high percentages of residents identifying as Chinese or the remaining hot spots have low percentages of (or no) residents self-identifying in this way. Despite the lack of overlapping hot spots in the NTA clusters of Brooklyn and Queens, we note that there are still some ginkgo trees distributed through those boroughs' neighborhoods. The lack of overlapping hot spots indicates that there are no concentrated areas within those neighborhoods where it is significantly more likely for residents to encounter ginkgo than elsewhere in the city.

While the NTA cluster in Brooklyn does not overlap with the nearby ginkgo hot spot, these neighborhoods have a collective 120 subway entrances that connect them to both the Brooklyn and Manhattan hot spots (Fig. 9.5a). The NTA cluster in Queens has 136

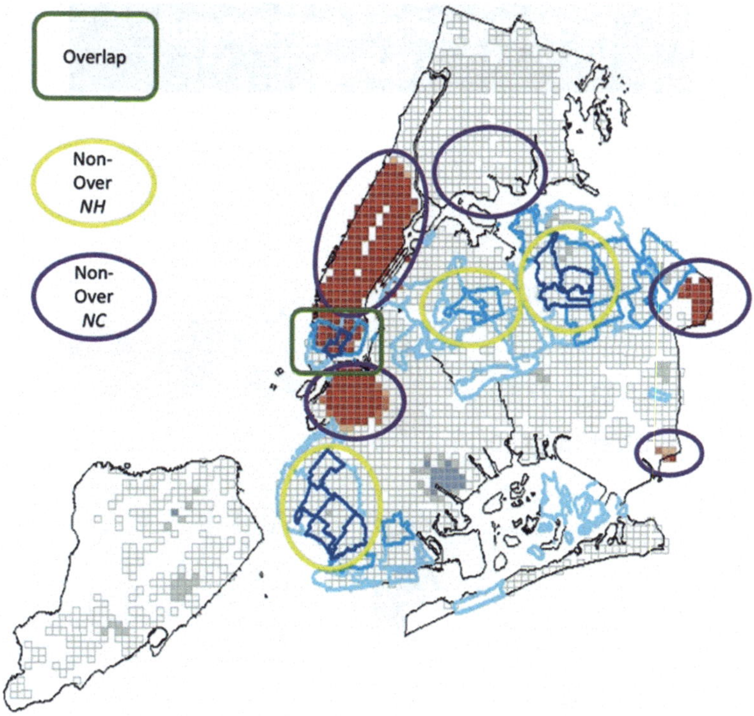

Fig. 9.4 Overlap of *Gingko* and mulberry hot spots with neighborhood tabulation areas that have high percentages of residents self-identifying as Chinese or Chinese-American

subway entrances that connect those neighborhoods to the two larger hot spots (Fig. 9.5a). However, these stations take more circuitous routes to eventually reach the hot spots, thereby reducing travel efficiency. There are no subway stations in the Queens hot spot, thereby limiting access to the ginkgo trees in that portion of the city. There may be other rail systems not mapped here that access that portion of Queens.

The system of bus stop shelters is more spatially extensive than the subway system, covering parts of the city otherwise unserved by the subway system and expanding potential access to a greater number of clusters. There are 81 bus stops that service the Manhattan NTA cluster, 212 that service the Brooklyn cluster, and 462 that service the cluster in Queens (Fig. 9.5b). Though our dataset does not have enough detail to map out actual routes and connections between buses, we reason that any of these stops will eventually connect with the bus stops in all three ginkgo hot spots. The Manhattan hot spot can be accessed by 599 bus stops, the Brooklyn hot spot by 103 stops, and the Queens hot spot by 11 stops. While the ginkgo hot spot in Queens is accessible neither by foot nor by subway, it is moderately accessible through bus routes.

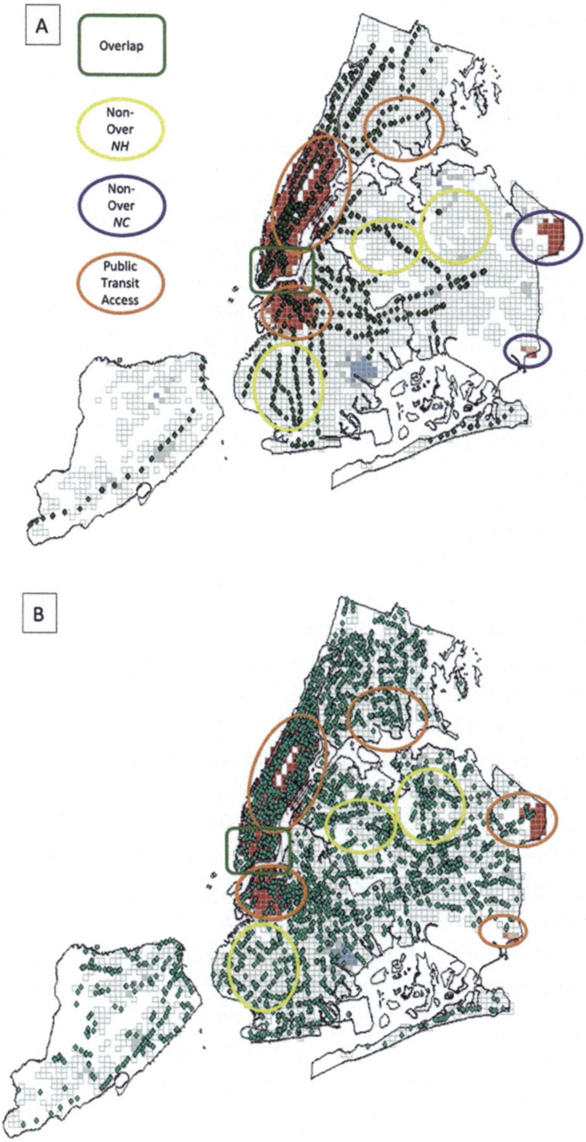

Fig. 9.5 Correspondence of *Gingko* and mulberry hot spots with neighborhoods with high percentages of residents who self-identify as Chinese or Chinese-American, along with street tree hot spots accessible by (**a**) the subway system and (**b**) existing bus routes

9.4 Discussion

In NYC, our study finds that gingko trees are much more abundant and widespread than mulberry trees as street trees. Still, there is direct correspondence of ginkgo hot spots with only one NTA in Lower Manhattan. Our analysis suggests that ginkgo street trees in these areas are accessible to foragers within their neighborhoods by means of walking alone. Other NTAs do not have direct access to areas with a high probability of providing desired materials through walking. The high accessibility of ginkgo street trees in this part of Manhattan, given the density of the borough's residential and commercial development, challenges assumptions about the availability of useful species in the city. The presence of ginkgo trees, however, given their popularity in cities as an ornamental, is perhaps not surprising in this area.

When combining insights from hot spot overlaps with public transit connectivity, a greater number of ginkgo street trees in Manhattan become accessible to Chinese and Chinese-American residents who forage. Thus, the street tree population in Manhattan is more well suited for foraging purposes than one might expect (Hurley and Emery 2018). Further consideration of overlaps and public transit connectivity reveal that ginkgo street trees in Brooklyn are moderately accessible with the use of either the subway or the buses, while those in the Queens hot spot are relatively inaccessible. Here, the difference hinges on which mode of public transportation is involved, with the Queens hot spot reachable only by bus.

By contrast, the lack of concentrated mulberry street trees indicates that this species is not highly accessible to residents in any part of the city. Due to this species' relatively limited density and abundance, foragers must look to other urban green spaces for harvests of this culturally important species. We note that as part of the fieldwork for this project, the authors visited illustrative examples of other urban green spaces reported as used by foragers. We observed wild mulberry trees in a number of locations, including in unused—older—sections of some cemeteries, reinforcing the importance that interstitial ecologies may have in providing access to culturally significant species (Hurley et al. 2008; Sardeshpande and Shackleton 2023).

Proximity and direct accessibility offer only partial considerations when assessing access. First, the accessibility of the remaining ginkgo street tree population relies on an understanding of the biological qualities of those trees (Biernacka et al. 2020; Jarvis et al. 2020). Due to the dioecious nature of ginkgo trees, we draw no conclusions about the accessibility of ginkgo trees elsewhere in the city that are not contained in one of the three hot spots. The presence of hot spots increases the likelihood of encountering fruiting female trees—the only trees that produce the harvestable fruit—within the perimeter. However, there is no means of guaranteeing this assumption when ginkgo density is low and managers are predisposed to planting male trees to avoid the "nuisance" caused by the fruit. It also bears mentioning that there may be other ginkgo populations elsewhere in the city that do not qualify as street trees (e.g., public parks) and therefore are not captured by this dataset. The failure to capture those additional trees through this approach means that we may overstate the barrier to trees that support cultural foraging practices for Chinese identifying residents who do not live near ginkgo hot spots.

Second, distance, or even travel time, to a species does not guarantee access to or actual use of green spaces or the desired natural features they contain. Our focus on this geographic aspect alone fails to recognize the other physical, economic, social, and cultural factors that also impact green space access and usage (Jarvis et al. 2020). Physical considerations, such as an individual's mobility and complications like steep slopes from the topography, may be more important than mere proximity to or the availability of public transit (Szaboova et al. 2019). While we attempt to approximate the role of subway and bus systems in improving accessibility, this analysis does not consider precise routes, travel time and distance, and the role of travel fees that complicate the usefulness of public transit. Long trips, either by means of walking or facilitated by public transportation, may introduce psychological barriers that prevent foraging. Even in cases where the trip is within their capabilities, poor health or a lack of fitness may create psychological barriers for some people that convinces them that those trips are unmanageable (Szaboova et al. 2019; Sardeshpande and Shackleton 2023).

Further barriers may be at play in limiting access. Psychosocial factors, which encompass individuals' social and cultural values and perceptions, and their influence on urban foraging practices are relatively unexplored in the literature. The ways in which individuals or social groups perceive urban forests have a direct influence on how they choose to interact with these landscapes. In a study examining the accessibility of ecosystem benefits as a whole, Szaboova et al. (2019) found that such perceptions, especially those concerning safety, are shaped by personal experiences. For instance, recent Russian-German and Balkan migrants to Berlin, Germany, perceive urban forests as safe spaces within which to forage because the forests bear similarities to the forests of their home countries (Jay and Schraml 2009). However, the public setting within which street trees grow lends itself to different perceptions regarding safety and the public's tolerance of foraging in that location. A more public setting, however, may also have the potential for greater public scrutiny and judgement and strong social norms that may dissuade individuals from foraging (Brandner and Schunko 2022; Sardeshpande and Shackleton 2023).

Our study provides a preliminary look at questions of accessibility, focusing on abundance and proximity. Beyond the psychosocial factors above, we are also quick to acknowledge that accessibility cannot be fully explored without also examining the role of both formal and informal access, including the interaction of rules and enforcement of those rules (Hurley et al. 2015), since street trees fall under the jurisdiction of the city's park system and are managed by the NYC Parks Department. Here, the powers and privileges exercised by the state to monitor and regulate the public's use of a given natural resource, in this context, encompass everything from selecting and planting species to policing their use (Peluso and Ribot 2020; Szaboova et al. 2019). Given the comparative mobility of humans versus trees, it is unlikely that these mature trees were planted to serve the current demographic makeup of the city. Moreover, the current urban forest landscape is likely due to the historic socio-economic composition of New York City (Locke and Baine 2014; Larondelle and

Strohbach 2016). The Parks Department has also chosen to use these rights to prohibit foraging on any of its property within the city. This is consistent with other managers' decisions across the country (McLain et al. 2014). Discussions with managers in New York suggest that many managers continue to have concerns regarding sustainability and the potential harms of overharvesting when justifying this policy (Emery and Hurley, unpublished data).

However, emerging theories surrounding management of natural resources call for incorporating more social-political dimensions into these decision-making processes (Jay and Schraml 2009; Kabisch and Haase 2014; Coles and Costa 2018; Peluso and Ribot 2020). Some argue specifically for the interdisciplinary incorporation of indigenous knowledge systems into inclusive urban green planning (Mata et al. 2020; Baumflek et al. 2021). Likewise, Calderón-Contreras and White (2020) argue that community resilience—supported by management systems that actively seek feedback from a diverse array of residents who engage with the natural resource in question—enhances ecological resilience efforts. Still others contend that sustainability alone is not reason enough to prohibit foraging, since sustainable foraging is feasible. In Berlin, Germany, researchers found that both native and nonnative species were harvested at levels consistent with their relative abundance, thus reducing the risk of overharvesting the sensitive species managers are most worried about protecting (Fischer and Kowarik 2020). However, according to preliminary evidence (Emery and Hurley, unpublished data), NYC Parks Department foraging policies are less rigorously enforced for street trees than they are for more formal urban green spaces. Though the act may not be legally sanctioned, a lack of enforcement removes a likely barrier for potential foragers in our study. Meanwhile, the discrepancy in enforcement means that this barrier to foraging still exists for park-based trees excluded from this study's dataset.

9.5 Conclusion

Some species of street trees have the potential to support biocultural dimensions of foraging practices, even for iconic and exotic species we might not think of in these utilitarian terms (Hurley and Emery 2018) and in some of the most densely built-up areas of the USA. Here, we recognize that some scholars may argue that the case of *Ginkgo* trees is an outlier, given this species' presence throughout the city in relation to the global plant trade (Pincetl et al. 2013), or for mulberry (Arrington 2021), and the particular aspects of *Ginkgo* biology that results in what are ultimately unwanted female trees and the ecosystem disservices from these messy fruits (Roman et al. 2020). Yet we are reminded of other studies in the literature on rural foraging by immigrants in Argentina (Kujawska and Łuczaj 2015) that point to the usage of these species, such as the queen palm (*Syagrus romanzoffiana*), that now are part of our urban forests in US cities (Nowak et al. 2011) and that are a function of the same global trade dynamics (Pincetl et al. 2013). Research on urban foraging, particularly biocultural aspects, should more closely consider the ways in which these global dynamics create connections between people, culture, and plants.

Data from this study suggest that some street tree species' abundance and widespread geographic distributions might possibly allay manager concerns about the impacts of foraging on particular species and could open up new opportunities for discussion about changes in regulations around foraging that take a species-specific approach, particularly to nonnative species which may be invasive in some parts of the city (Arrington 2021). A biocultural approach opens up the potential for recognition of the ways that species are viewed as producing ecosystem disservices maybe, in fact, simultaneously creating culturally significant benefits for some groups in the city, thereby providing a way to better connect resources and particularly groups. Greater attention is needed within urban foraging research on these culturally mediated ways of interacting with particular species (Marquina et al. 2022).

At the same time, our assessment suggests that street tree species, which are culturally significant to Chinese and Chinese-Americans, are available to resident foragers in highly uneven ways. First, ginkgo street trees are more accessible than mulberry street trees, due to both greater abundance and a wider distribution throughout the city, including within the five boroughs. This is not to say that ginkgo accessibility is the same throughout the city; ginkgo street trees in Manhattan are highly accessible by three modes of transportation, while the street trees in Brooklyn and Queens are moderately and poorly accessible, respectively. Second, mulberry street trees are virtually inaccessible throughout the city, both in terms of abundance and geographic distribution, although that does not indicate that the species itself is lacking in the city. Low accessibility here means that foragers are likely turning to other green spaces with mulberry trees, such as in parks or cemeteries, to meet their needs. Further ground-truthing and associated qualitative research is needed to better understand the dynamics that shape location and harvest of these materials from species of street trees that are of biocultural importance to residents.

Indeed, our findings offer a point of entry for considering the role that street tree species can play in supporting biocultural dimensions of urban foraging practices. Much more research is needed on this topic, including consideration of dimensions of access that go beyond proximity and direct accessibility. Greater attention is needed to the formal and informal practices of control, in terms of rules for harvesting, but also importantly how these are enforced (or not) for particular cultural groups in cities. Likewise, greater attention to the psychosocial aspects of biocultural foraging practices is needed: for example, what role might questions of social stigma play in dissuading harvests of these species or how might the strong and pungent smells of harvested fruits dissuade the usage of modes of public transportation to harvest these culturally important species. Finally, greater inclusion of the voices representing biocultural dimensions of foraging in the species selection of our urban forests is needed (McLain et al. 2014; Hurley et al. 2015; Poe et al. 2013; Nesbitt et al. 2020).

Acknowledgements We thank the interview participants who shared their harvesting practices with us and the U.S. Forest Service Urban Field Station, including staff from both the U.S. Forest

Service's New York City Urban Field Station and the New York City Parks Department, for their assistance with data collection.

Funding This research was funded by the Office of the Vice President for Academic Affairs at Ursinus College.

References

Adeyemi O, Shackleton CM (2023) Understanding foraging practices in Lagos Metropolis to redesign urban greenspaces in support of human-nature interactions. Urban For Urban Green 79(January):127805. https://doi.org/10.1016/j.ufug.2022.127805

Arrington A (2021) Urban foraging of five non-native plants in NYC: balancing ecosystem services and invasive species management. Urban For Urban Green 58(March):126896. https://doi.org/10.1016/j.ufug.2020.126896

Baumflek M, Degloria S, Kassam K-A (2015) Habitat modeling for health sovereignty: increasing indigenous access to medicinal plants in northern Maine, USA. Appl Geogr 56(January):83–94. https://doi.org/10.1016/j.apgeog.2014.10.012

Baumflek M, Kassam K-A, Ginger C, Emery M (2021) Incorporating biocultural approaches in Forest management: insights from a case study of indigenous plant stewardship in Maine, USA and New Brunswick, Canada. Soc Nat Resour 34(July):1–19. https://doi.org/10.1080/08941920.2021.1944411

Biernacka M, Kronenberg J, Łaszkiewicz E (2020) An integrated system of monitoring the availability, accessibility and attractiveness of urban parks and green squares. Appl Geogr 116(March):102152. https://doi.org/10.1016/j.apgeog.2020.102152

Brandner A, Schunko C (2022) Urban Wild food foraging locations: understanding selection criteria to inform green space planning and management. Urban For Urban Green 73(July):127596. https://doi.org/10.1016/j.ufug.2022.127596

Bunge A, Diemont S, Bunge J, Harris S (2019) Urban foraging for food security and sovereignty: quantifying edible Forest yield in Syracuse, New York using four common fruit- and nut-producing street tree species. J Urban Ecol 5(February). https://doi.org/10.1093/jue/juy028

Calderón-Contreras R, White CS (2020) Access as the means for understanding social-ecological resilience: bridging analytical frameworks. Soc Nat Resour 33(2):205–223. https://doi.org/10.1080/08941920.2019.1597233

Clark KH, Nicholas KA (2013) Introducing urban food forestry: a multifunctional approach to increase food security and provide ecosystem services. Landsc Ecol 28(9):1649–1669. https://doi.org/10.1007/s10980-013-9903-z

Coles R, Costa S (2018) Food growing in the City: exploring the productive urban landscape as a new paradigm for inclusive approaches to the design and planning of future urban open spaces. Landsc Urban Plan 170(February):1–5. https://doi.org/10.1016/j.landurbplan.2017.10.003

Dhyani S, Kadaverugu R (2020) Food security and cultural benefits from urban green spaces: exploring urban foraging as a silently growing global movement. Clim Change Environ Sustain 8(2):219–225. https://doi.org/10.5958/2320-642X.2020.00022.8

Fischer LK, Kowarik I (2020) Connecting people to biodiversity in cities of tomorrow: is urban foraging a powerful tool? Ecol Indic 112(May):106087. https://doi.org/10.1016/j.ecolind.2020.106087

Garekae H, Shackleton CM (2020) Foraging Wild food in urban spaces: the contribution of Wild foods to urban dietary diversity in South Africa. Sustainability 12(2). https://doi.org/10.3390/su12020678

Garekae H, Shackleton CM, Tsheboeng G (2022) The prevalence, composition and distribution of Forageable plant species in different urban spaces in two medium-sized towns in South Africa. Global Ecol Conserv 33(January):e01972. https://doi.org/10.1016/j.gecco.2021.e01972

Hurley PT, Becker S, Emery MR, Detweiler J (2022) Estimating the alignment of tree species composition with foraging practice in Philadelphia's urban Forest: toward a rapid assessment of provisioning services. Urban For Urban Green 68(February):127456. https://doi.org/10.1016/j.ufug.2021.127456

Hurley PT, Emery MR (2018) Locating provisioning ecosystem Services in Urban Forests: Forageable Woody species in new York City, USA. Landsc Urban Plan 170(February):266–275. https://doi.org/10.1016/j.landurbplan.2017.09.025

Hurley PT, Emery MR, McLain R, Poe M, Grabbatin B, Goetcheus CL (2015) Whose urban Forest? The political ecology of foraging urban nontimber Forest products. In: Isenhour C, McDonogh G, Checker M (eds) Sustainability in the Global City: myth and practice. Cambridge University Press, Cambridge, pp 187–212. New directions in sustainability and society. https://doi.org/10.1017/CBO9781139923316.014

Hurley, P., Emery, M. R., McLain, R., Poe, M., McGillis, K., Chan, J., & Rangel, M. L. Forthcoming. Hidden harvests: profiles in urban foraging. Taylor and Francis Publishing

Hurley PT, Grabbatin B, Goetcheus C, Halfacre A (2013) Gathering, buying, and growing Sweetgrass (Muhlenbergia Sericea): urbanization and social networking in the Sweetgrass basket-making industry of Lowcountry South Carolina. In: Voeks R, Rashford J (eds) African ethnobotany in the Americas. Springer New York, New York, NY, pp 153–173. https://doi.org/10.1007/978-1-4614-0836-9_6

Hurley PT, Halfacre AC, Levine NS, Burke MK (2008) Finding a 'disappearing' nontimber Forest resource: using grounded visualization to explore urbanization impacts on Sweetgrass Basketmaking in greater Mt. Pleasant, South Carolina. Prof Geogr 60(4):556–578. https://doi.org/10.1080/00330120802288941

Jarvis I, Gergel S, Koehoorn M, van den Bosch M (2020) Greenspace access does not correspond to nature exposure: measures of urban natural space with implications for Health Research. Landsc Urban Plan 194(February):103686. https://doi.org/10.1016/j.landurbplan.2019.103686

Jay M, Schraml U (2009) Understanding the role of urban forests for migrants–uses, perception and integrative potential. Urban For Urban Green 8(4):283–294. https://doi.org/10.1016/j.ufug.2009.07.003

Kabisch N, Haase D (2014) Green justice or just green? Provision of urban green spaces in Berlin, Germany. Landsc Urban Plan 122(February):129–139. https://doi.org/10.1016/j.landurbplan.2013.11.016

Kaoma H, Shackleton C (2015) The direct-use value of urban tree non-timber Forest products to household income in poorer suburbs of south African towns. Forest Policy Econ 61(November):104–112. https://doi.org/10.1016/j.forpol.2015.08.005

Knigge LD, Cope M (2006) Grounded visualization: integrating the analysis of qualitative and quantitative data through grounded theory and visualization. Environ Plan A: Econ Space 38(11):2021–2037. https://doi.org/10.1068/a37327

Kowalski JM, Conway TM (2019) Branching out: the inclusion of urban food trees in Canadian urban Forest management plans. Urban Food Forestry: Current State and Future Perspectives 45(October):126142. https://doi.org/10.1016/j.ufug.2018.05.012

Kujawska M, Łuczaj Ł (2015) Wild edible plants used by the Polish community in Misiones, Argentina. Human Ecology 43:855–869

Landor-Yamagata JL, Kowarik I, Fischer LK (2018) Urban foraging in Berlin: people, plants and practices within the metropolitan green infrastructure. Sustainability 10(6). https://doi.org/10.3390/su10061873

Larondelle N, Strohbach M (2016) A murmur in the trees to note: urban legacy effects on fruit trees in Berlin, Germany. Urban For Urban Green 17(March):11. https://doi.org/10.1016/j.ufug.2016.03.005

Locke D, Baine G (2014) The good, the bad, and the interested: how historical demographics explain present-day tree canopy, vacant lot and tree request spatial variability in New Haven, CT. Urban Ecosyst 18(September):391. https://doi.org/10.1007/s11252-014-0409-5

Marquina T, Emery M, Hurley P, Gould R (2022) The 'quiet hunt': the significance of mushroom foraging among Russian-speaking immigrants in New York City. Ecosyst People 18(December):226–240. https://doi.org/10.1080/26395916.2022.2055148

Mata L, Ramalho CE, Kennedy J, Parris KM, Valentine L, Miller M, Bekessy S, Hurley S, Cumpston Z (2020) Bringing nature Back into cities. People Nat 2(2):350–368. https://doi.org/10.1002/pan3.10088

McLain RJ, Hurley PT, Emery MR, Poe MR (2014) Gathering 'Wild' food in the City: rethinking the role of foraging in urban ecosystem planning and management. Local Environ 19(2):220–240. https://doi.org/10.1080/13549839.2013.841659

Nesbitt L, Konijnendijk C, Lauster N, Park H (2020) Intercultural learning in contested space: the biocultural realities of global cities through the lens of Vancouver, Canada

[dataset] New York City Open Data (2018a) 2015 street tree census: tree data. Accessed 28 Nov 2019]

[dataset] New York City Open Data (2018b) Bus stop shelters. Retrieved from https://data.cityofnewyork.us/Transportation/Bus-Stop-Shelters/qafz-7myz. Accessed 18 Sep 2019

[dataset] New York City Open Data (2018c) Community Parks Initiative Zone Boundaries. Retrieved from https://data.cityofnewyork.us/City-Government/Community-Parks-Initiative-Zone-Boundaries/vvdx-b56i. Accessed 15 Nov 2019

[dataset] New York City Open Data (2018d) Subway entrances. Retrieved from https://data.cityofnewyork.us/Transportation/Subway-Entrances/drex-xx56. Accessed 18 Sep 2019

[dataset] New York City Open Data (2020) Neighborhood tabulation areas (NTA). Retrieved from https://data.cityofnewyork.us/City-Government/Neighborhood-Tabulation-Areas-NTA-/cpf4-rkhq Accessed 15 Nov 2019

New York City Parks (2017) 2015 street tree census report: census overview. Retrieved from http://media.nycgovparks.org/images/web/TreesCount/Index.html. Accessed 19 Mar 2020

Nowak DJ, Hoehn R, Crane DE, Weller L, Davila A (2011) Assessing urban forest effects and values: Los Angeles' urban forest. USDA Forest Service, Northern Resource Bulletin NRS-47, Newtown Square, PA, 30

Peluso NL, Ribot J (2020) Postscript: a theory of access revisited. Soc Nat Resour 33(2):300–306. https://doi.org/10.1080/08941920.2019.1709929

Pincetl S, Prabhu S, Gillespie T, Jenerette D, Pataki D (2013) The evolution of tree nursery offerings in Los Angeles County over the last 110 years. Landsc Urban Plan 118(June):10–17. https://doi.org/10.1016/j.landurbplan.2013.05.002

Poe MR, LeCompte J, McLain R, Hurley P (2014) Urban foraging and the relational ecologies of belonging. Soc Cult Geogr 15(8):901–919. https://doi.org/10.1080/14649365.2014.908232

Poe MR, McLain RJ, Emery M, Hurley PT (2013) Urban Forest justice and the rights to Wild foods, medicines, and materials in the City. Hum Ecol 41(3):409–422. https://doi.org/10.1007/s10745-013-9572-1

Roman LA, Conway TM, Eisenman TS, Koeser AK, Barona CO, Locke DH, Darrel Jenerette G, Östberg J, Vogt J (2020) Beyond 'trees are good': disservices, management costs, and tradeoffs in urban forestry. Ambio 50(3):615–630. https://doi.org/10.1007/s13280-020-01396-8

Russo A, Escobedo FJ, Cirella GT, Zerbe S (2017) Edible green infrastructure: an approach and review of provisioning ecosystem services and disservices in urban environments. Agric Ecosyst Environ 242(May):53–66. https://doi.org/10.1016/j.agee.2017.03.026

Sardeshpande M, Hurley PT, Mollee E, Garekae H, Dahlberg AC, Emery MR, Shackleton C (2021) How people foraging in urban greenspace can mobilize social–ecological resilience during Covid-19 and beyond. Cities Front Sustain, 3. https://www.frontiersin.org/articles/10.3389/frsc.2021.686254

Sardeshpande M, Shackleton C (2023) Fruits of the City: the nature, nurture and future of urban foraging. People Nat 5(1):213–227. https://doi.org/10.1002/pan3.10428

Schunko C, Brandner A (2022) Urban nature at the fingertips: investigating Wild food foraging to enable nature interactions of urban dwellers. Ambio 51(5):1168–1178. https://doi.org/10.1007/s13280-021-01648-1

Schunko C, Wild A-S, Brandner A (2021) Exploring and limiting the ecological impacts of urban Wild food foraging in Vienna, Austria. Urban For Urban Green 62(July):127164. https://doi.org/10.1016/j.ufug.2021.127164

Shackleton CM, Hurley PT, Dahlberg AC, Emery MR, Nagendra H (2017) Urban foraging: a ubiquitous human practice overlooked by urban planners, policy, and research. Sustainability 9(10). https://doi.org/10.3390/su9101884

Shortly A, Kepe T (2021) Consuming the City: challenges and possibilities for foraging in Toronto's parks. For Trees Livelihood 30(2):75–89. https://doi.org/10.1080/14728028.2020.1863865

Stark PB, Miller D, Carlson TJ, Rasmussen K, de Vasquez. (2019) Open-source food: nutrition, toxicology, and availability of Wild edible greens in the East Bay. PLoS One 14(1):e0202450. https://doi.org/10.1371/journal.pone.0202450

Synk CM, Kim BF, Davis CA, Harding J, Rogers V, Hurley PT, Emery MR, Nachman KE (2017) Gathering Baltimore's bounty: characterizing behaviors, motivations, and barriers of foragers in an urban ecosystem. Urban For Urban Green 28(December):97–102. https://doi.org/10.1016/j.ufug.2017.10.007

Szaboova L, Brown K, Fisher J (2019) Access to ecosystem benefits: more than proximity. Soc Nat Resour 33(February):1–17. https://doi.org/10.1080/08941920.2018.1556759

Turner NJ, Cuerrier A, Joseph L (2022) Well grounded: indigenous peoples' knowledge, ethnobiology and sustainability. People Nat 4(3):627–651. https://doi.org/10.1002/pan3.10321

Sarah Becker (PhD) is an alumnus of the Ursinus Environmental Studies department and now works on sustainability issues in Boston, Massachusetts.

Patrick T. Hurley (PhD) is Professor and Chair of Environmental Studies at Ursinus College. When not researching urban foraging and its relation to urban forests in the USA, Dr. Hurley works with students to steward the College's food forest.

Marla R. Emery (PhD) is a Scientific Advisor with the Norwegian Institute for Nature Research (NINA). Dr. Emery is a globally recognized scholar on indigenous and local knowledge and served as the Co-Chair of the Assessment of Sustainable Use of Wild Species, a report of the Intergovernmental Science-Policy Platform on Biodiversity and Ecosystem Services (IPBES).

Joana Chan is an educator and environmental scholar centering social-ecological resilience and compassion in communities. Dr. Chan currently serves as a Mindfulness Teacher and Program Coordinator at Peace in Schools.

Designing Edible Cities: Exploring the Origins and Future of Urban Foraging and Wild Herbalism in the United Kingdom

10

Alessio Russo and Laura McCarthy

Abstract

Foraging and wild herbalism are important elements of the edible green infrastructure which could, by design, enhance the urban framework of the edible city. Despite recent attention to urban agriculture, landscape architecture has rarely focused on aspects of urban foraging and how to design sustainable urban green infrastructure for foragers.

A brief review of current literature precedes a historical context comparison with particular reference to the presence of physic gardens in towns and cities across the UK and Europe and hedgerow herbalism. The study examines three urban foraging case studies in England. Direct participant observations of three foraging groups in cities in Southern England enabled rich data collection and help to inform our recommendations for designing foraging-rich urban landscapes.

Whilst provisioning is at the heart of urban foraging, the research revealed insights into the relevance and importance of foraging at the individual, group and community levels.

At the individual level, urban foraging presents an opportunity for individuals not only to connect more deeply with nature in their local urban environment but also to increase social interactions. Guided groups enable knowledge transmission and collective learning between experts and participants. Groups facilitate the cross-disciplinary sharing of knowledge and information on a range of allied

A. Russo (✉)
School of Architecture and Built Environment, Faculty of Engineering, Queensland University of Technology, Brisbane, Australia
e-mail: alessio.russo@qut.edu.au

L. McCarthy
School of Arts, University of Gloucestershire, Francis Close Hall Campus, Gloucestershire, UK

S. Dhyani, M. Sardeshpande (eds.), *Urban Foraging in the Changing World*,
https://doi.org/10.1007/978-981-97-0345-6_10

activities including methods and recipes for culinary recipes and remedy preparations.

The research collates information gathered at the three urban locations. This data is compared and contrasted and used to inform urban foraging design recommendations, which are anchored in a place-specific historical context.

Keywords

Wild plants · Edible green infrastructure · Hedgerows · Food medicine

10.1 Introduction

Humans are by nature hunter-gatherers; Hill et al. (2011) state that for 95% of our species' historical timeline, we have relied on the direct provisioning services of nature, particularly wild flora, to provide remedies and to supplement our diets. Shaheen et al. (2017) refer to ethnobotanical studies, which suggest that approximately 7000 different plant species have been collected and eaten in communities worldwide. Foraging was a necessity in prehistoric times and continued in importance into mediaeval times to improve inadequate agricultural diets. The collecting and storing of wild foods were of particular importance during winter months (Chambers 2010).

Our hunter-gatherer history explains the continued appeal of collecting and using the wild and cultivated flora growing in our local environment. Foraging is a fundamental part of who we are. As a species, we seek meaningful connections to our natural world to provide us with a sense of belonging and to enhance place attachment (Wilson 1984).

As urbanisation and the corresponding LULC diminish our natural spaces within urban environments, the desire to connect with nature through foraging continues. Ian McHarg stressed that "we need nature as much in the city as in the countryside" (McHarg 1969) (pg. 5).

Urban foraging is defined as "the act of gathering wild plants, fungi, and animals from formal and informal urban green spaces" (Shackleton et al. 2017). Foraging opportunities exist in a variety of urban environments, including urban amenity forests, parks, abandoned and brownfield land, river corridors and coastal zones (Roe et al. 2016).

Schunko et al. (2021) refer to the extensive knowledge requirements of foraging, including plant identification, specific plant parts and their uses, where to find species in the wild as well as their timely harvesting (Schunko et al. 2021). Such information can be found in physic gardens. Physic gardens were originally established by monks and physicians to provide remedies and to enable the study of plant properties. They have now become places to educate and reconnect urban residents to plants and their uses. Many physic gardens became the botanical gardens we can still find in major cities worldwide.

There are still extensive knowledge gaps, and urban residents are increasingly looking for experts and enthusiasts to guide them in the field of urban foraging. This can be seen in the growth of urban foraging groups across the UK.

Indeed, wild plant gathering is a resurgent activity, globally (Frazee et al. 2016; Amato-Lourenco et al. 2020).

Research conducted in Kampala, Uganda, found that 47% of respondents used wild plants collected from their local urban environments for therapeutic and culinary purposes (Mollee et al. 2017). The results also show that the urban collector demographics are those individuals living in peri-urban areas with lower incomes and under 51 years of age (Mollee et al. 2017). Somesh et al. (2021), in Bengaluru, India, identified that foragers are primarily female, typically from lower socio-economic origins.

Urban foraging could provide significant contribution to household economies, supplementing home cooking and sustenance support systems such as food banks and community larders if urban green spaces are designed and managed with foraging in mind. Reliance on emergency food parcels in the UK has risen from around 26,000 in 2008/2009 to almost three million in 2022/2023 (Statista 2023). Increasing pressure on food security highlights an increased need for accessible, local food sources, particularly in areas which are 'sensitive to gentrification, pollution, and changes in land usage brought on by urbanisation' (Somesh et al. 2021). Urban landscape design which incorporates foraging design elements could increase ecosystem services (ESs). Foraging delivers provisioning ES through the direct practice of gathering flora for culinary and therapeutic purposes. Foraging has also been found to deliver cultural ES (Townsend 2020). Forage-rich landscapes support connectivity and improve a sense of belonging by increasing opportunities for people to interact with nature (Sardeshpande and Shackleton 2020). According to Fischer and Kowarik (2020), urban foraging could become a current strategy for bringing vast segments of society closer to urban nature.

However, if urban foragers are unfamiliar with the plant species they are collecting, they may compromise their safety and the integrity of public green spaces and negatively impact plant populations (Schunko et al. 2021).

Hundreds of wildflower species are protected in the UK under the 1981 Wildlife and Countryside Act, and the legalities concerning the collection of wild plants are often poorly understood, which creates a significant barrier to urban foraging (Landor-Yamagata et al. 2018). In the UK, the removal of roots is still prohibited; however, Section 4 (3) of the Theft Act (1968) states, "a person who picks mushrooms growing on any land, or who picks flowers, fruit or foliage from a plant growing wild on any land, does not steal what he picks, unless he does it for reward or for sale or other commercial purpose". This is a similar system under the Federal Conservation Act in Germany, which permits small-scale personal gathering in public spaces. However, there are often area-specific rules and regulations which override national policies (Linnekin 2018).

Urban foraging is thought to be a common practice throughout the establishment and growth of cities, old and new around the world (Shackleton et al. 2017), despite often confusing and conflicting legalities.

This chapter aims to provide background information on the historic foundations of urban foraging in the UK, to draw comparisons with foraging origins and traditions across Europe and to provide current insights into urban foraging groups actively collecting local flora in UK cities in Southern England. Three case studies, based in London, Bristol and Birmingham, are used to analyse current motivations and trends in the UK urban foraging. Research data is collated and used to provide urban design recommendations. Recommendations aim to enhance edible green infrastructure policy and contribute towards sustainable urban futures through enhanced local resilience.

10.2 Physic Gardens

Wild plants often have biologically active compounds, which can provide therapeutic value in addition to their nutritional significance (Ogle et al. 2003; Carvalho and Barata 2016). As a result, they can be regarded as functional foods or food medicine (Carvalho and Barata 2016). Ancient Eastern medicinal traditions have included medicinal foods; more recently, the fields of functional foods, nutraceuticals, and phytonutrients have drawn interest in the USA and Europe (Ogle et al. 2003).

Plants play a significant role in both early and modern societies (Carvalho and Barata 2016), particularly in the field of medicine. Chemicals and compounds found in plants, referred to as phytochemicals, can be potent and have both negative and positive effects on human health (Leitzmann 2016).

The majority of the early botanical investigations were conducted for medicinal purposes with apothecaries characterising plants and their compounds for medical concoctions (Chambers 2010). Private physic gardens, distinct from monastic herbaries, existed by the end of the fifteenth century, and some of these evolved into municipal gardens for the cultivation of 'simples' (Hill 1915). The Botanical Garden of Padua, which appears to be one of the earliest of these gardens, was established in 1545 (Hill 1915). The Padua Botanical Garden is distinctive because it was created as a 'Simples Garden', or garden of medicinal plants (Fig. 10.1). 'Simples' are remedies that have been taken straight from nature and are prepared for use without further blending (Hallett 2006). Dr. Francesco Bonafede requested the creation of the garden, which was initially known as '*Hortus simplicum*', designed to assist students in learning about the uses of medicinal plants (Hallett 2006).

Fig. 10.1 Medicinal plants at Padua's Botanic Garden, Italy

In England, the Worshipful Society of Apothecaries created the Chelsea Physic Garden, originally known as the Apothecaries' Garden, in London, England, in 1673 to cultivate plants for medicinal purposes. Today, the Chelsea Physic Garden (Fig. 10.2) contains a remarkable collection of about 5000 different edible, useful and medicinal living plants (Chelsea Physic Garden 2022).

A contemporary physic garden is situated at Dilston in Northumberland, England. Dilston Physic Garden (Fig. 10.3) was founded in the early 1990s, opened to the public in 2005. The garden was incorporated as a charity mainly for educational purposes in 2007 (Dilston Physic Garden 2022). Its origins are in neuroscience and neuropharmacology with links to Newcastle, Northumbria and Durham Universities. They are open to the general public and offer educational sessions to community groups, schools, colleges and universities on efficacious plant medicines and the possible health benefits derived from plants. They are actively involved in clinical trials and scientific research on plants to enhance memory and mind, and those to improve cognitive function and sleep. The research is supported by Royal Botanic Gardens, Kew, London.

In England, medicinal plants were also planted in traditional country houses and herb gardens. Cerney House Gardens (Fig. 10.4) is an example of a traditional country house, which features a herb garden split into two sections: a culinary area with herbs planted for use in teas and infusions, and a medical garden based on Culpeper's Herbal (Ross 2022).

Fig. 10.2 Chelsea Physic Garden in London

Fig. 10.3 Information board at Dilston Physic Garden, Northumberland, showing details of their active scientific study on the memory-enhancing properties of plants, with particular reference to Sage

10.2.1 A Historical Context of Hedges and Hedgerow Herbalism in the UK

In the UK, the use of medicinal plants was documented in the tenth-century Anglo-Saxon period (Rubin 1970; Thomas 2011).

The Leechbook of Bald,[1] the Lacnunga and the Old English Herbarium are three Old English medical texts compiled by the Anglo-Saxons (Rubin 1970; Thomas 2011). Thomas (2011) suggests these are amongst the earliest English medical writings and warrant further examination to provide a deeper understanding, of traditional herbal medicine (Thomas 2011). Oswald Cockayne collected and edited these

[1]The Leechbook of the Bald combined traditional herbal knowledge with paganism and superstition.

Fig. 10.4 Medicinal herb garden at Cerney House Gardens, North Cerney, Gloucestershire, England

writings to create Leechdoms, Wortcunning and Starcraft, and these three volumes were finally printed in the mid 1860's (Thomas 2011).

Prior to this period, most of Europe's population lived rurally. Only 8.3% of the UK population lived in towns and cities with over 10,000 inhabitants in 1650 (Barzun et al. 2023). Rural populations relied on small-scale agriculture and the use of common land to graze livestock.

Foraging for plants, nuts and fungi would have supplemented diets (Chambers 2010), and as pointed out by Fissell (2008), women were predominantly the home-makers, "Preparing food and providing bodily care was largely a female task and was gendered as such". Wild plants would have been used by women not only for culinary purposes but also for the creation of herbal remedies to tend sick members of their families and also people in their wider communities. However it is suggested by Fissell (2008) that remedies were also made in larger quantities and prepared through complex processes (Fissell 2008). Indeed, Fissell (2008) states that "women were central to health and healing in Europe before 1800".

However, the witch trials and the discreditation of local hedgerow herbalism may have contributed to the loss of confidence in wild herbalism. Across the UK, the number of people engaged with foraging plummeted, and by 1800, the percentage of people actively collecting wild foods had decreased to around 1% (Łuczaj et al. 2021). However, books documenting foraging activities in the early twentieth century show that the usage of wild foods had not fully disappeared (Łuczaj et al. 2021).

Hedges were commonly used to enclose estates, country houses and village settlements from the wider landscape (De Vries 2008); however, as part of the UK Enclosure Act of 1773, extensive hedges were established to delineate land, and ownership was transferred away from villagers to larger existing landowners (Wright 2016). Common land and open farmlands were reassigned, divided up and passed from rural communities to politically prominent families. They are some of the oldest remaining landscape features in the British countryside and have for that reason become a defining character of the English landscape (Sanderson and Prendergast 2002).

'Scrumping' became a way of subsidising low yields from the small plots of less fertile land the average resident received during the act of enclosure (McElroy 2012). 'Scrumping' is defined as 'stealing fruit' and was punishable by flogging, and assailants were referred to as 'prisoners' (News From The Past 2017). This may have had a long-standing impact on people's perceptions of foraging in the UK today with many people unsure of the legalities of foraging.

10.2.2 Hedge Species and Herbalism

Hedges continue to provide many cultural, supporting and regulating ecosystem services as well as food provisioning services (Staley et al. 2015). Hedges are crucial for wildlife and serve as vital archives of the area's past development. Additionally, they act as a wildlife corridor, a shelter for livestock, a way to manage livestock and a windbreak for crops and prevent soil erosion and water runoff (Sanderson and Prendergast 2002).

According to research by Gosling et al. (2016), urban and rural areas in England had significantly different woody plant species in hedges. In contrast to Blackthorn, Bramble, Dog Rose, Elder and Hawthorn, which were more frequently found in rural hedges, Beech, Holly, Ivy, Laurel, Privet and Yew were more frequently found in urban hedges. Urban species provide significantly lower provisioning services when compared with rural hedges (Gosling et al. 2016).

In their book 'Hedgerow Medicine: Harvest and Make Your Own Herbal Remedies', Bruton-Seal and Seal reported 50 wild plants that are abundant in British hedgerows, which contain medicinal properties (Bruton-Seal and Seal 2008).

For example, bramble (*Rubus species*) is the most commonly foraged fruit in the UK (Fig. 10.5). 'Going Blackberrying', also known as 'brambling', is an ancient

Fig. 10.5 Bramble is the most commonly foraged fruit in the UK

Fig. 10.6 *Sambucus nigra* berries

social activity that is practised around the world (Bruton-Seal and Seal 2008). The blackberry crop is known as ‘black gold’ in the US Ozark Mountains, where communal picking is still practised (Bruton-Seal and Seal 2008). Bramble has a reputation for promoting health, which is backed up by traditional folk remedies and, more recently, scientific evidence (Hummer 2010). In the past, brambles were used to treat dysentery according to the Leechbook of Bald. The root extract of *Rubus* is cathartic and emetic, and its leaf infusions are approved for use as a gargle for sore mouth and throat cankers and as a wash for wounds (Hummer 2010). The tannin-rich bark of *Rubus* was also used as a tonic for diarrhoea. Studies have measured the high levels of ellagic acid, anthocyanin, total phenolics and total antioxidant content in *Rubus* fruits (Hummer 2010).

One of the most widely used medicinal plants in the world today is Elder (*Sambucus nigra* L.) (Fig. 10.6), which has a long ethnobotanical history as a remedy for viral infection across many different cultures (Porter and Bode 2017). In species-poor hedges, Elder is relatively common, and its representation rarely increases in species-rich hedges (Atkinson and Atkinson 2002). This is consistent with both its rapid ability to colonise disturbed habitats and its short lifespan (Atkinson and Atkinson 2002). *Sambucus*’ flowers are still a fever remedy and delicious in drinks or desserts (Bruton-Seal and Seal 2008). The berries help with cough relief and are effective against colds and the flu. The leaves work well as an ointment for bruises (Bruton-Seal and Seal 2008).

10.3 The Difference in Foraging Traditions Between the UK and Europe

Hedgerow foraging can be seen as a significant part of Britain’s foraging past and present; however, it is interesting to compare foraging traditions in Britain with those in mainland Europe. UK tree cover is under 10% of the total land area; across Europe, a much greater percentage of forest tree cover remains. In 2021, tree cover in France was 31% of its total land area, tree cover in Germany was 36% of the total land area and in Italy 31% of its total land area. These figures increase in Eastern Europe with 41% tree cover in the Czech Republic (Global Forest Watch 2023). Forests in Europe provide extensive foraging opportunities from nuts to fruits to fungi. As wild fungi are an integral part of forest ecosystems, the collecting of mushrooms is found to be an important part of European cultural heritage dating back to Palaeolithic Age (Procházka et al. 2023). A lack of forest or woodland foraging opportunities in the UK, due to the significantly reduced tree cover, may explain why UK populations are often characterised as 'mycophobic'. The reduced opportunity for woodland foraging in the UK can partly explain the lack of knowledge and reduced cultural significance; however, it is documented that individuals foraged for edible wild mushrooms during the Victorian era (Łuczaj et al. 2021).

Procházka et al. (2023) highlight the significance of fungi as a 'staple' for European diets, "Wild mushrooms are a source of fibre, proteins, vitamins, and minerals. Some mushrooms are beneficial for their medicinal properties or dietary properties, such as their content of cholesterol-lowering substances, improving diabetes conditions, and even cancer prevention effects" (Procházka et al. 2023, pg. 1). The nutritional and perceived medicinal properties of wild fungi help to explain why this continued recreational activity, has increased in popularity across Europe in recent years (Kotowski 2016; Wright 2016; Jehlička et al. 2020; Šiftová 2020; Procházka et al. 2023). Kotowski (2016) highlights that wild mushrooms are increasingly considered to be a valuable commercial product and are often regarded as a delicacy. In turn this could be helping to fuel the growing popularity of recreational mushroom hunting (Kotowski 2016). However, the misidentification of wild mushroom species has resulted in an increasing number of poisoning cases (Schenk-Jaeger et al. 2012; Kotowski 2016; Cervellin et al. 2018). National and local regulations for access and mushroom hunting vary greatly across Europe (Turtiainen and Nuutinen 2012). For example, in all forest areas in Finland and Sweden, people have the freedom to gather mushrooms and wild berries, and this privilege extends to both residents and tourists (Turtiainen and Nuutinen 2012). On the other hand, mushroom picking is regulated by regional laws in Italy. For example, residents of the Campania region are permitted to pick mushrooms after passing a qualification interview designed to assess their knowledge of the species and varieties of mushrooms; the methods of research, collection and marketing permitted by the current regulations; as well as general concepts of mycology and forestry (Regione Campania 2022).

10.3.1 Contaminants

It is noted by Dowalti et al. (2021) that heavy metals often present in urban environments may accumulate in the fruiting bodies of edible mushrooms. The transferral through digestion may cause damage to kidneys or heart and impacts one or more of the body's vital systems (Dowlati et al. 2021).

Concerns around other food crop contamination are presented as a barrier to participation in wild urban food collection with several studies focused on the absorption of pollutant particles from both air- and soil-borne contaminants (Leake et al. 2009; Amato-Lourenco et al. 2020; Romanova and Lovell 2021).

Romanova and Lovell (2021) stress that pollutants from the built environment, industries and vehicles can be trapped within urban environments, cycling from air to soil to water. A study of London-based urban allotment gardens, used for the production of edible crops, found higher than recommended levels of lead concentrations in roots and leafy vegetables (Leake et al. 2009). However, the higher-than-normal concentrations of lead were found not to be transferred upon consumption. Blood samples were taken and tested from a number of allotment holders and found to be within normal range (Leake et al. 2009). Several studies looking at contaminants present in apples grown in urban centres found high levels of heavy metals such as zinc, copper and lead (Scherz and Kirchhoff 2006; Imeri et al. 2019; Medvedeva et al. 2021). However, as long as fruits are adequately washed, the toxic load from the average human consumption would be insignificant enough to impact people's health (von Hoffen and Säumel 2014). The type of fruit, nuts, pome, stone fruits and berries may influence the amount of trace metals present (von Hoffen and Säumel 2014). Säumel et al. (2012) suggest that additional protection may be obtained through the use of woody vegetation, this could filter out airborne particles and therefore help to prevent contamination of urban edibles. Overall, results suggest that the negative effects of urban pollutants on edible plants are outweighed by the provisioning and cultural ecosystem services derived from the inclusion of foraging elements designed into edible green infrastructure (Russo et al. 2017).

10.3.2 The Resurgence of Urban Foraging in the UK

Despite historical hang-ups and current contamination concerns, the collecting of wild food in urban areas continues. The number of urban foraging groups across UK towns and cities has been growing steadily since the early 2000s, but a sharp increase has been noted since the pandemic (Lawrence 2020). The pandemic created a renewed appreciation for green spaces as people took the time to explore their local areas at a pace that allowed them to connect with the natural world around them and take note of the flora in their immediate locality (Clouse 2022; Crossley and Russo 2022). Waitrose & Partners Food report of 2021 noted "a UK wide 89% increase in foraging interest on social media". The promotion of wild food gathering on popular media platforms such as YouTube and Instagram and the use of foraged elements in UK cooking programmes such as MasterChef have led to the rise of demand and interest in foraged foods. Locally and wild-sourced foods are appearing on Michelin-starred menus across UK cities.[2] However, this

[2] See *Best UK Restaurants using foraged and locally sourced food – https://oracleoftime.com/best-restaurants-uk-foraged-and-locally-sourced-food/*

Ten of our Favourite Farm-to-Table Restaurants – https://guide.michelin.com/gb/en/article/features/ten-of-our-favourite-farm-to-table-restaurants

does raise the issue of foraging for commercial benefit. A rural resident from Oxfordshire expressed concerns about commercial foraging stripping hedgerows of sloes, and Kenneth Greenway, Tower Hamlets Cemetery Manager and Foraging Leader, said that he had encountered conflict with "commercial foragers, who should be sourcing their wild plants ethically and by buying their plants from farms that grow them commercially".

Foraging and herbalism have also been recognised in several RHS show gardens in recent years (Fig. 10.7). The 2023 Chelsea show garden, Centre for Mental Health's The Balance Garden designed by Jon Davies and Steve Williams of Wild

Fig. 10.7 The Centre for Mental Health—The Balance Garden 2023 (top left and right) (Image courtesy: RHS/Neil Hepworth). Alder Hey Urban Foraging Station Garden (bottom left) (Image courtesy: RHS/Sarah Vivienne). The Wellbeing of Women Garden (bottom right) (Image courtesy: RHS/Sarah Cuttle)

City Studio, included a designated space for mushroom cultivation and incorporated so-called Edimentals,[3] perennial vegetables, herbs and forest food species.

In 2022, Howard and Hugh Miller designed the Alder Hey Urban Foraging Station Garden for the Chelsea Flower Show. The garden featured species-rich hedgerow planting, an orchard meadow and bog-ditch habitats with herbs and marginal plants for foraging.

The RHS Sanctuary Garden, featured at Hampton Court in 2019, designed by Ula Maria, former 2017 winner of the RHS Young Designer award, featured an inner courtyard full of herbs and medicinal plants.

The Health and Wellbeing Garden designed by Alexandra Noble for the 2018 Hampton Court featured a Chamomile seat amongst other medicinal and herbal planting.

The RHS Wellbeing of Women Garden, designed by Wendy von Buren, Claire Moreno and Amy Robertson, featured at Hampton Court RHS Show in 2015. The planting included medicinal and aromatic planting including *Echinacea*, *Salvia officinalis* 'Purpurascens' and *Salvia rosmarinus*.

The presence of foraging in mainstream media and high-profile public UK events such as the RHS shows may be driving an interest and desire to engage in foraging activities amongst more affluent urban communities as well as more marginalised less food-secure segments of UK society. An increased reliance on food banks can be found in all urban centres across the UK, which reflects food insecurities (Brown et al. 2022; Clouse 2022). Kenneth Greenway, suggests this is due to "austerity measures, the effects of Brexit on the UK food market as well as disruptions to long international supply chains which has meant that UK customers are currently faced with lower food availability". There are notable feelings of food vulnerability in the UK. The Office for National Statistics states that "UK food price inflation was amongst the highest across G7 economies in March 2023". As the UK cost of living crisis and increasing food insecurities remain, foraging presents itself as a way to mitigate pressures and, as Kenneth Greenway suggests, foraging also satisfies a 'desire for self-sufficiency and provides a degree of autonomy away from a reliance on supermarkets'.

Foraging in the UK reflects it's many cultural communities (Townsend 2020). Foraging facilitates nature connectedness and encourages green space stewardship whilst also enabling access to food which is "nutritious, gourmet and sustainable" (Townsend, 2020).

10.4 UK Urban Foraging Case Studies: Findings

Foraging is said to have a universal appeal for all people (Nyman 2019).

For the purposes of this research, we identified and participated in three urban foraging groups in major UK cities in the Southern England. Participant observation was selected as the main research method to gain insights into urban foraging group

[3] Plants which are both aesthetically pleasing and edible.

experiences, participant motivations, preferred locations for foraging, and conflicts and concerns around foraging as well as to observe how individuals interacted with each other and their local environment. During informal discussions, participants in all groups were asked about their views on foraging and the opportunities to collect flora in urban landscapes. We also observed basic participant demographics. The groups took place in London, Birmingham and Bristol and varied in participant numbers, group formality and cost (Table 10.1).

The Tower Hamlets Cemetery became a designated nature reserve in 2000, after local opposition to the demolition of the site, and saw the ownership pass to London Borough of Tower Hamlets. In 1990, its management was taken over by the Friends of Tower Hamlets Cemetery. The nature reserve is currently tended by 3000 volunteers and invites 7000 school children to visit annually and hosts 170 walks and talks with various public and private groups including corporate events. The knowledge transfer and educational value of the reserve are increased by providing supporting materials such as online resources, digital handouts and species lists.

The Birmingham foraging group led by Stephanie Bale, Sustainable Life Birmingham, visited two cemeteries, Warstone Lane and Key Hill, both situated in Birmingham's Jewellery Quarter, in the heart of the city.

In Bristol, the Frenchay Common provided little in the way of foraging due to the large expanse of mown amenity grass. However, Frenchay Moor is a National Trust-owned conservation area with woodland and lowland meadow habitats and improved and unimproved grasslands managed by grazing livestock. Hedgerows and veteran trees provide additional foraging opportunities for people and wildlife.

10.4.1 General Observations

The UK urban foraging sessions we attended attracted predominantly female participants. Sixty-seven per cent of attendees across all three groups were women under the age of 60. London had the most diverse attendance with international participants and a mix of ethnicities. Bristol had the youngest participants with 100% under the age of 50. Birmingham had a more even split of male and female attendees. Couples and friends attended the foraging groups together, as well as several male and female individuals who attended alone.

The size of urban foraging groups in the UK varies from 5 to 20 participants, and the cost of group participation has been found to range from a voluntary contribution up to £65+ for complete foraging and cooking experiences.[4]

The cost appears to correlate with the formality of leadership style. The less formal groups attended in Bristol and Birmingham provided more social benefits with an increase in repeat participation found on a weekly or monthly basis. A male participant in Birmingham who attended the group alone said that he had 'attended several groups, as well as foraging alone, as he is still learning'.

[4] See https://www.deliciousmagazine.co.uk/foragingdaysoutpen/

Table 10.1 Characteristics of foraging groups in London, Birmingham and Bristol

City	Location	Cost	Duration	Formality	Group leaders	Leader expertise	Group size	Gender split	Foraging group established	Frequency of walks
London	Tower Hamlets Cemetery	£42.89	2 h	Formal	2	Environmental biology, horticulture and cemetery manager	15	6 men 9 women	2009/2010	Monthly
Birmingham	Key Hill Cemetery	£10.00	2 h	Informal	1	10 Years' experience. Charity leader for Sustainable Life Birmingham	5	2 men 3 women	2016	Monthly
Bristol	Frenchay Common and Moor	£6.00	1.5 h plus Social	Informal	1	Studying herbal medicine	16	4 men 12 women	2020	Weekly

Foraging group leaders at all three locations gave clear identification methods, discussing stem and leaf shapes, colour and texture. Leaders and participants discussed plant edibility and shared information on the medicinal properties and household use of the plants collected. Collaborative learning approach was observed as participants shared recipe ideas for soups and salads and suggested which plants could be eaten with what to enhance flavour and nutrition. Processing techniques such as syrups, teas, jams, and wine were also discussed during the foraging walks, as well as less mainstream methods for creating coffee, soap, gelatine and flour substitutes were also discussed along with herbal remedies and the creation of infusions and tinctures.

Table 10.2 details an extensive list of the species collected by each of the case study groups during our urban foraging sessions. The properties listed in the table were provided by group leaders and draw on the work of Robin Harford's, "Edible and Medicinal Wild Plants of Britain and Ireland" , published in 2019.

10.4.2 Motivations and Access to Knowledge

It was suggested by Kenneth Greenway at the Tower Hamlets Cemetery, London, that foraging starts with 'Gateway Plants'. These include the aforementioned blackberries (*Rubus fruticosus*) and plants such as wild garlic (*Allium ursinum*), flowers and berries from the Elder (*Sambucus nigra*) and sloes from the blackthorn (*Prunus spinosa*). According to Kenneth Greenway, learning to identify these common UK hedgerow species is thought to build confidence with foraging and act as stepping stones to deeper interactions with, and understanding of, nature in our urban environments.

Urban residents are, as we have seen through the growth of foraging groups online and in cities across the UK, eager for greater knowledge and understanding, and are looking to professionals and enthusiasts for expert advice and ethnobotanical know-how. This was confirmed by 53% of participants in London. Additionally, the fear of poisonous plants was mentioned as a reason to look to experts for guidance on how to interact with the natural world, safely. Collaborative learning was common across all the case study groups, with all three leaders inviting participants to share their own knowledge and experience, especially where cultural differences led to greater plant use and culinary insights. In London, Kenneth Greenway spoke of social events where foraging facilitated inclusive, nutritional lunch opportunities for local elderly residents at risk of isolation. Overarchingly, people are looking to foraging to enable them to interact, and connect in a three-way process, to nature, to each other and to the places in which they live. Kenneth Greenway renamed a botanical tour, in London's Tower Hamlet Cemetery, as a foraging tour and found participant numbers increased significantly. This can be interpreted as demonstrating urban residents' desire for interaction and a deeper connection with nature. All foragers were observed to be keen to enhance their observations and improve connections to nature in their local environment. A male and female couple attending

Table 10.2 Species collected during urban foraging sessions in London, Birmingham and Bristol

Scientific Name	Common Name	Uses	Properties	British Native	Habit	London	Bristol	Birmingham
Achillea millefolium	Yarrow	Leaves - teas, poultice, tinctures	Anti-inflammatory, healing wounds, cramps	Yes	Herb	Yes	Yes	Yes
Aegopodium podagraria	Ground Elder	Salad, cook like spinach	Anti-inflamatory, treats Gout	No	Herb	No	No	Yes
Alliaria petiolata	Jack in the Hedge	Leaves - cooked like spinach	Vitamin A, respiratory system benefits, antiseptic, induce sweating	Yes	Herb	Yes	Yes	No
Allium triquetrum	Three cornered Leek	Bulbs & leaves - soups and stews	Insect repellent, used as onions	No	Herb	Yes	No	No
Anthriscus sylvestris	Cow Parsley	Stems & leaves	Insect repellent, anti-oxidant	Yes	Herb	Yes	No	No
Artemisia argyi	Chinese Mugwort	Leaves - soups, stews, teas	Vitamin A, K, E, Potassium, antioxidant, improves sleep, reduces cramps	No	Herb	Yes	No	No
Bellis perennis	Daisy	Leaves & flowers	Colds and upper respiratory tract, digestive aid	Yes	Herb	Yes	No	No
Betula pendula	Silver Birch	Sap, leaves, bark and seeds	Rehydrates quickly, fires	Yes	Tree	Yes	No	Yes
Cardamine pratensis	Lady's Smock	Leaves - wasabi or watercress substitute	Vitamin C, digestive aid	Yes	Herb	Yes	No	No
Centranthus ruber	Red Valerian	Flowers & leaves	Unknown	No	Herb	Yes	No	No
Cichorium intybus	Chickory	Flowers, leaves & stem	Proteins, vitamin A, B, C & K, calcium, potassium and iron	Yes	Herb	Yes	No	No
Chamaenerion angustifolium	Rosebay Willowherb	Leaves, shoots, stems & flowers - tea, boiled or steamed	Vitamin A & C	Yes	Herb	Yes	No	No

Corylus avellana	Hazel	Nuts	Protein, vitamin E, digestive aid, regulates blood sugars, protects cells	Yes	Tree	Yes	Yes	Yes
Crateagus	Hawthorne	Berries & young leaves - fruit leather, salads	Antioxidants and anti-inflammatory, stress reliever, reduces cholesterol	Yes	Tree	Yes	No	Yes
Daucus carota	Wild Carrot	Root	Digestive benefits	Yes	Herb	No	No	Yes
Diplotaxis tenuifolia	Wild Rocket	Leaves & flowers	Vitamin C, fiber, antioxidant,	Yes	Herb	Yes	No	No
Galium aparine	Goosegrass	Leaves & Stem	Glandular fever, insect bites, skin conditions, immune boosting	Yes	Herb	Yes	No	Yes
Galium mollugo	Hedge Bedstraw	Leaves	Anti-inflamatory, skin conditions,	No	Herb	Yes	No	No
Geranium robertianum	Herb Robert	Flowers & leaves	Antiseptic, anti-inflammatory, anti-bacterial, anti-fungal, anti-oxidants	Yes	Herb	No	No	Yes
Geum urbanum	Herb Bennet	Root only	Spice	Yes	Herb	No	No	Yes
Hedera helix	Ivy	Leaves - soap	Soap and detergent	Yes	Woody Climber/ Creeper	No	Yes	Yes
Juniperus communis	Juniper	Seeds	Antiseptic, anti-inflammatory, anti-bacterial, anti-fungal, anti-oxidants	Yes	Tree	Yes	No	No
Lamium album	Dead Nettles	Flowers & leaves - wine, cooked leaves like spinach	Mushroom flavouring, protein, vitamin C & A	Yes	Herb	Yes	Yes	Yes
Lycoperdon perlatum	Puffball mushroom	Fruit	Protein, carbohydrates and fats	Yes	Fungi	No	No	Yes
Mahonia	Mahonia	Berries	Skin conditions, ant-oxidant	No	Shrub	No	Yes	Yes
Malus sylvestris	Crab apple	Fruit - jams, cider	Lowers cholesterol, vitamin A and C, antioxidant	Yes	Tree	Yes	Yes	Yes

(continued)

Table 10.2 (continued)

Scientific Name	Common Name	Uses	Properties	British Native	Habit	London	Bristol	Birmingham
Malus domestica	Apple	Fruit - eaten fresh	Fiber, potassium and vitamin C, antioxoidant	No	Tree	No	No	Yes
Oenothera biennis	Evening Primrose	Seeds	Oil to treat hormone imbalances, improves skin conditions	No	Herb	No	Yes	No
Pinus sylvestris	Scots pine	Pollen & Needles - tea	Aphrodisiac, antioxidants, vitamin C	Yes	Tree	No	No	Yes
Plantago	Plantain	Leaves, seeds & flower buds	Salads, mushroom flavour flower buds, flour, vitamins A,B, C and K, protein, trace elements, anti-histimine	Yes	Herb	Yes	Yes	Yes
Potentilla anserina	Silver Weed	Leaves - teas, salads, roots - dried and ground	Anti-spasmodic, carbohydrates, starch	Yes	Herb	Yes	Yes	Yes
Quercus robur	Oak	Acorns - soak and rinse 7 times, Leaves - wine	Protein, carbohydrates and fats, calcium, potassium	Yes	Tree	Yes	No	Yes
Rosa canina	Rosehips	Hips	Face oil, Vitamin C, immune system booster	Yes	Shrub	Yes	Yes	Yes
Rumex sorrel	Lemon Sorrel	Leaves - stews, salads	Vitamin C,	Yes	Herb	No	Yes	No
Sambuccus nigra	Elder	Blossom & Berries	Antioxidants, anti-inflammatory, anti-bacterial, Vitamin C	Yes	Tree	Yes	Yes	Yes
Sanguisorba minor	Salad Burnet	Leaves	Digestive aid, burn treatment, aphrodisiac, oral health, anti-inflamatory	No	Herb	Yes	Yes	Yes
Sorbus	Rowan	Berries - syrup, tea, juice	Anti-bacterial, anti-fungal, helps fight intestine inflections, boosts immune system	Yes	Tree	No	No	Yes

Stellaria media	Chickweed	Leaves & stem - soups and salads, lilac dye	Vitamin C, A, B, minerals and fatty acids, anti-inflammatory, eczema	Yes	Herb	No	No	Yes
Tanacetum parthenium	Feverfew	Leaves & flowers	Anti-inflammatory	No	Herb	No	No	Yes
Taraxacum	Dandelion	Leaves & roots - salads, tea, tincture, poultice, coffee	Anti-inflammatory, Vitamin A, C, K and E, folate, reduce blood pressure and cholesterol, decrease blood sugars	Yes	Herb	Yes	No	Yes
Tillia cordata	Lime	Leaves, seeds & buds - tea, chocolate substitute,	Mild sedative, reduces stress, antispasmodic	Yes	Tree	Yes	No	No
Urtica dioica	Nettles	Leaves & seeds - soup, tea, pesto	Anti-inflammatory, adaptagenic, vitamins A, C and K, Potassium, Calcium, stimulates circulation	Yes	Herb	Yes	Yes	Yes
Verbascum	Mullein	Leaves - tea	Chest infections	Yes	Herb	No	No	Yes

the Birmingham group said that they had never previously foraged but "go for walks and want to engage with what's around more'" People are seeking greater knowledge of plants in their locality but want context and meaning through direct interaction with their green spaces; the opportunity to experience their urban green space from a provisioning ecosystem service perspective. Direct provisioning services obtained from foraging activities appeared to delight participants across all three urban foraging groups attended, with participants ready to touch, taste and smell flora identified by the experts.

Fourteen per cent of participants across all three groups recognised foraging as a way to strengthen their place attachment, saying that it improved their observation of their immediate natural and built environment, and noted how green spaces change throughout the seasons. Foraging has been found by this study to increase cultural ES derived from urban green spaces. Walking, talking and exploring new areas were aspects of foraging noted as beneficial for mental and physical health (Figs. 10.8, 10.9, 10.10, 10.11, 10.12 and 10.13).

10.4.3 Foraging Locations and Terrain

Two out of three groups relied on cemeteries predominantly to run their urban foraging groups. Cemeteries were said by one leader to "offer increased plant variety due to shading effects from trees as well as gravestones".

Fig. 10.8 Urban foraging group discussing species in Tower Hamlets Cemetery Park, London

Fig. 10.9 Urban foraging group collecting Sorrel on Frenchay Common, Bristol

Fig. 10.10 Urban foraging group collecting apples and *Sorbus* berries in Brookfield Cemetery, Birmingham

Fig. 10.11 Red valerian (*Centranthus ruber*) in Tower Hamlets Cemetery

Fig. 10.12 Collecting rose hips on roadside in Frenchay, Bristol

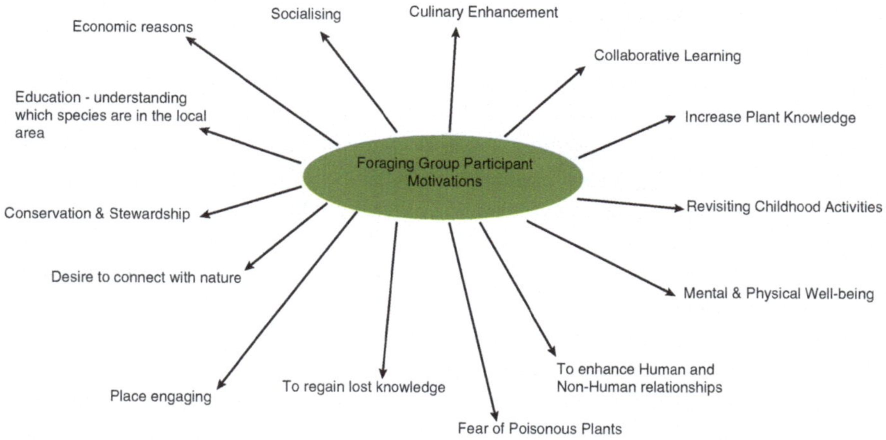

Fig. 10.13 Chart showing the motivations of participants in joining an urban foraging course

It was noted that areas with shape and undulations were more varied in terms of species and were therefore favoured over open areas of mown amenity grass. Some species such as daisy (*Bellis perennis*) and yarrow (*Achillea millefolium*) continued to thrive even in areas subjected to regular mowing. Both species were discussed in length on Frenchay Common, particularly the use of yarrow for stemming bleeding during WW1 and WW2. Such areas, including many public parks, are seen to be more managed, less diverse and more likely to implement chemical management programmes.

During the group sessions in Bristol and Birmingham, additional foraging was carried out along roadsides and path edges. Dry, compact ground was noted by Maria Fernandez Garcia, Bristol, as often beneficial for finding species such as plantain (*Plantago* sp.) and more boggy ground recognised as better for finding lady's smock (*Cardamine pratensis*).

Riversides and waterways were identified by 100% of foraging leaders as great foraging locations due to an increased variety of marginal, aquatic and bog planting in addition to hedges and grasslands. The Bristol leader preferred foraging in central locations for ease of access for participants, but urban fringes for woodland edge and hedgerow foraging were also favoured above woodland foraging, which was considered 'repetitive'.

Interestingly, the London and Birmingham group leaders both referred to sites where the biodiversity is actively enhanced and managed. Stephanie Bale referred to the Ten Acres in Stirchley where she had spent 'the last 4 years working to improve its native biodiversity'. They had included 'rare and endangered plant species' based on past ecological records. Kenneth Greenway referred to the Tower Hamlets Cemetery as a very-well-managed site with a lot of diversity for its (relatively) small size. It was noted that volunteers actively propagated some species of popular plants and thinned out seedlings of species that were over-abundant.

Diversity of landscape and species are key to foraging

10.4.4 Conflict and Contamination

All leaders collected along roadsides and path edges with little concern for contamination. Stephanie Bale, Sustainable Life Birmingham, advised against collecting fungi in cemeteries due to their high absorption rate of lead from lead-lined coffins.[5] Stephanie Bale also expressed concerns, particularly in graveyards, of Weil's disease[6] from rats. Weil's disease can also be spread by infected dogs. Twelve per cent of attendees in Bristol discussed 'dog wee' contamination. However, rinsing with water was deemed sufficient by all groups to protect against urine contamination from dogs and urban wildlife.

Airborne pollutants from traffic and industry were also noted to be removed through the process of washing and where relevant cooking. The London expert stated that "plants don't absorb air pollutants". However, all group leaders referred to chemical management programmes in urban green spaces. Tower Hamlets has a no-spray policy across the whole district including in parks and public spaces. Wild and less manicured spaces were identified as areas less likely to have chemical contamination. The group leader from Healing Weeds said, 'if there are weeds growing then the area hasn't been sprayed'; she recommended avoiding areas which are 'too manicured'. The Sustainable Life Birmingham leader also highlighted that the spraying of areas which included the UK invasive Japanese knotweed (*Fallopia japonica*) was likely and should be avoided.

However, overall concerns about chemical green space management programmes are lower than expected and concern about such schemes lower than anticipated with foragers finding the benefits to outweigh any negative effects of plant consumption or applications. This was perhaps due to the local knowledge and understanding of the chosen foraging sites and trust in the foraging leader experts. These findings are in concordance with research by Scarrow (2018), which found that 94% of urban gardeners in Daejeon, Korea, were unconcerned about food safety.

Overall results suggest that the negative effects of urban pollutants on edible plants are outweighed by the extensive ES derived from edible green infrastructure (Russo et al. 2017).

Urban foragers, in the cases we observed, generally did not experience conflicts between different users of the urban green spaces, with the foraging experts saying that most people they encountered were interested in what they were collecting and why. Two additional people joined the London foraging walk as it was in progress. Birmingham leader Stephanie Bale found that more conflict arose with members of the public when foraging in more affluent areas.

[5] In the late nineteenth and early twentieth centuries, coffins were often lined with lead to prevent external disturbances, reduce the spread of disease and slow down the decay process.

[6] Weil's disease is also known as leptospirosis and is spread through the urine of infected animals. This is usually rats and mice but can also be spread by dogs, pigs and cows. It is a bacterial infection that can lead to kidney or liver failure or meningitis and in severe cases death.

10.4.5 Foraging Etiquette and the Impact of Foraging on Biodiversity

Foraging etiquette is a common theme discussed during sessions. Leaders set out that foraging should be from a mix of locations to maintain plant communities. Phrases such as 'leaving no trace' and 'slow, kind and patient' were used to ensure the safety of foragers and to protect and maintain biodiversity.

Foragers were also referred to by the London foraging leader as 'gardeners', and participants were encouraged to consider thinning out of seedlings of certain plant communities such as Jack-by-the-hedge (*Alliaria petiolata*) and wild rocket (*Diplotaxis tenuifolia*), as well as monitoring more unwanted invasive species such as Japanese knotweed (*Reynoutria japonica* syn. *Fallopia japonica*). Foraging therefore can encourage and create urban green space stewards. Stewardship can enhance and support urban green space management, which is both cost-effective and beneficial to the local community in terms of increased ecosystem services.

The scattering of seeds and leaves during foraging walks provides regulating service support through supporting the increase of plant communities. At the Tower Hamlets Cemetery, London, active propagation and planting of species to support foraging and wildlife are also undertaken.

Specifically, the propagation of plants to support butterfly and moth species was discussed in London, Birmingham and Bristol foraging groups.

For example, Jack-by-the-hedge (*Alliaria petiolata*) supports orange-tip butterfly, and hedge bedstraw (*Galium album*) is propagated and planted in London for the hummingbird hawk-moth caterpillars to feed on. Seeds from the Devil's-bit scabious in Bristol Frenchay Common were scattered to encourage fritillary butterflies. Foragers, therefore, become stewards of both local flora and fauna.

Signage can play an important role in supporting foraging activities and guiding the stewardship of UGS (Fig. 10.14).

The Key Hill Cemetery in Birmingham is maintained by the Friends of Key Hill Cemetery group, and signage was present to inform visitors of species, planting aims and maintenance schedules. A QR code was also provided for additional information on species of flora and their locations in the cemetery.

Signage is noted, through case study observations, to increase connectivity, making landscapes more interactive and accessible to local communities and visitors alike.

Additional information, similar to the signs used in physic gardens, could be provided to assist with the identification and uses of species of plants especially those included in more complex urban planting schemes.

Foragers also benefit greatly from a growing digital support network. Vital resources can be found on online foraging groups, social media platforms, forums and online communities; for example, in London, Raynes Park resident Solomon Dolphin, who began foraging 10 years ago, is currently the administrator of the Facebook group Foraging SW London in which foraging participants share photos, recipe suggestions and identification tips for various plants and fungi (Cole 2021).

Fig. 10.14 Wayfinding signage in Tower Hamlets Cemetery (top left). Wildflower planting signage in Birmingham (top right). Plant properties and usage signage in Dilston Physic Garden (bottom left). Informative signage in Oxford Botanic Garden (bottom right)

Plant identification apps for smartphones as well as specific foraging apps such as Seek by iNaturalist and Forage—Wild Free Food—are also important tools in enabling greater urban green space interaction.

10.4.6 Foraging by Design

Foraging leaders were asked the following questions about foraging design implementation:

1. How do you think urban green space design could be amended to support foraging?
2. What features would you include and why?

Urban green space management programmes were key to answers given. A reduction in mowing and weeding was presented by Kenneth Greenway. Stephanie Bale of Sustainable Life Birmingham stated that "management and maintenance needs to be written in from the start, and a varying scope of environments needs to be included". It was suggested that a reduction in dense tree canopy and a 'gentle guiding hand' to shape green spaces rather than areas being 'left alone to rewild' would support more foraging opportunities.

Diverse planting was mentioned by all three foraging leaders in the cases studied with planting 'below and around trees' and ground cover planting rather than bare ground between planting in formal beds and borders. A reduction in 'weeding' allows wild areas of natural seed banks to establish rather than over-reliance on amenity grass. However, it is apparent that there continues to be a lack of tolerance and understanding to reductions in mowing with areas left for wildflowers described by one West Oxfordshire town resident as an 'unkempt mess'.

Experts supported the inclusion of more urban hedgerows for foraging and to support wildlife. Native trees and orchards were suggested design elements and multifunctional planting to "create aesthetic appeal, seasonal interest and extended flowering times".

We are already seeing foraging entering the design of public spaces. The recent Grey to Green regeneration of Sheffield's industrial quarter designed by Nigel Dunnett of Sheffield University includes wildflowers and edible non-native and native cultivars. The apples (*Malus*) included in Sheffield's Grey to Green city regeneration scheme (Fig. 10.15) are the most accessible to urban foragers as a dietary supplement; however, several other plant species included in the scheme are of potential foraging interest when considering herbal medicine plant properties. These include UK natives *Achillea* and *Betula pendula* as well as non-natives with herbal properties such as *Agastache* and *Salvia yangii*.

The planting scheme also includes *Eupatorium and Kniphofia*. It is claimed that yarrow (*Achillea millefolium*) was used to staunch bleeding in both World War I and II as well as the Trojan War. It can also act as an anti-inflammatory and anti-spasmodic (Benedek et al. 2007; Akram 2013). There are studies on the medicinal properties of *Eupatorium*, which is said to have antimicrobial, anti-inflammatory, antioxidant, antitumor as well as antibacterial and antifungal properties amongst others (Sharma et al. 1998; El-Seedi et al. 2002). Studies found *Kniphofia* to contain anti-malarial properties (Wube et al. 2005; Alebachew et al. 2021). *Salvia yangii* was found in a study by Ashraf et al. (2014) to have antimicrobial and antioxidant properties. *Agastache mexicana* is stated in a study by Palma-Tenango et al. (2021) to have potentially beneficially bioactive compounds providing anti-inflammatory, antioxidant and anxiolytic effects.

Diverse planting, with the inclusion of native and non-native species, maximises the culinary and herbal remedy foraging opportunities. Diverse planting could also support an extended period of flowering for pollinators and urban aesthetics, creating additional provisioning, cultural and supporting ES. In turn, this contributes towards urban resilience in the face of climate change.

Fig. 10.15 Malus in Sheffield Grey to Green scheme (left). *Eupatorium*, *Achillea* and *Salvia yangii* in Sheffield Grey to Green scheme (above, right)

10.5 Conclusion and Design Recommendations

In conclusion, our exploration into urban foraging and wild herbalism in the UK reveals their potential to provide social and ecological benefits. Forage-rich urban landscapes can provide social and ecological benefits and support edible and resilient city models (Sardeshpande et al. 2021).

Using nature-based solutions (NBSs) in cities, in the form of blue and green infrastructure elements, which promote urban food production offer numerous advantages for the development of sustainable cities in terms of enhancing ES (Artmann et al. 2020; Sardeshpande et al. 2021).

In comparison to other NBSs, edible city projects have demonstrated significantly more successful inclusive urban regeneration and emancipative citizen commitment (Scharf et al. 2019). Foraging connects urban residents to their natural spaces, and promotes additional engagement and stewardship of our UGS and therefore contributes to overall community resilience.

Incorporating and managing edible urban areas for foraging, recreation and nature observation, edible cities can be seen as a strategy for boosting urban attraction and place identity, fostering social cohesion and an experiential human-food connection that supports urban biodiversity (Artmann, Sartison and Vávra, 2020).

10.5.1 Ecological

Recommendations for an increase in diverse urban planting schemes include incorporating a wider range of 'wild' or native species of trees, shrubs and wildflowers. ES derived from urban planting can also be enhanced through the inclusion of non-native species of woody and herbaceous perennials, herbs and vegetables, which could increase flowering times as well as offer a greater breadth of plants for food and food medicine. Some such species as *Salvia yangii* and *Kniphofia* are more suited to hotter, drier, poor soil conditions often found in urban environments.

Hedges have been found to have place-specific social and ecological significance, particularly in the UK. The use of rural hedgerow species is found in the UK such as *Prunus spinosa*, *Sambucus nigra*, *Corylus avellana*, *Rubus fruticosus*, *Crataegus monogyna*, *Sorbus aucuparia*, *Malus sylvestris* and *Rosa canina*. Traditionally planted in double rows, hedgerows provide the perfect green corridors for wildlife (Wright 2016). A network of urban hedges could provide habitat space and seasonal food sources for people and wildlife increasing in ES of urban green infrastructure. The hedging species mentioned above are all pioneer species, which are quick to establish. Bare-root hedging whips are a low-cost and low-maintenance alternative to large specimen, pot-grown trees.

With minimal intervention and higher success rates, hedging whips could establish into interactive multifunctional green corridors supporting foraging and urban biodiversity. Many of the plants identified through the foraging case study evaluation establish communities in hedges as the trees provide protection and nutrients through the process of facilitation (Schulze et al. 2019).

Urban hedges could also be viewed as riparian buffers and used to filter particle contamination in areas of poor air quality especially if species are selected for their ability to thrive without high levels of pollutants amassing in their fruits or leaves (Romanova and Lovell 2021).

Designing and planning species to suit urban conditions are key; encouraging species tolerant of wet and waterlogged ground into areas prone to urban flooding such as *Salix*, *Cardamine pratensis*, *Matteuccia struthiopteris* and members of the *Mentha* genus may be beneficial in providing regulating ES.

Species tolerant of shade may be beneficial in urban environments where building and infrastructure create reduced hours of sunlight. These could include *Allium triquetrum*, *Allium ursinum*, *Stellaria media* and *Juniper communis*, all of which can tolerate semi-shade conditions.

Pioneer species, which often self-seed into post-industrial, brownfield sites, such as *Betula pendula*, *Taraxacum*, *Sambucus nigra* and *Rubus* fruticosus, could be incorporated into urban planting schemes to quickly transform areas of urban wasteland into foraging and wildlife-rich urban pocket gardens. Vacant, derelict and underutilised commercial and industrial lands have been identified as areas that could support additional forage-rich woodlands or hedgerows (Perry and Handley 2000).

10.5.2 Existing and New UGSs

The design of new UGSs must consider incorporating a range of terrains incorporating blue infrastructure elements as well as green infrastructure elements such as woodland, scrub, hedging, formal planting, meadows and grasslands. Undulating landscapes present opportunities for varied planting as exposure and protection from direct sun and wind varies. Careful planning and place-specific spatial arrangements of new UGSs enable diversity and can create multifunctional spaces with degrees of 'wildness' to meet the UGS requirements of urban residents whilst maximising ES.

10.5.3 Social

Foraging-rich landscapes have the potential to offend or delight depending on the views and opinions held by urban residents. A balance in the degrees of maintenance could help to gain acceptance for more NBSs in our urban environments. A top-down approach is required for programmes such as no-spray maintenance policies regarding pesticides and fungicides. Our recommendation, based on this study, is that public consultations and access to information could help understand and support 'wild' forage-rich urban landscapes. Education, volunteering and vocational training for landscape maintenance professionals may enable a 'gentler' approach to UGS stewardship. This requires input and commitment from local councils as well as cross-disciplinary experts from local charities and community organisations, enthusiasts, academics and commercial enterprises.

Physic gardens are currently a valuable resource, providing an opportunity for the general public, and would help foragers to acquire a deeper understanding of the culinary and medicinal uses of the plants which can be found in their local urban environments. However, these are not found in all towns and cities. Introducing smaller accessible 'educational' gardens into urban centres could promote the use of foraging and food medicines. The use of QR code labelling and information boards, as found across the Chelsea and Dilson Physic Gardens, could be utilised in urban green spaces to guide foragers and provide guidelines to mitigate negative ecological impacts.

Urban nature reserves such as the Tower Hamlets Cemetery appear particularly suited to urban cemeteries. Designating cemeteries as urban nature reserves across all UK towns and cities could increase a sense of place and belonging and encourage greater public use and facilitate learning. The success of the Tower Hamlets Cemetery in educating 700 school children each year in botany and the uses of plants coupled with the recruitment of volunteers to maintain such spaces could enable extensive knowledge transfer and a valuable increase in UGS stewardship.

The design of compact cities is required to ensure the effective use of available space; therefore, edibles should be integrated into the fabric of our cities (Russo et al. 2017). Retrofitting developed urban areas with edible green infrastructure at suitable locations poses a challenge for landscape architects and urban planners (Aparicio Uribe et al. 2022).

References

Akram M (2013) Minireview on Achillea millefolium Linn. J Membr Biol 246(9):661–663. https://doi.org/10.1007/s00232-013-9588-x

Alebachew Y et al (2021) In vivo anti-malarial activity of the hydroalcoholic extract of rhizomes of Kniphofia foliosa and its constituents. Malar J 20(1):3. https://doi.org/10.1186/s12936-020-03552-7

Amato-Lourenco LF et al (2020) Edible weeds: are urban environments fit for foraging? Sci Total Environ 698:133967. https://doi.org/10.1016/j.scitotenv.2019.133967

Aparicio Uribe CH, Bonilla Brenes R, Hack J (2022) Potential of retrofitted urban green infrastructure to reduce runoff–a model implementation with site-specific constraints at neighborhood scale. Urban For Urban Green 69:127499. https://doi.org/10.1016/j.ufug.2022.127499

Artmann M, Sartison K, Vávra J (2020) The role of edible cities supporting sustainability transformation–a conceptual multi-dimensional framework tested on a case study in Germany. J Clean Prod 255:120220. https://doi.org/10.1016/j.jclepro.2020.120220

Ashraf SN et al (2014) Compositional studies and biological activities of Perovskia abrotanoides Kar. Oils. Biol Res 47(1):12. https://doi.org/10.1186/0717-6287-47-12

Atkinson MD, Atkinson E (2002) Sambucus nigra L. J Ecol 90(5):895–923. Available at: http://www.jstor.org/stable/3072258

Barzun J et al. (2023) History of Europe, Encyclopedia Britannica. Available at: https://www.britannica.com/topic/history-of-Europe

Benedek B, Kopp B, Melzig MF (2007) Achillea millefolium L. s.l.–is the anti-inflammatory activity mediated by protease inhibition? J Ethnopharmacol 113(2):312–317. https://doi.org/10.1016/j.jep.2007.06.014

Brown H, Mills S, Albani V (2022) Socioeconomic risks of food insecurity during the Covid-19 pandemic in the UK: findings from the understanding society Covid survey. BMC Public Health 22(1):590. https://doi.org/10.1186/s12889-022-12964-w

Bruton-Seal J, Seal M (2008) Hedgerow medicine: harvest and make your own herbal remedies. Merlin Unwin Books, Ludlow, Shropshire

Carvalho AM, Barata AM (2016) The consumption of Wild edible plants. In: Wild plants, mushrooms and nuts. John Wiley & Sons, Ltd, Chichester, UK, pp 159–198. https://doi.org/10.1002/9781118944653.ch6

Cervellin G et al (2018) Epidemiology and clinics of mushroom poisoning in northern Italy: a 21-year retrospective analysis. Hum Exp Toxicol 37(7):697–703. https://doi.org/10.1177/0960327117730882

Chambers P (2010) Foraging: discover free food from fields, streets, gardens and the coast. Pen & Sword, Barnsley, South Yorkshire, UK

Chelsea Physic Garden (2022) History of the garden. Available at: https://www.chelseaphysicgarden.co.uk/about/history/. Accessed 30 Sep 2022

Clouse C (2022) The resurgence of urban foraging under COVID-19. Landsc Res 47(3):285–299. https://doi.org/10.1080/01426397.2022.2047911

Cole, E. (2021). The growing popularity of foraging in London. https://www.swlondoner.co.uk/news/20062021-foraging-growing-popularity-london (Accessed: 30 September 2022).

Crossley AJ, Russo A (2022) Has the pandemic altered public perception of how local green spaces affect quality of life in the United Kingdom? Sustainability 14(13):7946. https://doi.org/10.3390/su14137946

Dilston Physic Garden (2022) About us. Available at: https://dilstonphysicgarden.com/about-us/. Accessed 30 Sep 2022

Dowlati M, Sobhi HR, Esrafili A, FarzadKia M, Yeganeh M (2021) Heavy metals content in edible mushrooms: A systematic review, meta-analysis and health risk assessment. Trends in Food Science & Technology 109:527–535. https://doi.org/10.1016/j.tifs.2021.01.064

El-Seedi H et al (2002) Antimicrobial diterpenoids from Eupatorium glutinosum (Asteraceae). J Ethnopharmacol 81(2):293–296. https://doi.org/10.1016/S0378-8741(02)00101-0

Fischer LK, Kowarik I (2020) Connecting people to biodiversity in cities of tomorrow: is urban foraging a powerful tool? Ecol Indic 112:106087. https://doi.org/10.1016/j.ecolind.2020.106087

Fissell ME (2008) Introduction: women, health, and healing in early modern Europe. Bull His Med 82(1):1–17. https://doi.org/10.1353/bhm.2008.0024

Frazee LJ et al (2016) Eat your weeds: edible and Wild plants in urban environmental education and outreach. Plant Sci Bull 62(2):72–84. https://doi.org/10.3732/psb.1500003

Global Forest Watch (2023) Forest monitoring designed for action. Available at: https://www.globalforestwatch.org/. Accessed 30 June 2023
Gosling L et al (2016) Differences between urban and rural hedges in England revealed by a citizen science project. BMC Ecol 16(S1):15. https://doi.org/10.1186/s12898-016-0064-1
Hallett S (2006) World's first botanical garden has roots in medicine. Can Med Assoc J 175(2):177–177. https://doi.org/10.1503/cmaj.060649
Hill AW (1915) The history and functions of botanic gardens. Ann Mo Bot Gard 2(1):185–240
Hill KR et al (2011) Co-residence patterns in hunter-gatherer societies show unique human social structure. Science 331(6022):1286–1289. https://doi.org/10.1126/science.1199071
von Hoffen LP, Säumel I (2014) Orchards for edible cities: cadmium and lead content in nuts, berries, pome and stone fruits harvested within the inner city neighbourhoods in Berlin, Germany. Ecotoxicol Environ Safety 101:233–239. https://doi.org/10.1016/j.ecoenv.2013.11.023
Hummer KE (2010) Rubus pharmacology: antiquity to the present. HortScience 45(11):1587–1591. https://doi.org/10.21273/HORTSCI.45.11.1587
Imeri R, Kullaj E, Millaku L (2019) Distribution of heavy metals in apple tissues grown in the soils of industrial area. J Ecol Engg 20(3):57–66. https://doi.org/10.12911/22998993/99733
Jehlička P et al (2020) Thinking food like an east European: a critical reflection on the framing of food systems. J Rural Stud 76:286–295. https://doi.org/10.1016/j.jrurstud.2020.04.015
Kotowski M (2016) Differences between European regulations on Wild mushroom commerce and actual trends in Wild mushroom picking. Slovak Ethnol 64(2):169–178
Landor-Yamagata J, Kowarik I, Fischer L (2018) Urban foraging in Berlin: people, plants and practices within the metropolitan green infrastructure. Sustainability. Multidisciplinary Digital Publishing Institute 10(6):1873. https://doi.org/10.3390/su10061873
Lawrence W (2020) How urban foraging became the new way to explore a city, National Geographic. Available at: https://www.nationalgeographic.com/travel/article/how-urban-foraging-became-the-new-way-to-explore-a-city. Accessed 30 June 2023
Leake JR, Adam-Bradford A, Rigby JE (2009) Health benefits of "grow your own" food in urban areas: implications for contaminated land risk assessment and risk management? Environ Health 8(Suppl 1):S6. https://doi.org/10.1186/1476-069X-8-S1-S6
Leitzmann C (2016) Characteristics and health benefits of phytochemicals. Complement Med Res 23(2):69–74. https://doi.org/10.1159/000444063
Linnekin BJ (2018) Food law gone Wild: the law of foraging. Fordham Urban Law J 45(4):995. Available at: https://ir.lawnet.fordham.edu/ulj/vol45/iss4/3
Łuczaj Ł, Wilde M, Townsend L (2021) The ethnobiology of contemporary British foragers: foods they teach, their sources of inspiration and impact. Sustainability 13(6):3478. https://doi.org/10.3390/su13063478
McElroy W (2012) The enclosure acts and the industrial revolution. Available at: https://www.fff.org/explore-freedom/article/enclosure-acts-industrial-revolution/. Accessed 30 June 2023
McHarg IL (1969) Design with nature. American Museum of Natural History
Medvedeva Y et al (2021) Human health risk assessment on the consumption of apples growing in urbanized areas: case of Kharkiv, Ukraine. Int J Environ Res Public Health 18(4):1504. https://doi.org/10.3390/ijerph18041504
Mollee E, Pouliot M, McDonald MA (2017) Into the urban wild: collection of wild urban plants for food and medicine in KampalaUganda. Land Use Policy 63:67–77. https://doi.org/10.1016/j.landusepol.2017.01.020
News From The Past (2017) Harsh punishment for scrumping. Available at: https://newsfromthepastblog.wordpress.com/2017/04/25/harsh-punishment-for-scrumping/. Accessed 30 Sep 2022
Nyman M (2019) Food, meaning-making and ontological uncertainty: exploring "urban foraging" and productive landscapes in London. Geoforum 99:170–180. https://doi.org/10.1016/j.geoforum.2018.10.009
Ogle BM et al (2003) Food, feed or medicine: the multiple functions of edible wild plants in Vietnam. Econ Bot 57(1):103–117. https://doi.org/10.1663/0013-0001(2003)057[0103:FFOMTM]2.0.CO;2

Palma-Tenango M, Sánchez-Fernández RE, Soto-Hernández M (2021) A systematic approach to Agastache mexicana research: biology, agronomy, Phytochemistry, and bioactivity. Molecules 26(12):3751. https://doi.org/10.3390/molecules26123751

Perry D, Handley J (2000) The potential for woodland on urban and industrial wasteland in England and Wales, forestry commission technical paper 29. Edinburgh. Available at: https://www.forestry.gov.uk/pdf/FCTP029.pdf/$FILE/FCTP029.pdf

Porter RS, Bode RF (2017) A review of the antiviral properties of black elder (Sambucus nigra L.) products. Phytother Res 31(4):533–554. https://doi.org/10.1002/ptr.5782

Procházka P et al (2023) Climatic factors affecting Wild mushroom foraging in Central Europe. Forests 14(2):382. https://doi.org/10.3390/f14020382

Regione Campania (2022) *Informazioni utili per la raccolta di funghi epigei spontanei commestibili in Campania*. Available at: http://www.agricoltura.regione.campania.it/tartufo/funghi_info.html#:~:text=L%27autorizzazione è personale e,– PagoPA–codice tariffa 1159. Accessed 29 Sep 2022

Roe M, Sarlöv Herlin I, Speak S (2016) Identity, food and landscape character in the urban context. Landsc Res 41(7):757–772. https://doi.org/10.1080/01426397.2016.1212324

Romanova O, Lovell S (2021) Food safety considerations of urban agroforestry systems grown in contaminated environments. Urban Agri Region Food Sys 6(1). https://doi.org/10.1002/uar2.20008

Ross D (2022) Cerney House Gardens. Available at: https://www.britainexpress.com/attractions.htm?attraction=2469. Accessed 11 Sep 2022

Rubin S (1970) The medical practitioner in Anglo-Saxon England. J R Coll Gen Pract 20(97):63–71

Russo A et al (2017) Edible green infrastructure: an approach and review of provisioning ecosystem services and disservices in urban environments. Agri Ecosys Environ 242:53–66. https://doi.org/10.1016/j.agee.2017.03.026

Sanderson H, Prendergast HDV (2002) Commercial uses of wild and traditionally managed plants in England and Scotland. Kew Richmond Surrey

Sardeshpande M et al (2021) How people Foraging in Urban greenspace can mobilize social–ecological resilience during Covid-19 and beyond. Front Sustain Cities 3. https://doi.org/10.3389/frsc.2021.686254

Sardeshpande M, Shackleton C (2020) Urban foraging: land management policy, perspectives, and potential. PLOS ONE 15(4):e0230693. https://doi.org/10.1371/journal.pone.0230693

Säumel I et al (2012) How healthy is urban horticulture in high traffic areas? Trace metal concentrations in vegetable crops from plantings within inner city neighbourhoods in Berlin, Germany. Environ Pollut 165:124–132. https://doi.org/10.1016/j.envpol.2012.02.019

Scarrow R (2018) Lead contamination in leafy greens. Nature Plants 4(7):402. https://doi.org/10.1038/s41477-018-0206-x

Scharf N et al (2019) Urban commons for the Edible City-first insights for future sustainable urban food systems from Berlin, Germany. Sustainability (Switzerland) 11(4):1–17. https://doi.org/10.3390/su11040966

Schenk-Jaeger KM et al (2012) Mushroom poisoning: a study on circumstances of exposure and patterns of toxicity. Eur J Intern Med 23(4):e85–e91. https://doi.org/10.1016/j.ejim.2012.03.014

Scherz H, Kirchhoff E (2006) Trace elements in foods: zinc contents of raw foods—a comparison of data originating from different geographical regions of the world. J Food Compos Anal 19(5):420–433. https://doi.org/10.1016/j.jfca.2005.10.004

Schulze E-D et al (2019) Plant ecology. Springer Berlin Heidelberg, Berlin, Heidelberg. https://doi.org/10.1007/978-3-662-56233-8

Schunko C, Wild A-S, Brandner A (2021) Exploring and limiting the ecological impacts of urban wild food foraging in Vienna, Austria. Urban Forestry Urban Green 62:127164. https://doi.org/10.1016/j.ufug.2021.127164

Shackleton CM et al (2017) Urban foraging: a ubiquitous human practice overlooked by urban planners, policy, and research. Sustainability (Switzerland) 9(10):1–18. https://doi.org/10.3390/su9101884

Shaheen S, Ahmad M, Haroon N (2017) Edible Wild plants: an alternative approach to food security. Springer International Publishing, Cham. https://doi.org/10.1007/978-3-319-63037-3

Sharma OP et al (1998) A review of the toxicosis and biological properties of the genus eupatorium. Nat Toxins 6(1):1–14. https://doi.org/10.1002/(SICI)1522-7189(199802)6:1<1::AID-NT3>3.0.CO;2-E

Šiftová J (2020) Foraging in Czechia: the nation's precious hobby. Norsk Geografisk Tidsskrift–Norwegian Journal of Geography 74(5):310–320. https://doi.org/10.1080/00291951.2020.1851757

Somesh D et al (2021) Patterns of urban foraging in Bengaluru city. Urban For Urban Green 57:126940. https://doi.org/10.1016/j.ufug.2020.126940

Staley JT et al (2015) Re-structuring hedges: rejuvenation management can improve the long term quality of hedgerow habitats for wildlife in the UK. Biol Conserv 186:187–196. https://doi.org/10.1016/j.biocon.2015.03.002

Statista (2023) UK foodbank users 2023. Available at: https://www.statista.com/statistics/382695/uk-foodbank-users/. Accessed 30 June 2023

Thomas V (2011) Do modern-day medical herbalists have anything to learn from Anglo-Saxon medical writings? J Herbal Med 1(2):42–52. https://doi.org/10.1016/j.hermed.2011.08.003

Townsend L (2020) The professional wild food community and Covid-19: the use of online platforms in supporting people to access alternative food sources. Local Develop Soc 1(2):160–165. https://doi.org/10.1080/26883597.2020.1797441

Turtiainen M, Nuutinen T (2012) Evaluation of information on Wild berry and mushroom Markets in European Countries. Small-scale Forestry 11(1):131–145. https://doi.org/10.1007/s11842-011-9173-z

De Vries E (2008) Hedge-rider: witches and the underworld. Pendraig Publishing

Wilson EO (1984) Biophilia. Harvard University Press

Wright J (2016) A natural history of the hedgerow: and ditches, dykes and dry stone walls. Profile Books, London

Wube AA et al (2005) Antimalarial compounds from Kniphofia foliosa roots. Phytother Res 19(6):472–476. https://doi.org/10.1002/ptr.1635

Alessio Russo (PhD) is a Senior Lecturer in Landscape Architecture at the Queensland University of Technology (QUT) in Brisbane, Australia. Before joining QUT, he served as a Senior Lecturer in Landscape Architecture and Academic Course Leader for the master's in landscape architecture at the University of Gloucestershire in England (2019-2024). Prior to that, he was an Associate Professor (2018) at RUDN University in Moscow, Russia. He has worked as a Professor and Head of Laboratory of Urban and Landscape Design (2016–2018) at Far Eastern Federal University in Vladivostok, Russia. Outside of academia, Dr Russo has worked as a Landscape Architect in the UK, Italy, and the UAE, dealing with sustainable design and planning.

Laura McCarthy (MA, MCIHort) is a knowledgeable horticulturalist with years of experience in developing and maintaining private gardens and estates across Southern England. She is a member of the Chartered Institute of Horticulture, the Organic Growers Alliance, and an associate member of the Landscape Institute. Her specialization is social and ecological urban landscapes, and her work aims to maximize ecosystem services derived from the use of NBS in urban centers.

11 Encountering the Hidden Bounty of the Urban Forest: Community Foraging Practices and Policies in Canada

Sivajanani Sivarajah, Md Abdul Halim, Sadia Butt, and Imrul Kayes

Abstract

Urban foraging research has been growing in the past two decades. Likewise, trends in social practice are moving towards connecting with nature through activities that include urban foraging. While the global urban population increases, it brings into focus on food security issues and practices such as foraging are seen as one possible mitigation strategy to provide local, accessible food. In this chapter, we highlight the growing phenomenon of urban foraging globally and provide an overview of Canada's current status and practices. Firstly, using grey literature, we unpack and synthesize existing urban foraging practices, and the status of legislative frameworks and by-laws that restrict the use of urban foraging across 16 Canadian municipalities (larger than 100,000 population). Then, we use a bibliometric analysis to review and synthesize the scholarly "urban foraging" literature to identify themes and trends in publication. The latter part of the chapter dives into a case study highlighting the historical importance and current status of urban fruit trees in the City of Toronto. Finally, we conclude the chapter by comparing urban foraging practices between the global north and south. We found Canadian urban foraging research is still in its early stages, while practices grow, conservation policies restrict and limit the activity on public lands.

S. Sivarajah (✉)
Département des sciences du bois et de la forêt, Faculté de foresterie, de géographie, et de géomatique, Université Laval, Québec, QC, Canada
e-mail: sivajanani.sivarajah@sbf.ulaval.ca

M. A. Halim · S. Butt · I. Kayes
Daniels Institute of Forestry and Conservation, University of Toronto, Toronto, ON, Canada

S. Dhyani, M. Sardeshpande (eds.), *Urban Foraging in the Changing World*,
https://doi.org/10.1007/978-981-97-0345-6_11

Keywords

Urban foraging · Community · Ecosystem services · Edible plants · Canada · Gathering · Non-timber products · Wild edibles · Urban fruit trees · Legislation · By-laws · Municipalities · Cities

11.1 Overview of Urban and Peri-Urban Foraging

Since our livelihoods have always been linked to nature, human survival has depended on foraging non-timber and timber products, such as plants, fungi, mosses, and tree components. In the ancient world, foraged products have always been part of the human diet. Past hunter-gatherer societies made more use of wild plants than our modern societies, which eventually led to the domestication of food crops (e.g. agriculture) (Diamond 2002; King 1994; Leonti et al. 2006; Zohary and Hopf 2000). These foraged products often provided the necessary ingredients to build fires and were used as medicine and food. However, as more than 56% of the world's population now resides in cities, this figure is about 80% in Canada (World Bank 2022; Statistics Canada 2021), humans are becoming isolated from the benefits of nature, as explained by the *biophilia hypothesis*. The *biophilia hypothesis* explains that humans have an innate need to stay connected to nature for their overall physical and mental well-being (Kellert and Wilson 1993). Based on this need for nature, the act of gathering non-timber and timber products has sociocultural, psychological, and physical fitness benefits (Shortly and Kepe 2021).

The advent of agriculture and technology has socially distanced humans from rudimentary tasks of gathering their own food and resources. In the global north (United States, Canada, and Europe), over the past few centuries, as the population moves away from rural areas to cities, away from natural areas, foraging in cities has been used to connect back with nature. This practice has recently gained popularity as more people are becoming interested in sustainable living, healthy eating, and reducing their overall environmental impact and carbon footprint (Martin 2018). Furthermore, this practice has become more of a hobby than a necessity in the global north. Urban and peri-urban foraging is the practice of harvesting edible timber and non-timber products from urban green spaces, including both public and private areas, for personal use or for sale. In this context, we define "*urban foraging*" as the practice of identifying and gathering (for personal use or sale) native and introduced plants and timber and non-timber products that grow naturally and are planted (not deliberately cultivated for human use) in urban and peri-urban areas. Foraging and gathering of wild plants can occur in parks, along trails, roads, waterways, front lawns, and backyards (Shackleton et al. 2017; Nyman 2019; Shortly and Kepe 2021; McLain et al. 2014). In this chapter, we argue that urban foraging differs from harvesting in community-based or private urban gardens, where crops and edible products are planted intentionally for harvesting and used as food, medicinal, or other sources (e.g. crafts). Urban foraging practices have been widely growing, persisted in the global south, and are even accepted in many countries. However, urban foraging has its fair share of political and social challenges and limitations in the global north (Shortly and Kepe 2021; Poe et al. 2013; Sardeshpande and Shackleton 2020). While challenges to foraging and food collection can be found to be universal, we highlight those that apply to the north.

One of the critical benefits of urban foraging is the ability to access fresh, healthy foods collected locally and sustainably. This practice further can reduce our carbon footprint, as it reduces the food system's environmental impact, as locally collected food requires less energy to transport and has a smaller carbon footprint. Additionally, urban foraging can provide people with access to a diverse range of edible products and products for other uses (e.g. arts, crafts, and campfires) that may not be readily available in their local grocery stores. Traditionally, wild edible fruits are among the mostly used non-timber forest products and important sources of nutrition, medicine, and income around the world (Sardeshpande and Shackleton 2019). Despite the many potential benefits of urban foraging, the practice has some challenges and drawbacks. One of the main challenges is that urban areas are often contaminated with pollutants such as heavy metals and pesticides, which can be absorbed by plants and considered very harmful to human health and well-being. For example, heavy metal concentrations (e.g., chromium, lead) have been found on plants and fungi located along high traffic urbanized zones. Some wild edible plants have been shown to have captured heavy lead (Pb) concentrations that exceed safe human consumption levels (Amato-Lourenco et al. 2020; Kokkoris et al. 2019). Another challenge is that urban foraging can be difficult for beginners to learn, as it requires an elevated knowledge about the safety of local plants and non-timber products. Some can be quite poisonous and toxic. This can be particularly challenging for people who may not readily access resources, field guides, and/or knowledgeable community members to share knowledge. The other major hindrances to urban foraging, particularly in Canada are the strict laws and regulations (refer to Table 11.1) governing the use of public and private green spaces, making it difficult for people to engage in urban foraging practices. This will be later discussed in more depth in the upcoming sections.

In Canada, urban foraging has a long history. Indigenous communities have long harvested wild plants for food, medicine, and other uses, and these land-based practices continue to exist today (Kuhnlein and Turner 2020). Settlers learned to also harvest from the wild along with establishing agricultural practices. In recent years, in Canada, urban foraging has become more mainstream, as people of all ethnicities and ages started participating in this practice. Given that Canada is a multicultural nation, with more than 250 different ethnicities, with 23% born outside of Canada with recent immigrants (2016–2021) making up 15% of all immigrants (Statistics Canada 2021). Foraging involves gathering wild mushrooms and plants, allows immigrants to connect with nature and appease their nostalgia for their native homelands. For example, some non-timber and timber products (e.g. fruits, nuts, flowers, berries, leaves, and medicinal plants) can be similar to the vegetation found in their homelands and help new and established immigrants to develop a sense of place and strengthen their attachments to their new surroundings. This act of gathering also provides ample opportunities for humans to make new long-term connections with their surroundings and form

Table 11.1 Municipal park legislation and by-laws for Canadian cities

City	Foraging allowed	Exceptions	By-laws
Toronto	No	No	*"BY-LAW No. 854-2004, § 608-6. Injury and damage No person shall in a park: (1) Climb a building, structure or equipment, unless it is equipment designed for climbing; (2) Break, injure, deface, destroy, move or remove the whole or any part of a flower, plant material, fungus, tree or other vegetation or a building, structure, equipment or other property of the City*" (City of Toronto 2004)
Montreal	No	No	*"BY-LAW RCG 10-016-2, 12.2. To protect the environment, in a park, no person may: (1) kill, injure or trap an animal; (2) abandon or release any animal; (3) feed a wild animal; (4) destroy the habitat, nest, eggs or nesting box of an animal; (5) damage, alter or modify any site, watercourse, spring or other natural formation; (6) damage, cut or remove rocks, trees, shrubs or any other dead or living plants, as well as any fossils, or write, paint or engrave inscriptions thereon; (7) enter areas designated for reforestation, ecological restoration or renaturalization"* (City of Montreal 2018)
Calgary	No	No	*"BYLAW NUMBER 11M2019, 2) A person must not remove any park vegetation, including weeds, without a permit*" (City of Calgary 2019)
Ottawa	No	No	*"By-law No. 2004-276, 9. Asset protection (1) No person shall, in a park, cut, climb, break, injure, deface, disturb or remove any property including: (a) a tree, shrub, bush, flower, plant, grass, wood, soil, sand, rock or gravel"* (City of Ottawa 2016)
Edmonton	No	No	*"BYLAW 2202, Preservation of natural areas, 8 While on Parkland no person shall: (1) enter into any undeveloped or natural area other than on an Improved Trail or an Unimproved Trail; (2) remove any rock, gravel, sand or soil; (3) move, remove, cut or damage any tree, shrub, flower, other plant or deadfall"* (City of Edmonton 2021)
Mississauga	No	No	*"By-law 0197-2020, C. Damage to Property 15. While in a Park, no person shall:* (1) *Plant, prune, remove or damage any Tree, Shrub, Plant, flower, flowerbed, bush, sod, grass, or any other vegetation unless otherwise authorized by the Commissioner; or* (2) *Plant, prune, climb, remove, damage, or deface any Tree, Tree roots or parts of a Tree; or (3) remove, damage, disturb or deface any Park Amenity; or (4) climb any Park Amenity, unless the amenity is designated for climbing; or (5) remove any soil, sand, gravel, stone, rocks, wood, or any other material located in the Park"*. (City of Mississauga 2020)

(continued)

Table 11.1 (continued)

City	Foraging allowed	Exceptions	By-laws
Winnipeg	No	No	*"By-law 85/2009, Damage to park 17) Unless authorized by the Chief Administrative Officer, a person must not (1) disturb, remove or damage a tree, branch, flower, plant or other vegetation in a park; (2) disturb, remove or damage soil or turf in a park, including embedding stakes in soil or turf"*. (City of Winnipeg 2017)
Vancouver	No	Yes-indigenous groups in some locations	*"PARKS CONTROL BY-LAW, No person shall cut, break, injure, remove or in any way destroy or damage any rock, soil, tree, shrub, plant, turf or flower"*. (City of Vancouver 2021)
Brampton	No	No	*"By-law 161-83, 3. Conduct of persons using parks or properties: (1) No person shall, within parkland, unless otherwise permitted (42–92) (a) remove or damage any plant, shrub or flower; (b) climb, remove or damage any tree or fence"* (City of Brampton 1983)
Hamilton	No	No	*"By-law No. 01-219, 8. No person shall, in any park, unless expressly authorized by permit: (b) break, injure, deface, move or remove the whole or any part of any flowers, plant material, trees or other vegetation or any building, structure, equipment or other property of the City; (c) move, remove, excavate, disturb, alter, damage, or cause the deterioration or erosion of, the whole or any part of rocks, boulders, rock faces, soil, sand, wood, or like substance, or any other property or thing; (e) conduct archaeological research, or remove any relic, artifact or natural object"* (City of Hamilton 2022)
Surrey	No	No	*"BY-LAW NO. 13480, Damage, 15. No person shall within a park: (a) remove, cut, break, injure or in any way destroy or damage any animal, tree, shrub, plant, turf, sod, or flower; (b) cut or remove any tree, timber or firewood"* (City of Surrey 1998)
Halifax	No	No	*"BY-LAW P-600, Protection of Park, 3. (1) While in any park, no person shall do, cause or permit any of the following: (a) add to, remove, destroy, defile, or damage any fauna or flora, or any park facility, structure, equipment, or sign"* (City of Halifax 2018)
London	No	No	*"BY-LAW NUMBER, PR-2, 3.1. Activities - prohibited In any park or recreation area within the City, no person shall:(19) climb, break, peel bark from, cut, remove, burn, deface or otherwise damage any tree, shrub or bush, or pick, destroy or remove any flower, plant, roots, grass or rock"* (City of London 2021)

(continued)

Table 11.1 (continued)

City	Foraging allowed	Exceptions	By-laws
Markham	No	No	*"BY-LAW 167-92, 4. No person shall: (a) in any park, destroy or cut, mark, break, dig, tear up, burn or in any way damage, injure or deface: (i) any tree, flower bed, shrub, plant or sod"* (City of Markham 2007)
Vaughan	No	No	*"BY-LAW NUMBER 134-95, Injury and Damage, No person shall in any park: (a) climb any building, structure or equipment, unless it is equipment designed for climbing;, (b) break, injure, deface or remove the whole or any part of any flowers, plant material, trees or other vegetation or any building, structure, equipment or other property of the municipality; (c) unless authorized by permit, climb or remove the whole or any part of rocks, boulders, rock faces or remove any soil, sand or wood"* (City of Vaughan 1995)
Burnaby	No	No	*BURNABY PARKS REGULATION BYLAW 1979 "5. No person shall cut, break, injure, damage, deface, destroy, foul or pollute any personal property or any tree, shrub, plant, turf or flower in or on any park"* (City of Burnaby 1979)

attachments to particular plants, fungi, and gathering sites (Poe et al. 2013). The frequency of gathering and engaging in nature can also help Canadian residents to develop local knowledge of landscape and ecology (e.g. growing habitats and growth periods) (Butt et al. 2021), which can provide mutualistic benefits for both humans and land (Shortly and Kepe 2021, Poe et al. 2013). They can also connect with individuals who share an interest in sustainable living, and help build a sense of community, while providing opportunities for education and skill-building (Poe et al. 2013).

The primary purpose of the chapter is twofold where we capture the current status of how Canada fits within the global existing urban foraging research, and examine the available research directions in urban foraging. Using grey literature, we unpack current practices, emerging trends, and the status of legislative framework and by-laws of urban foraging in Canada. Then we synthesize "*urban foraging*" research using bibliometric analyses and identify themes emerging from the scholarly literature. The latter part of the chapter dives into a case study to illustrate a portrait of urban fruit trees in Toronto, Canada, highlighting the importance of urban food trees. Finally, we conclude the chapter by comparing global north and south and how they differ in terms of urban foraging practices.

11.2 Canadian Policies and Governance of Urban Foraging

In Canada, urban governance varies across municipalities, whereby policy concerning urban forests (Morzillo et al. 2022), park planning, and policy for urban agriculture are locally developed albeit influenced by provincial official plans. However, when it comes to urban foraging, there is a consensus across Canada that urban foraging, the collection of natural materials on public lands, for personal, informal livelihood, or commercial purposes, be it live or dead trees, plants, non-timber products, and wildlife are not permitted to be removed. This edict stems from the use of commons from the *Old World* before North American colonization to ensure that the land was protected for the owner and that public lands were not to be violated. The *Old World* property right approaches influence how the public is permitted to interact with nature in parks and public spaces in the *New World* (Greer 2012). Canada, a British commonwealth country maintains these English common property law-based governance approaches along with international-based conservation and protection edicts through international conventions. In this section, we review the various policy tools at the federal, provincial, and local levels that govern how public land is used in Canada in the context of urban foraging for plant and non-timber materials.

In Canada, 89% of Canada's land is Crown land, with 48% provincial and 41% federally managed. These are public lands and are governed by the federal and provincial governments. In contrast, municipalities govern municipal-owned lands and establish by-laws that establish authority over privately owned lands with respect to such issues as building footprints, property standards, and tree removals. Public lands can be defined as land under the jurisdiction of a government body. In Canada, in 1992, the Real Property Act was implemented to replace previous Crown land Acts (Secretariat TB of C 1996). This Act replaces "*land*" with the term "*Real property*" and defines the term as "*land, mines, minerals, buildings, and fixtures on, above or below the surface, and any interest therein, both in Canada and abroad. The definition includes both legal interests in land, such as estates, and physical interests in land, such as mines and minerals*" (Secretariat TB of C 1996). This land includes Crown land across Canada or abroad.

The provincial governments work with the federal government to administer activity on Crown land according to their Ministry. For example in Ontario, Crown land is managed by the Ontario Ministry of Natural Resources and Forestry (OMNRF) under the Crown Land Use (CLU) policy. Provincial Acts also influence what can be done on Crown lands, for example, the Wild Rice Harvesting Act, the Endangered Species Act, or Public Lands Act. The Wild Rice Harvesting Act prohibits the foraging of wild rice on public land without a license, and only residents in designated areas are permitted to apply for a license (Wild Rice Harvesting Act RSO 1990). In addition, there are permissions to collect timber from crown land with limits and for personal use only (Ministry of Natural Resource and Forestry 2023). Though timber is permitted, there are limitations and authorization is not required for "*collecting a maximum of 10 cubic metres or four full stacked cords of downed wood (dead wood that is on the ground), cutting one Christmas tree that is a maximum of 2.5 metres high (north of the French and Mattawa rivers only), transplanting up to five trees that are a maximum of 1.4 metres high, cutting branches from standing trees, if the branches are a maximum of 7.5 centimetres in diameter and where the removal of the branches will not kill the tree, collecting wood for a campfire (downed wood or cut standing dead trees) while lawfully camping or recreating on Crown land, for immediate use*" (Ministry of Natural Resources 2023).

Furthermore, legislation for the Species at Risk Act (SARA) also protects plants and trees by creating "*prohibitions to protect listed threatened and endangered species and their critical habitat…*" (Government of Canada 2022). In the Species at Risk Act under the General Prohibitions section, wildlife species which include plants are protected from harm and collection, whereby under "***Killing, harming, etc., listed wildlife species: 32 (1)*** *No person shall kill, harm, harass, capture or take an individual of a wildlife species that is listed as an extirpated species, an endangered species or a threatened species*" and under "***Possession, collection, etc. (2)*** *No person shall possess, collect, buy, sell or trade an individual of a wildlife species that is listed as an extirpated species, an endangered species or a threatened species, or any part or derivative of such an individual*" (Species at Risk Act, SC 2002; Government of Canada 2022). There are also prohibitions regarding the destruction of the habitat for endangered and extirpated species for which contraventions against SARA can incur hefty fines. For example, fines for an individual if found guilty of committing any of the prohibitions can face a summary conviction of $50,000 or 1 year imprisonment while if guilty of an indictable offense can be fined $250,000 or 5 years of imprisonment (Species at Risk Act, SC 2002). These fines are prohibitive yet, if offenses are repeated, the fines are doubled and if they are continuous, then the fines are incredibly punitive.

11.2.1 Foraging in Provincial Parks

Here we discuss the status of policy and legislature in five Canadian provinces as case studies to understand the various structures and language that governs removal of wildlife from parks in the context of urban foraging. They range in the complexity of the language and the number of departments or ministries that oversee the conservation of parks. In all cases, park conservation officers have legislated authority to implement enforcement of any contraventions against the provincial park.

In Ontario, urban foraging is prohibited in provincial parks. Provincial parks are managed by the Ministry of the Environment, Conservation and Parks (MECP). Visitors are not permitted to remove any natural object from a provincial park and can be fined $125 (Ontario Parks 2023). The province specifically uses the term in their outreach materials and states in one of their blogs in April 2022 that they have noticed an increase in urban foraging behaviour (Ontario Parks 2023). They provide many reasons why urban foraging is not sustainable for the protection and conservation of provincial parks. They cite that the plants are needed for other wildlife, are part of the ecosystem, and are important to biodiversity to create necessary feedback loops (e.g. food chains). Ontario Parks states that unregulated and increased foraging is unsustainable and removal will disturb the habitat characteristics (Ontario Parks 2023). Similarly, in Alberta, provincial Crown lands are governed by the Public Lands Act or Special Areas Act, parks under the provincial parks. Activities, such as tree harvesting and recreational use on these lands, fall under statutory consents for permission to do certain activities on public land.

In the Province of Quebec, the removal of any plant, tree, or mushroom materials is also prohibited in Quebec's Parks Regulation Act, Division V, Article 20 States that, "*No person who travels, stays or engages in an activity in a park for purposes other than scientific or management purposes may: (1) fell, damage, remove or bring in a tree, a shrub, a plant, a mushroom or a part thereof; however, the picking of edible vegetable products is allowed, except for commercial purposes or in a preservation zone or maximum preservation zone*". The permission to pick a vegetable assumes that it is not a native product and may introduce ambiguity among residents who may not have local knowledge between native and non-native products. This type of language is found in various provincial parks acts across the provinces.

In Eastern Canada in Newfoundland provincial parks, visitors have no right to any animate or inanimate park objects and are made aware of the **Prohibitions** under section "**3.** (1) A person shall not, except in accordance with a permit for management or scientific purposes, remove, harm, hunt, chase, destroy or cause damage to any object whether animate or inanimate that is contained within the boundaries of a provincial park" (*Provincial Parks Regulations* under the *Provincial Parks Act* (O.C. 97–510)) (Newfoundland and Labrador 1997). In British Columbia in their Park, Conservancy and Recreation Area Regulation under Prohibited activity, article 32 "*(1)Except as authorized by a park officer, a person must not, (a) damage or destroy any natural resource or property in a park, conservancy or recreation area, (b)possess any natural resource or property of a park, conservancy or recreation area, (c)remove any natural resource or property from a park, conservancy or recreation area, (d)engage in any research or collection activity in a park, conservancy or recreation area, (e)remove water from a park, conservancy or recreation area for purposes other than personal consumption, …[am. B.C. Regs. 215/2006, Sch. s. 8; 141/2009, Sch. s. 1 (e).]*". Fines are imposed for this behaviour ranging from \$50 to \$500 and can include jail time.

11.2.2 Municipal Urban Foraging Policies (Prohibitions for Protection)

We look at the by-laws in 16 municipalities in Canada with over 100,000 people (Table 11.1). All of Canada's major cities prohibit the removal of natural materials from public municipal parks, and none has established urban foraging locations for the public. The language varies from general decrees where no "*natural object live or dead is allowed to be removed or injured*" to language that specifically states the items. Furthermore, the by-law "Article" headers under which the prohibition for foraging is stated are "*Damage*", "*Visitor Behaviour*", and "*Asset Protection*". In some cases, they preface that no removal is permitted by explaining that it will damage the environment. A few by-laws also include the trampling and disturbance caused by walking in park areas protected from disturbance. In Canada's Capital City Ottawa, under the Parks and Facilities By-law No. **2004-276** even grass is excluded**,** "*No person shall, in a park, cut, climb, break, injure, deface, disturb or* *remove any property including: a) a tree, shrub, bush, flower, plant, grass, wood, soil, sand, rock or gravel*" (City of Ottawa 2016). Often, these by-laws targeted at "urban foraging" are sandwiched under other by-laws to protect their natural assets (Table 11.1).

People who forage must resort to finding nearby crown land that are not provincial parks or find private landowners who will permit them to forage on their land. To our knowledge, the City of Vancouver is the only large municipality to date that is creating foraging opportunities in certain parks for indigenous groups, such as the Musqueam, Squamish and Tsleil-Waututh Nations and other indigenous groups where they have engaged in public consultations to determine how and where these activities can be carried out in a sustainable manner (see Table 11.1).

11.2.3 Urban Foraging Organizations in Canada

The number of foraging organizations has increased over the years (Shortly and Kepe 2021; Waygood 2019; McLain et al. 2014) and their premise that nature can provide food, medicines, and products for human use are at odds with municipal land managers who feel a responsibility to protect and conserve urban and peri-urban nature for ecosystem benefits (Shortly and Kepe 2021; Kowalski and Conway 2023; McLain et al. 2014). Though urban foraging groups advocate for sustainable picking, usually advising that one should only pick one-third of any population of forageable materials, that is predicated on an assumption only one visitor visits the patch of a desirable plant. This trend in growing forager groups (Shortly and Kepe 2021) has led to municipalities making it known in social media that urban foraging is not acceptable as the land area of natural spaces is already greatly reduced and their health is at a risk of decline should people disturb the wild plants and/or their habitats. The fact that natural areas and the plants that it consists of are at a premium and at risk of degradation by human activity, municipal land managers are concerned about the dire consequences. "*Many plants that foragers collect grow in these sensitive habitats*", the city [Toronto] says. "*Trampling soft soils and other plants negatively affects ecosystems*" (CTV News 2015). Despite the overwhelming prohibition across Canadian municipalities and provincial parks, foraging groups exist and promote the activity (see Sect. 7.3).

Apart from the obvious benefits of getting food free, local access, and purported health benefits, the groups promote a life that goes back to nature where they can connect with nature as they learn about what is edible, immersing themselves in knowledge building. The words used to describe their activities are hunting, scavenging, frugal foraging, gathering, and even surviving (Kowalski 2014; McLain et al. 2017). These groups always promote safety and encourage members to ensure that they are picking only what they can 100 percent identify. Most will also encourage checking one's local regulations and by-laws on collecting items from parks and public lands.

11.3 Case Study of Urban Fruit Trees in Toronto, Canada

The City of Toronto, one of the largest metropolitan cities in Canada, is home to 6,313,000 people (Macrotrends 2023). During the historical amalgamation, on January 1, 1998, Toronto merged with five different boroughs, Scarborough, North York, Etobicoke, York, East York, and developed into a large unified "megacity" (Schwartz 2009). This historic amalgamation also led to major restructuring of municipal governments and landscapes across the city. Even before the historic amalgamation, when we look back at Toronto's heritage, remaining farmlands with orchards (e.g. apple (*Malus* sp.)) were converted to streets, residential neighbourhoods, and permanent lots. This is one of the reasons why we still see rows and rows of crabapples (*Malus* sp.) in certain parts of Toronto. Many of the existing urban fruit trees in Toronto were imported from the original homelands of first-generation Torontonians of the city. They were planted because these trees had cultural significance and reminded the immigrants of their homeland (Heritage Toronto 2020). The few existing fruit orchards give us a glimpse into the past. Some Toronto neighbourhoods such as St. Clair West (see Fig. 11.1) have a distribution of peach and apricot trees which are attributed to early Torontonians who had roots in Italy and Portugal. Another introduced species is the garlic mustard (*Alliaria petiolata*), although not a tree, but an edible herb is considered an invasive species. This invasive species was likely introduced in the 1800s by European settlers (Heritage Toronto 2020).

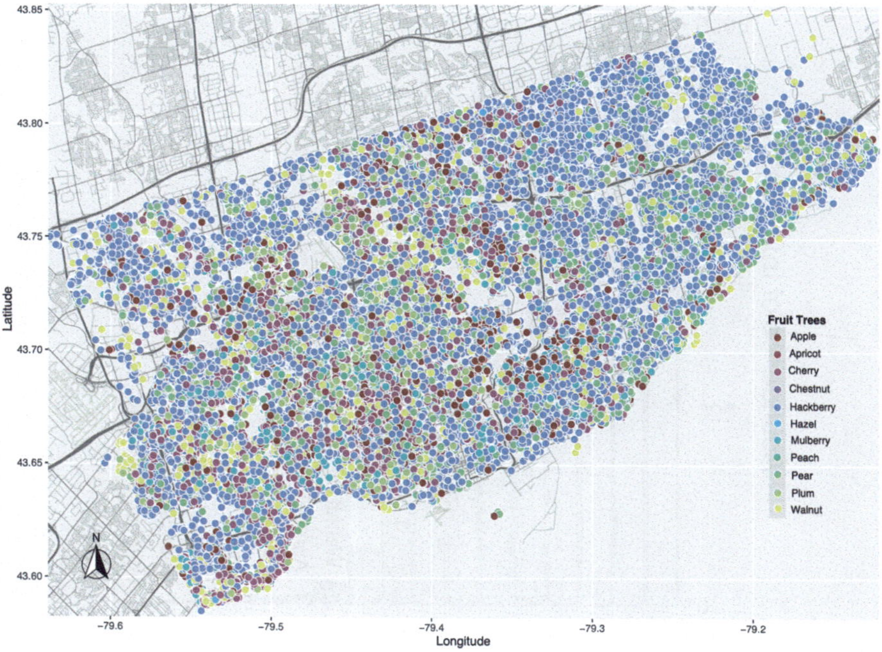

Fig. 11.1 Distribution map of fruit trees located on public lands of the City of Toronto (Data: https://open.toronto.ca/dataset/street-tree-data/)

Likewise, Sakura (*Prunus avium*) was first introduced in 1959 within High Park, Toronto by first-generation Japanese Torontonians (High Park Nature Centre 2023). Historically, this act of importing urban fruit trees has largely shaped Toronto's landscape. In this section, we will highlight the current portrait of urban fruit trees in the City of Toronto.

Urban forests comprising a network of street trees, park trees, and privately owned trees can include fruiting tree species that might not be fully intended for harvesting. Furthermore, municipalities and private landowners often avoid planting fruit bearing trees in parking lots where they can be thought to be messy, while black walnut (*Juglans nigra*) and other nut-based trees are often avoided in schoolyards due to safety and allergy concerns. By introducing fruit trees in urban environments, biodiversity of the urban forest can increase, along with its resilience (Clark and Nicholas 2013). Many municipalities in Canada including the City of Toronto often plant fruit trees to diversify their canopy and to support biodiversity initiatives (e.g. Toronto biodiversity Strategy 2019).

Using Toronto's open data, we have highlighted a portrait of the City of Toronto's fruit trees. In total, there are 569,412 trees on public streets and an estimated 9.7 million more on private land (City of Toronto 2022). Approximately, 3.8% of public trees produce edibles for urban wildlife and human use. The dominant fruiting species include cherries, apples, pears, peaches, plums, and various nuts and berries (Fig. 11.2). According to Fig. 11.2, hackberries (*Celtis occidentalis*) account for

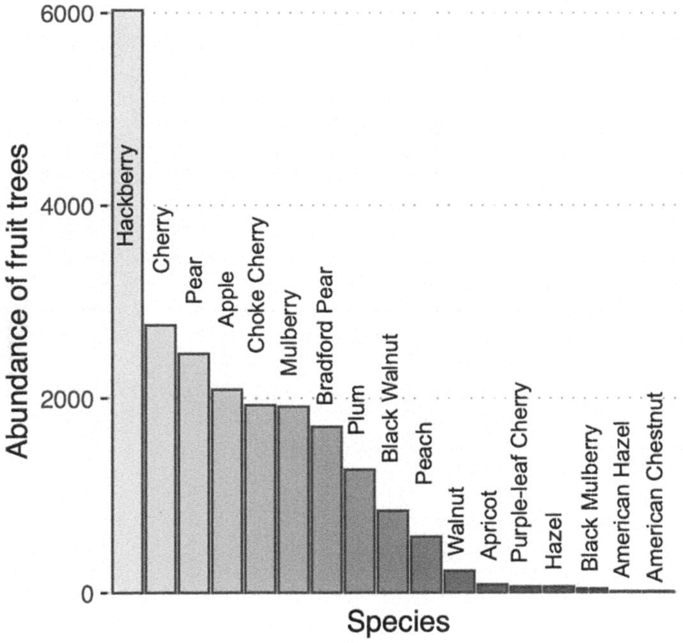

Fig. 11.2 Abundance of fruit trees by species in the public land of the City of Toronto (Data source: https://open.toronto.ca/dataset/street-tree-data/)

27% (6046) of the fruiting trees in the city, accounting for over a quarter of the total fruit tree species (City of Toronto 2022). This spatial distribution of hackberries can be explained by the fact that they are known to be a resilient urban species, and they can grow and thrive in a variety of environmental and soil conditions. Therefore, they were often planted as street trees. Spatial distribution of fruit trees per hectare for all Toronto neighbourhoods, based on City of Toronto's open data, is presented in Fig. 11.1.

In the global north, fruit trees have been successfully introduced in urban forestry community projects to help improve food security, enhance wildlife conservation, and strengthen community networks. Some organizations that are active in community urban fruit tree projects include the *Chicago Rarities Orchard Project* (www.chicagorarities.org), in Chicago; *Cityfruit* (www.cityfruit.org) and the *Beacon Food Forest* (www.beaconfoodforest.org), in Seattle; *Falling Fruit* (www.fallingfruit.org), in Los Angeles, *The Orchard Project* (www.theorchardproject.org.uk/), in Glasgow; *Just food* (www.justfood.ca), in Ottawa; and *Not Far From The Tree* (www.notfarfromthetree.org), in Toronto; *Mundraub* (https://mundraub.org/), in Germany. Organizations such as Toronto's "Not far from the tree" have been successful in making foraging fruit trees acceptable to municipal decision-makers, as they address "food waste" on private lands while contributing to "food security". The program allows private fruit tree owners to register their tree for the "Not far from the tree" program. After initial registration, the organization sends volunteers to collect fruits from the newly registered tree, where one-third of the fruits gathered gets donated to local community food shelters, one-third goes to the volunteers engaged in the program, and the rest goes back to the owner of the tree (Not far from the tree 2023).

Overall, foraging planted trees is considered a safe "*urban foraging practice*", while in wild or natural areas, foraging can be considered detrimental to the urban ecosystem. Considering urban areas has low greenspace, low species availability of rare and edible urban species (e.g. ginseng and wild leeks), popularizing urban foraging of wild species may eventually put species at risk. As a result of strict by-laws and policies, organizations often support and conduct urban foraging programs on privately owned lands and avoid having to deal with the legislative complexity that exists surrounding public lands.

11.4 Current Status of Literature in Urban Foraging

11.4.1 Selection Criteria and Rationale

In this section, we will outline the current status of the scholarly literature centred around "*urban foraging*". When we performed a quick search using the keyword "*foraging*", we often captured a wide range of topics, including rural and animal foraging. However, to keep our search aligned with the topic discussed in this chapter, we were specifically interested in capturing literature on "*urban foraging by humans*". Web of Science was the selected database for this study because it is an extensive interdisciplinary database, inclusive of all scientific domains, and frequently used in bibliometric analyses across various disciplines (Li et al. 2018; Polinko and Coupland 2021).

After some preliminary tests, we used a comprehensive search query on title, abstract, and keywords to identify and retrieve all published studies related to urban foraging. Using the search criteria illustrated in Fig. 11.3, bibliometric data were retrieved on 3 December 2022. We found 85 research items from the Web of Science core database. As a primary step, we manually screened these 85 candidate items by thoroughly reading them, considering the definition of "*urban foraging*" outlined above in this chapter's introduction. Performing this step resulted in 69 peer-reviewed journal articles, with 180 authors contributing from around the globe from 1978 to 2022. Therefore, we did not include grey literature, such as newspapers, popular books, or blog posts, in this search.

11.4.2 Results from Bibliometric Analysis

We performed bibliometric analyses on the selected research items using the "*bibliometrix*" package (Aria and Cuccurullo 2017) of R (R Core Team 2022) that expedites comprehensive bibliometric analyses using various techniques to perform both bibliometric and scientometric quantitative research. Our bibliometric analysis results showed that from 1978 until 2022, the number of research items published on urban foraging had an annual growth rate of 4.84%, with increasing interest and publications in the field of urban foraging after 2010 (Fig. 11.4). This increasing interest in urban foraging is likely to be influenced by a variety of factors, such as changes in the field of study, new research methods and technologies, and shifts in social and cultural attitudes towards urban foraging (de Jong and Varley 2018). Over this period (1978–2022), corresponding authors from 25 countries published at least one research item (Fig. 11.5), with the USA publishing 26%, Canada publishing 2.9%, and North American (including USA and Canada) countries publishing 32% of the total research items considered. Overall, 64% of the research items were published by the corresponding authors from the global north, and the rest was from the authors of the global south. Although there exists some urban foraging literature from the cities around the world, this research domain is a wildly understudied topic (Fischer and Kowarik 2020), particularly from the global south (Russo and Escobedo 2022).

Two published studies from Canada (Kowalski and Conway 2019; Shortly and Kepe 2021) matched our search criteria. One of these contributions focuses solely on Toronto (Shortly and Kepe 2021), while the other examines 47 municipalities across Canada (Kowalski and Conway 2019). Even though urban foraging is illegal (Table 11.1) in parks and other urban forests of Canada, these studies indicate that urban foraging for berries, mushrooms, weeds, etc. is gaining popularity in Canadian cities. In both of these studies, the researchers observed that individuals forage in urban green spaces to connect with nature, retain cultural traditions, and safeguard the environment (Shortly and Kepe 2021; Kowalski and Conway 2019). The authors

WOS Search Criteria	Number of Research Items
(TI = ((foraging OR gathering OR wild OR edibles OR OR non-timber) AND urban))	528
(TI=((foraging OR gathering OR edibles OR non-timber OR wild OR) AND urban) NOT TS = (wildlife OR bat OR gull OR mammal OR insect OR bird OR Merlin OR cottontails OR ant OR termite OR fish OR bee OR bear OR boar OR cat OR *flies* OR dogs OR callout OR rat OR rodent OR vertebrates OR carnivores OR OR monkey OR *c?ows* OR cattle))	244
(TI=((foraging OR gathering OR edibles OR non-timber OR wild OR) AND urban) NOT TS = (wildlife OR bat OR gull OR mammal OR insect OR bird OR Merlin OR cottontails OR ant OR termite OR fish OR bee OR bear OR boar OR cat OR *flies* OR dogs OR callout OR rat OR rodent OR vertebrates OR carnivores OR OR monkey OR *c?ows* OR cattle) NOT TS = (DNA OR pathogen OR coli OR bacteria OR virus OR parasite OR infection OR gut))	224
(TI=((foraging OR gathering OR edibles OR non-timber OR wild OR) AND urban) NOT TS = (wildlife OR bat OR gull OR mammal OR insect OR bird OR Merlin OR cottontails OR ant OR termite OR fish OR bee OR bear OR boar OR cat OR *flies* OR dogs OR callout OR rat OR rodent OR vertebrates OR carnivores OR OR monkey OR *c?ows* OR cattle) NOT TS = (DNA OR pathogen OR coli OR bacteria OR virus OR parasite OR infection OR gut) NOT TS = (seller OR horticulture OR germination OR transit OR sea OR rivers OR *estuary* OR wastewater OR pond OR coast OR lake OR geospatial))	206
(TI=((foraging OR gathering OR edibles OR wild OR non-timber) AND urban) NOT TS = (wildlife OR mammal OR insect OR bird OR bat OR gull OR Merlin OR ant OR termite OR fish OR bee OR bear OR boar OR cottontails OR *flies* OR dogs OR cat OR callout OR vertebrates OR carnivores OR rodent OR rat OR monkey OR *c?ows* OR cattle) NOT TS = (DNA OR pathogen OR coli OR bacteria OR virus OR parasite OR infection OR gut) NOT TS = (seller OR horticulture OR germination OR transit OR sea OR rivers OR *estuary* OR wastewater OR pond OR coast OR lake OR geospatial) NOT TS = (network OR films OR photography OR "agent-based" OR Wi-Fi OR computing OR mobile OR "high-tech"))	178
(TI=((foraging OR gathering OR edibles OR wild OR non-timber) AND urban) NOT TS = (wildlife OR mammal OR insect OR bird OR bat OR gull OR Merlin OR ant OR termite OR fish OR bee OR bear OR boar OR cottontails OR *flies* OR dogs OR cat OR callout OR vertebrates OR carnivores OR rodent OR rat OR monkey OR *c?ows* OR cattle) NOT TS = (DNA OR pathogen OR coli OR bacteria OR virus OR parasite OR infection OR gut) NOT TS = (seller OR horticulture OR germination OR transit OR sea OR rivers OR *estuary* OR wastewater OR pond OR coast OR lake OR geospatial) NOT TS = (network OR films OR photography OR "agent-based" OR Wi-Fi OR computing OR mobile OR "high-tech") NOT DT=("BOOK REVIEW" OR "NOTE" OR "CORRECTION" OR "MEETING ABSTRACT"))	140
(TI=((foraging OR gathering OR edibles OR wild OR non-timber) AND urban) NOT TS = (wildlife OR mammal OR insect OR bird OR bat OR gull OR Merlin OR ant OR termite OR fish OR bee OR bear OR boar OR cottontails OR *flies* OR dogs OR cat OR callout OR vertebrates OR carnivores OR rodent OR rat OR monkey OR *c?ows* OR cattle) NOT TS = (DNA OR pathogen OR coli OR bacteria OR virus OR parasite OR infection OR gut) NOT TS = (seller OR horticulture OR germination OR transit OR sea OR rivers OR *estuary* OR wastewater OR pond OR coast OR lake OR geospatial) NOT TS = (network OR films OR photography OR "agent-based" OR Wi-Fi OR computing OR mobile OR "high-tech") NOT DT=("BOOK REVIEW" OR "NOTE" OR "CORRECTION" OR "MEETING ABSTRACT") NOT WC = ("Veterinary Sciences" OR Theater OR Religion OR "Public Administration" OR "Psychology Clinical" OR "Physics Applied" OR "Meteorology Atmospheric Sciences" OR "Materials Science Ceramics" OR "Marine Freshwater Biology" OR Management OR "History Of Social Sciences" OR "Engineering Geological" OR "Engineering Chemical" OR "Education Scientific Disciplines" OR "Criminology Penology" OR "Construction Building Technology" OR "Energy Fuels" OR "Computer Science Theory Methods" OR "Computer Science Artificial Intelligence" OR "Business Finance" OR "Biotechnology Applied Microbiology" OR "Behavioral Sciences" OR "Asian Studies" OR History OR Agronomy OR "Social Sciences Interdisciplinary" OR "Humanities Multidisciplinary" OR Literature OR "Geography Physical" OR "Public Environmental Occupational Health" OR "Psychology Multidisciplinary" OR Architecture OR "Agriculture Multidisciplinary" OR Ecology))	85

Fig. 11.3 Systematic search criteria for the Web of Science core database

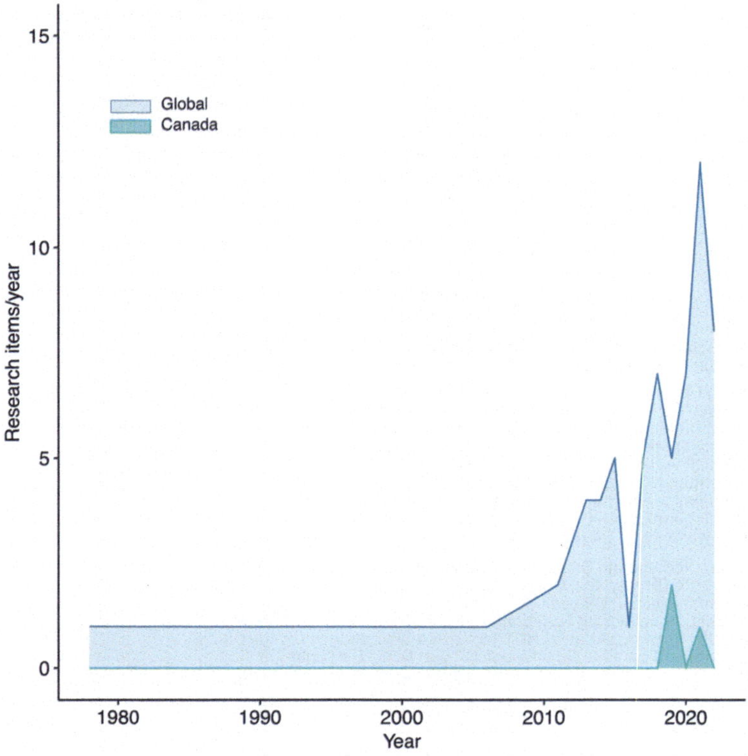

Fig. 11.4 Number of research items on urban foraging published globally (blue) and in Canada (turquoise) each year during 1978 to 2022

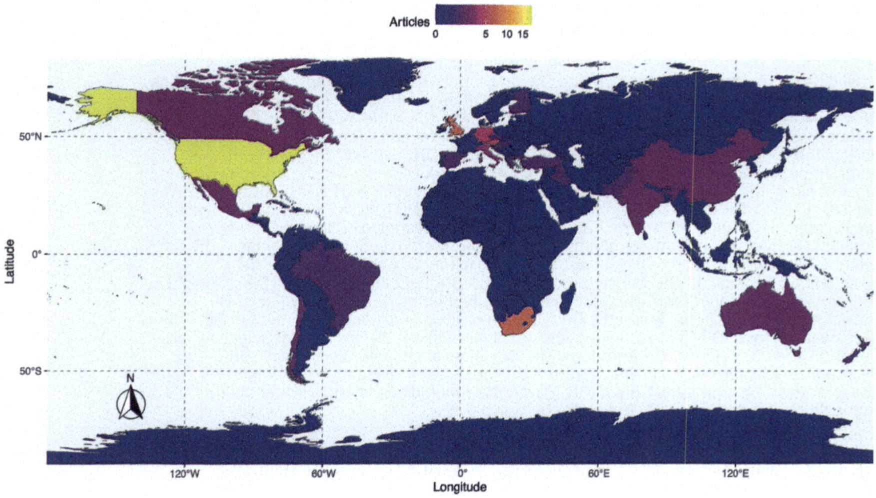

Fig. 11.5 World map showing total number of publications on urban foraging per (corresponding author's) country from 1978 to 2022

argued that wild foraging in urban environments could improve human-nature connections and promote equitable access and control over urban forests and green space resources. These studies also recommended that communal orchards and food forests can be a potentially safe and sustainable way to integrate foraging into public green spaces. Despite the global prevalence of varying attitudes among public authorities towards urban foraging, ranging from full restriction to laissez-faire approaches, and occasionally even tacit encouragement (Schunko and Brandner 2022; Sardeshpande and Shackleton 2020), researchers and policymakers must look through a socio-ecological lens and recognize the knowledge and ethical considerations of foragers (Rombach and Dean 2023) when integrating urban foraging into urban management plans for cities.

11.4.3 Emerging Themes from Bibliometric Analyses on "Urban Foraging"

Our findings reveal three conceptual themes related to urban foraging emerged from our bibliometric analyses of the 69 peer-reviewed journal articles: i) natural conservation and ecosystem services, ii) social justice and accessibility, iii) possible threats to human health (Fig. 11.6). According to our metric multidimensional scaling (MDS) analysis, the largest cluster represents literature surrounding nature conservation and ecosystem services (59% of articles retrieved). The second largest cluster represents social justice and accessibility (29% of articles retrieved), and the third largest encompasses possible threats to human health (12% of articles retrieved) (Fig. 11.6). We will outline the various sub-themes in the following sections.

11.4.3.1 Nature Conservation and Ecosystem Services

Meaningful interactions between nature and humans are crucial to enhancing our connection with nature, which counteracts the "extinction of experience"[1] (Soga and Gaston 2016) and encourages nature conservation. With the rapid growth of the urban population, opportunities for residents to experience nature are becoming increasingly limited, posing severe challenges for sustainable urban development planning. Within our bibliometric database, peer-reviewed articles (e.g. Synk et al. 2017; McLain et al. 2017; Threlfall and Kendal 2018) addressing the "natural conservation and ecosystem services" theme focus on the benefits of urban wild foraging for supporting human well-being (physical and psychological), while fostering a positive attitude towards biodiversity conservation without endangering native biodiversity (Fischer and Kowarik 2020). Urban foragers have been reported to collect endangered plant species (Landor-Yamagata et al. 2018) and engage in unsustainable harvesting practices (McLain et al. 2017), primarily due to the ambivalent

[1] It is the ongoing estrangement of people from the natural world, which Robert M. Pyle described as the "extinction of experience" in his 1993 memoir, The Thunder Tree (Pyle 1993). Soga and Gaston (2016) went into great detail about this and made the point that the "extinction of experience" can lead to a vicious cycle of negativity towards nature.

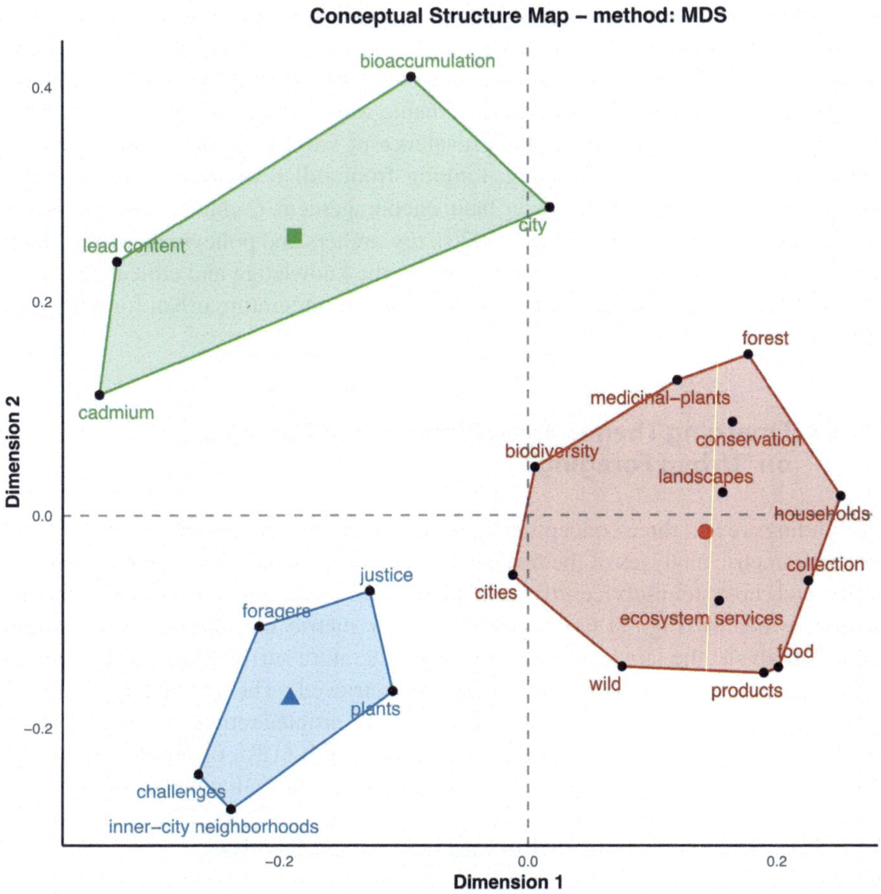

Fig. 11.6 A conceptual structure map of the Keywords Plus® (Web of Knowledge) using the multidimensional scaling method showing the major clusters of the concepts. Out of 69 selected articles, 41 encompass natural conservation and ecosystem services, 20 on social justice and accessibility, and 8 on possible threats to human health related to urban foraging. The clear clustering (no overlapping) of themes indicates articles published precisely focused on each theme related to urban foraging

nature of government imposed regulatory frameworks. However, Schunko and Brandner (2022) suggested that these can be counterbalanced by the stewardship practices performed by urban foragers that contribute to preserving and enhancing urban vegetation. Whereas, another study by Threlfall and Kendal (2018) emphasized that maintaining urban wild spaces can serve as refuges for species in the absence of other managed or planned green spaces.

Moreover, urban wild spaces are distinct in composition and structure compared to other managed urban spaces that provide critical ecosystem services (e.g. recreational parks). McLain et al. (2017) observed that urban foraging reduces undesirable species and their habitats similarly to organized volunteer restoration efforts. In addition to nature conservation-related benefits, urban vegetation can offer other direct and indirect ecosystem services such as carbon sequestration, pollution reduction, habitat for wildlife, and stormwater management (Shackleton et al. 2015; Poe et al. 2013). However, urban non-woody vegetation ecosystem services are understudied (Hurley and Emery 2018). Overall, these studies concluded that although foraging can provide a significant portion of food and cultural needs, can maintain biodiversity, and provides a variety of ecosystem services, it is unclear how foraging may be accommodated into urban green space design.

11.4.3.2 Social Justice and Accessibility

In the context of urban foraging, social justice and accessibility are examined in terms of who has access to resources and benefits, forager participation in resource management decisions, and acknowledging how cultural identities, livelihoods, and social relationships are connected to urban foraging (Poe et al. 2013). In the past two decades, research based in North American cites has found that disparities exist in access to green spaces and urban forests with respect to socio-economic class, race, and education (Nesbitt et al. 2018; Conway et al. 2011; Heynen et al. 2006). As a result, by connecting people with their local environment and resources, urban foraging offers a unique way to rethink social justice. Urban foraging involves more than just collecting edible plants from urban environments; it also entails developing a fair and open system that allows everyone access to green spaces and their resources. Within our filtered bibliometric database, the key articles (e.g. Poe et al. 2014; Sardeshpande and Shackleton 2020; Brandner and Schunko 2022) addressing the social justice and accessibility theme regard urban foraging as a way to make cities more resilient, inclusive, safe, and sustainable by eradicating hunger, ensuring food security, and enhancing nutrition, all while upholding urban forest justice. They contend that urban foraging provides for increased human–nature interactions, which contributes to the expansion of urban forest justice (Poe et al. 2013). Urban foraging also ensures the rights of local people to govern their culturally appropriate diets and health systems, as well as their access to natural resources and decision-making processes impacting them. Institutional dynamics and marginalizing gatherers are critical challenges in implementing social justice and accessibility in urban foraging. Recognizing urban foraging is a critical first step to promoting active community participation in managing and creating urban green spaces (Garekae and Shackleton 2021).

11.4.3.3 Possible Threats to Human Health

From our analyses, the third conceptual cluster in urban foraging research relates to possible threats of the behaviour to human health (Fig. 11.6). Our database illustrates that the articles published on this theme, primarily arise from concerns related to bioaccumulation and deposits of urban pollutants (e.g. heavy metals such as lead and cadmium) on commonly foraged plants and fungi (e.g. mushrooms). Heavy metals such as Zn, Pb, and Cd are typically higher in urban soils and air compared to urban and peri-urban areas, primarily originating from vehicle tyres, motor oil, workshops, and emissions from vehicles and industries (Nadgórska-Socha et al.

2017). These heavy metals can either directly leach into the soil or enter the soil by deposition of particulate matter (adsorption) from the air. The main pathways by which these heavy metals and trace elements get into plants are through their roots and leaves. Plants uptake a bioavailable amount of these elements and store them in their tissues (Norouzi et al. 2015). Sometimes, this mechanism, plants' ability to absorb nutrients through roots and leaves, can be viewed positively, as plants help reduce air and soil pollution in heavily urbanized areas. However, it can be viewed negatively and also be harmful if the plants are consumed, and the concentration of heavy metals or rare earth elements is higher than the permissible limit for human consumption and use. As such, eating such plants can lead to the accumulation of harmful concentrations of heavy metals in human bodies, leading to serious health issues.

Many of the studies within our 69 peer-reviewed articles (e.g. Schlecht and Säumel 2015; Kokkoris et al. 2019; Ivanic et al. 2021) compared the concentration of heavy metals and rare earth elements in edible urban foraged plants and fungi found in urban areas, compared to the ones gathered from rural or peri-urban areas. They observed that wild edibles could have bioavailable heavy metal concentrations higher than the safe upper limit in urban areas, with significant variation between species and locations. Because bioaccumulation is highly spatially dependent and has been reported to decrease by 10% for about 50 m away from the primary traffic (Yin et al. 2011), foraging in properly planned locations can help to minimize or avoid possible contamination. Studies suggested that policy makers integrate urban foraging into biodiversity- and wilderness-friendly greenspace management with access to foraging sites, codes of conduct, and comprehensive information about the bioavailability of contaminants to foragers (Brandner and Schunko 2022). These recommendations are expected to help improve urban foraging, increase urban biodiversity, and ensure the multifunctional advantages of public urban green areas for residents.

11.5 Challenges and Opportunities of Urban Foraging for Canada

Patterns of urban foraging vary significantly between the global north and south, primarily due to social, cultural, and political factors, and offer an excellent opportunity to learn from one another to develop resilient food systems and self-sufficient communities. Urban foraging is typically prevalent and accepted in the developing world, where access to affordable produce may be limited. It is a culturally accepted practice in the global south, where traditional knowledge of urban foraging is passed down through the generations (Kumar et al. 2018). In addition, urban foraging in the global south is subject to less stringent legal restrictions, as barriers to urban foraging are more in line with a lack of knowledge, access, interest, or health concerns (Somesh et al. 2021).

Researchers have found that foragers are quite diverse in their economic status and motivations (Galt et al. 2014; Martin 2018). In Canada, in large metropolitan cities, access to a wide variety of food and other resources is generally readily available. Urban foraging has experienced an upsurge in the global north, where it is often regarded as a novelty, trend, or form of environmental education (Shortly and Kepe 2021). As highlighted in this chapter, municipal land managers and provincial parks are supported by the establishment of clear anti-foraging by-laws in Canada (Table 11.1).

The key challenges faced by urban foragers in the global north are as follows:

1. A lack of understanding of what might be safe to eat and how to prepare them (Gaither et al. 2020)
2. Identifying and harvesting wild plants due to factors such as limited availability, lack of accessibility, and uncertainty about legal status (Poe et al. 2014)
3. The fact that urban foraging is often considered illegal or a violation of private property rights, and many cities have by-laws and regulations prohibiting the collection of wild food on public or private lands (Shackleton et al. 2017)
4. Urban foragers may also face opposition from property owners who perceive them as trespassers (Shackleton et al. 2017; McLain et al. 2012)

Despite the differences, there are several potential ways in which urban foragers in the global north and south can benefit from one another's experiences and perspectives. For example, the global north can learn from the global south about the importance of urban foraging for food security and sustainability. In contrast, the global south can learn from the global north about ways to promote urban foraging as a form of environmental education and recreation (Shortly and Kepe 2021). Urban foragers in the global north can learn from the traditional knowledge and practices of foragers in the global south, who have often developed sophisticated systems for identifying, harvesting, and using wild foods (Kumar et al. 2018). Additionally, research and knowledge exchange between the global north and south can provide insights into best practices for urban foraging, such as identifying, harvesting, and preparing wild foods safely and sustainably.

As food security issues grow globally, including Canada (Galt et al. 2014; Widener et al. 2017) and Canada moves to create more National Urban Parks (Parks Canada 2023), the understanding of urban foraging as a growing activity will need to be assessed and incorporated into strategic park planning. If urban foraging is to be permitted, park planning practices must address what steps are likely needed to ensure that urban foraging by-laws are implemented, and that access is equitable, safe, and sustainable. Whether the motivations for urban foraging are increasing the connection of people with nature or to address food security access issues (Widener et al. 2017), local consultation using a participatory approach to develop urban foraging would benefit the implementation of urban foraging opportunities as exemplified in Vancouver's initiative to create urban foraging opportunities for local indigenous groups (Table 11.1). Shortly and Kepe (2021) state that the consultation process can be difficult with urban foragers as they are a diverse group and not easy to contact. Consulting with diverse groups established and new forager groups can help provide the networks and organizational capacity to engage and shape a sustainable future of urban foraging.

11.6 Conclusion and Future Research Directions

As far as research directions for the emerging field of urban foraging will be necessary to consider gaps in a cross-disciplinary approach. Galt et al. (2014) have already used the political ecology approach to analyse the socio-environmental transformations that characterize urban foraging practices. It is important to consider a framework that will address:

1. The gaps of the spatial extent of where and how frequently urban foraging occur across major urban centres and smaller municipalities alike. This can be done in a case study approach or an extended spatial mapping endeavour.
2. How to connect with urban foraging communities and even more complex how to connect with individuals (Shortly and Kepe 2021) and explore the existing networks.
3. How motivations vary across Canada, perhaps exploring the diversity in population and geographies. There is very little work in this field. Although the literature in the United States has some reference to motivations, we need in a Canada wide study to enable policy makers to better understand the needs of the urban foraging community.
4. Research needs to build a framework to understand the impacts of current urban foraging practices on ecosystems.
5. The characterization of who forages, their status of knowledge and their capacity building behaviours. We cannot romanticize that foragers have the knowledge to do the right thing.
6. The governance and policy changes in municipalities affect the rules towards urban foraging, penalties, and messaging provided to people on how they may use natural areas. A framework would ideally include this important aspect to gauge changing views on urban foraging and that access is equitable.
7. Mapping hotspots for urban foraging for managers to understand the biophysical extent and impacts. This will help identify feasible locations for urban foraging by creating a risk and vulnerability index for those locations where people forage (e.g. sidewalk and hard surface harvesting vs provincial significant areas).
8. The relationships and patterns between regulations and enforcement by documenting and summarizing enforcement incidents related to urban foraging contraventions in major municipalities in Canada.
9. Comparison of provincial and municipal urban foraging practices.

Very little research has been done on urban foraging (less than 70 peer-reviewed journal articles according to our bibliometric analyses). Urban foraging research has much to grow and addressing the above gaps will allow policy makers to use science-based evidence to re-asses current legislation and incorporate the existing needs of the urban foraging community in Canada.

References

Amato-Lourenco LF, Ranieri GR, de Oliveira Souza VC et al (2020) Edible weeds: are urban environments fit for foraging? Sci Total Environ 698:133967. https://doi.org/10.1016/j.scitotenv.2019.133967

Aria M, Cuccurullo C (2017) Bibliometrix: an R-tool for comprehensive science mapping analysis. J Informet 11:959–975. https://doi.org/10.1016/j.joi.2017.08.007

Brandner A, Schunko C (2022) Urban wild food foraging locations: understanding selection criteria to inform green space planning and management. Urban For Urban Green 73:127596. https://doi.org/10.1016/j.ufug.2022.127596

Butt S, Smith S, Moola F, Conway TM (2021) The relationship between knowledge and community engagement in local urban forest governance: a case study examining the role of resident association members in Mississauga, Canada. Urban For Urban Green 60:127054. https://doi.org/10.1016/j.ufug.2021.127054

City of Brampton (1983) Park lands by-law 161–83. https://www.brampton.ca/EN/City-Hall/Bylaws/All%20Bylaws/Parkland.PDF. Accessed 13 Jan 2023

City of Burnaby (1979) The Corporation of the District of Burnaby Bylaw No. 7331. https://bylaws.burnaby.ca/media/Consolidated/7331C.pdf. Accessed 13 Jan 2023

City of Calgary (2019) Bylaws related to parks and pathways. https://www.calgary.ca/bylaws/parks-pathways.html. Accessed 13 Jan 2023

City of Edmonton (2021) City of Edmonton Bylaw 2202. Parkland Bylaw. https://www.edmonton.ca/sites/default/files/public-files/C2202.pdf?cb=1673399089. Accessed 13 Jan 2023

City of Halifax (2018) Halifax Regional Municipality By-Law P-600. https://www.halifax.ca/sites/default/files/documents/city-hall/legislation-by-laws/By-law%20P-600.pdf. Accessed 13 Jan 2023

City of Hamilton (2022) A By-law (No. 01-219) To manage and regulate municipal parks. https://www.hamilton.ca/sites/default/files/2022-06/01-219-OfficeConsolidation06.15.2022.pdf. Accessed 13 Jan 2023

City of London (2021) Parks and recreation area by-law—PR-2. https://london.ca/by-laws/parks-recreation-area-law-pr-2. Accessed 13 Jan 2023

City of Markham (2007) By-law for the general management and regulation of parks within the Town of Markham. https://www.markham.ca/wps/wcm/connect/markham/2defe9c8-eee6-4ce2-aaa3-88b374c5e991/Bylaw-167-92.pdf?MOD=AJPERES&CONVERT_TO=url&CACHEID=ROOTWORKSPACE.Z18_2QD4H901OGV160QC8BLCRJ1001-2defe9c8-eee6-4ce2-aaa3-88b374c5e991-mrL-qy. Accessed 13 Jan 2023

City of Mississauga (2020) Parks by-law 0197-2020. https://www.mississauga.ca/wp-content/uploads/2022/08/09103049/Parks-By-law-0197-2020-1.pdf. Accessed 13 Jan 2023

City of Montreal (2018) Cleanliness regulations—City of Montreal. https://ville.montreal.qc.ca/pls/portal/docs/PAGE/ARROND_OUT_FR/MEDIA/DOCUMENTS/R_AO-78_201801.PDF. Accessed 13 Jan 2023

City of Ottawa (2016) Parks and facilities by-law. By-law No 2004-276. https://documents.ottawa.ca/sites/documents/files/2004_276_en.pdf. Accessed 13 Jan 2023

City of Surrey (1998) Surrey parks, recreation and cultural facilities regulation by-law, 1998. https://www.surrey.ca/sites/default/files/bylaws/BYL_reg_13480.pdf. Accessed 13 Jan 2023

City of Toronto (2004) City of Toronto BY-LAW No. 854-2004. https://www.toronto.ca/legdocs/bylaws/2004/law0854.pdf. Accessed 13 Jan 2023

City of Toronto (2022) Street Tree Data (shapefile). https://www.toronto.ca/city-government/data-research-maps/open-data/. Accessed 4 Dec 2022

City of Vancouver (2021) Parks By-Law City of Vancouver. https://parkboardmeetings.vancouver.ca/files/BYLAW-ParksBylawsConsolidated-20210621.pdf. Accessed 13 Jan 2023

City of Vaughan (1995) Parks By-Law City of Vancouver. By-Law Number 134-95. https://www.vaughan.ca/cityhall/by_laws/Bylaws/134-95.pdf. Accessed 13 Jan 2023

City of Winnipeg (2017) The City of Winnipeg Park By-Law NO 85/2009. https://clkapps.winnipeg.ca/dmis/docext/viewdoc.asp?documenttypeid=1&docid=4751. Accessed 13 Jan 2023
Clark KH, Nicholas KA (2013) Introducing urban food forestry: a multifunctional approach to increase food security and provide ecosystem services. Landsc Ecol 28:1649–1669. https://doi.org/10.1007/s10980-013-9903-z
Conway TM, Shakeel T, Atallah J (2011) Community groups and urban forestry activity: drivers of uneven canopy cover? Landsc Urban Plan 101(4):321–329
CTV News (2015) Foraging prohibited by Toronto bylaw, city warns. Toronto. https://toronto.ctvnews.ca/foraging-prohibited-by-toronto-bylaw-city-warns-1.2497105?cache=trfslrtptoe%3FclipId%3D104056. Accessed 13 Jan 2023
de Jong A, Varley P (2018) Foraging tourism: critical moments in sustainable consumption. J Sustain Tour 26:685–701. https://doi.org/10.1080/09669582.2017.1384831
Diamond J (2002) Evolution, consequences and future of plant and animal domestication. Nature 418:700–707. https://doi.org/10.1038/nature01019
Fischer LK, Kowarik I (2020) Connecting people to biodiversity in cities of tomorrow: is urban foraging a powerful tool? Ecol Indic:112
Gaither CJ, Aragón A, Madden M et al (2020) "Black folks do forage": examining wild food gathering in Southeast Atlanta Communities. Urban For Urban Green 56:126860. https://doi.org/10.1016/j.ufug.2020.126860
Galt RE, Gray LC, Hurley P (2014) Subversive and interstitial food spaces: transforming selves, societies, and society–environment relations through urban agriculture and foraging. Local Environ 19(2):133–146. https://doi.org/10.1080/13549839.2013.832554
Garekae H, Shackleton CM (2021) Knowledge of formal and informal regulations affecting wild plant foraging practices in urban spaces in South Africa. Soc Nat Resour 34:1546–1565
Government of Canada (2022) Consolidated federal laws of Canada, Species at Risk Act. https://laws-lois.justice.gc.ca/eng/acts/s-15.3/FullText.html. Accessed 13 Jan 2023
Greer A (2012) Commons and enclosure in the colonization of North America. Am Hist Rev 117:365–386. https://doi.org/10.1086/ahr.117.2.365
Heritage Toronto (2020) Toronto's Urban Orchard. https://www.heritagetoronto.org/explore-learn/toronto-urban-orchard-heritage/. Accessed 13 Jan 2023
Heynen N, Perkins HA, Roy P (2006) The political ecology of uneven urban green space—the impact of political economy on race and ethnicity in producing environmental inequality in Milwaukee. Urban Aff Rev 42(1):3–25
High Park Nature Centre (2023). https://www.highparknaturecentre.com/. Accessed 13 Jan 2023
Hurley PT, Emery MR (2018) Locating provisioning ecosystem services in urban forests: Forageable woody species in New York City, USA. Landsc Urban Plan 170:266–275. https://doi.org/10.1016/j.landurbplan.2017.09.025
Ivanic M, Furdek Turk M, Tkalcec Z et al (2021) Distribution and origin of major, trace and rare earth elements in wild edible mushrooms: urban vs. forest areas. J Fungi 7
Kellert SR, Wilson EO (1993) The biophilia hypothesis. Island Press
King FB (1994) Interpreting wild plant foods in the archaeological record. In: Eating on the wild side: the pharmacologic, ecologic, and social implications of using noncultigens. University of Arizona Press, Tucson, pp 185–209
Kokkoris V, Massas I, Polemis E et al (2019) Accumulation of heavy metals by wild edible mushrooms with respect to soil substrates in the Athens metropolitan area (Greece). Sci Total Environ 685:280–296. https://doi.org/10.1016/j.scitotenv.2019.05.447
Kowalski J (2014) Gathering abundance: an exploration of urban foraging practices in Toronto. Dissertation, York University
Kowalski JM, Conway TM (2019) Branching out: the inclusion of urban food trees in Canadian urban forest management plans. Urban For Urban Green 45:126142. https://doi.org/10.1016/j.ufug.2018.05.012
Kowalski JM, Conway TM (2023) The routes to fruit: Governance of urban food trees in Canada. Urban For Urban Green 8:28045

Kuhnlein H, Turner N (2020) Traditional plant foods of Canadian indigenous peoples: nutrition, botany and use. Routledge, London

Kumar S, Das G, Shin H-S et al (2018) Diversity of plant species in the steel city of Odisha, India: ethnobotany and implications for conservation of urban bio-resources. Braz Arch Biol Technol 61

Landor-Yamagata JL, Kowarik I, Fischer LK (2018) Urban foraging in Berlin: people, plants and practices within the metropolitan green infrastructure. Sustainability 10(6):1873. https://doi.org/10.3390/su10061873

Leonti M, Nebel S, Rivera D, Heinrich M (2006) Wild gathered food plants in the European mediterranean: a comparative analysis. Econ Bot 60:130–142. https://doi.org/10.1663/0013-0001(2006)60[130:WGFPIT]2.0.CO;2

Li K, Rollins J, Yan E (2018) Web of science use in published research and review papers 1997–2017: a selective, dynamic, cross-domain, content-based analysis. Scientometrics 115:1–20. https://doi.org/10.1007/s11192-017-2622-5

Macrotrends (2023) Toronto, Canada Metro Area Population 1950–2023. https://www.macrotrends.net/cities/20402/toronto/population. Accessed 10 Jan 2023

Martin M (2018) Urban foraging: rethinking the human-nature connection in cities. eTropic 17:149–163. https://doi.org/10.25120/etropic.17.1.2018.3647

McLain R, Poe M, Hurley PT et al (2012) Producing edible landscapes in Seattle's urban forest. Urban For Urban Green 11:187–194

McLain RJ, Hurley PT, Emery MR, Poe MR (2014) Gathering "wild" food in the city: rethinking the role of foraging in urban ecosystem planning and management. Local Environ 19:220–240

McLain RJ, Poe MR, Urgenson LS et al (2017) Urban non-timber forest products stewardship practices among foragers in Seattle, Washington (USA). Urban For Urban Green 28:36–42

Ministry of Natural Resources and Forestry (2023) Using wood from Crown land for personal use. https://www.ontario.ca/page/using-wood-crown-land-personal-use. Accessed 6 Jan 2023

Morzillo AT, Campbell LK, King KL et al (2022) A tale of urban forest patch governance in four eastern US cities. Urban For Urban Green 75:127693

Nadgórska-Socha A, Kandziora-Ciupa M, Trzęsicki M, Barczyk G (2017) Air pollution tolerance index and heavy metal bioaccumulation in selected plant species from urban biotopes. Chemosphere 183:471–482. https://doi.org/10.1016/j.chemosphere.2017.05.128

Nesbitt L, Meitner MJ, Sheppard SR, Girling C (2018) The dimensions of urban green equity: a framework for analysis. Urban For Urban Green 34:240–248

Newfoundland and Labrador (1997) NLR Regulations 91/97—Provincial Parks Regulations under the Provincial Parks Act. https://www.assembly.nl.ca/legislation/sr/regulations/rc970091.htm#4_. Accessed 13 Jan 2023

Norouzi S, Khademi H, Faz Cano A, Acosta JA (2015) Using plane tree leaves for biomonitoring of dust borne heavy metals: a case study from Isfahan, Central Iran. Ecol Indic 57:64–73. https://doi.org/10.1016/j.ecolind.2015.04.011

Not Far From the Tree (2023) Not far from the tree—we pick fruit and share it. https://notfarfromthetree.org/. Accessed 13 Jan 2023

Nyman M (2019) Food, meaning-making and ontological uncertainty: exploring "urban foraging" and productive landscapes in London. Geoforum J Phys Hum Reg Geosci 99:170–180

Ontario Parks (2023) The effects of foraging at Ontario Parks. Ontario Parks Blog. https://www.ontarioparks.com/parksblog/foraging/. Accessed 6 Jan 2023

Parks Canada (2023) Government of Canada invests $130 million to work with partners to create a network of national urban parks. https://www.canada.ca/en/parks-canada/news/2021/08/government-of-canada-invests-130-million-to-work-with-partners-to-create-a-network-of-national-urban-parks.html. Accessed 6 Jan 2023

Poe MR, McLain RJ, Emery M, Hurley PT (2013) Urban forest justice and the rights to wild foods, medicines, and materials in the city. Hum Ecol 41:409–422

Poe MR, LeCompte J, McLain R, Hurley P (2014) Urban foraging and the relational ecologies of belonging. Soc Cult Geogr 15:901–919. https://doi.org/10.1080/14649365.2014.908232

Polinko AD, Coupland K (2021) Paradigm shifts in forestry and forest research: a bibliometric analysis. Can J For Res 51:154–162. https://doi.org/10.1139/cjfr-2020-0311

R Core Team (2022) R: a language and environment for statistical computing. R Foundation for Statistical Computing, Vienna, Austria. https://www.R-project.org/

Rombach M, Dean DL (2023) Exploring key factors driving urban foraging behavior in garden and non-garden locations. Foods 12:1032. https://doi.org/10.3390/foods12051032

Russo A, Escobedo FJ (2022) From smart urban forests to edible cities: new approaches in urban planning and design. Urban Plan 7:131–134

Sardeshpande M, Shackleton C (2019) Wild edible fruits: a systematic review of an under-researched multifunctional NTFP (non-timber forest product). Forests 10:467. https://doi.org/10.3390/f10060467

Sardeshpande M, Shackleton C (2020) Urban foraging: land management policy, perspectives, and potential. PloS One 15(4):e0230693. https://doi.org/10.1371/journal.pone.0230693

Schlecht MT, Säumel I (2015) Wild growing mushrooms for the Edible City? Cadmium and lead content in edible mushrooms harvested within the urban agglomeration of Berlin, Germany. Environ Pollut 204:298–305. https://doi.org/10.1016/j.envpol.2015.05.018

Schunko C, Brandner A (2022) Urban nature at the fingertips: investigating wild food foraging to enable nature interactions of urban dwellers. Ambio 51:1168–1178

Schwartz H (2009) Ten years after amalgamation. Can J Reg Sci XXXII:483–494

Secretariat TB of C (1996) Guide to the Federal Real Property Act and Federal Real Property Regulation. https://www.canada.ca/en/treasury-board-secretariat/services/federal-real-property-management/guide-federal-real-property-act-regulation.html. Accessed 13 Jan 2023

Shackleton S, Chinyimba A, Hebinck P et al (2015) Multiple benefits and values of trees in urban landscapes in two towns in northern South Africa. Landsc Urban Plan 136:76–86. https://doi.org/10.1016/j.landurbplan.2014.12.004

Shackleton CM, Hurley PT, Dahlberg AC et al (2017) Urban foraging: a ubiquitous human practice overlooked by urban planners, policy, and research. Sustainability 9(10):1884. https://doi.org/10.3390/su9101884

Shortly A, Kepe T (2021) Consuming the city: challenges and possibilities for foraging in Toronto's parks. For Trees Livelihoods 30:75–89

Soga M, Gaston KJ (2016) Extinction of experience: the loss of human–nature interactions. Front Ecol Environ 14:94–101. https://doi.org/10.1002/fee.1225

Somesh D, Rao R, Murali R, Nagendra H (2021) Patterns of urban foraging in Bengaluru city. Urban For Urban Green:57

Species at Risk, SC (2002) Species at Risk Act, SC 2002, c 29. https://canlii.ca/t/55cfs. Accessed 13 Jan 2023

Statistics Canada (2021) Population growth in Canada's rural areas, 2016 to 2021. https://www12.statcan.gc.ca/census-recensement/2021/as-sa/98-200-x/2021002/98-200-x2021002-eng.cfm. Accessed 13 Jan 2023

Synk CM, Kim BF, Davis CA et al (2017) Gathering Baltimore's bounty: characterizing behaviors, motivations, and barriers of foragers in an urban ecosystem. Urban For Urban Green 28:97–102

Threlfall CG, Kendal D (2018) The distinct ecological and social roles that wild spaces play in urban ecosystems. Urban For Urban Green 29:348–356

Toronto Biodiversity Strategy (2019) Wild, connected and diverse: a biodiversity strategy for Toronto. https://www.toronto.ca/legdocs/mmis/2019/ie/bgrd/backgroundfile-136906.pdf. Accessed 13 Jan 2023

Waygood U (2019) Urban foraging: its role in conservation and green space management in London. https://doi.org/10.13140/RG.2.2.15580.13445

Widener MJ, Minaker L, Farber S et al (2017) How do changes in the daily food and transportation environments affect grocery store accessibility? Appl Geogr 83:46–62. https://doi.org/10.1016/j.apgeog.2017.03.018

Wild Rice Harvesting Act (1990) Wild Rice Harvesting Act, RSO 1990, c W.7. https://canlii.ca/t/1jjq. Accessed 8 Jan 2023

World Bank (2022) Urban development. World Bank. https://www.worldbank.org/en/topic/urbandevelopment/overview. Accessed 13 Jan 2023

Yin S, Shen Z, Zhou P et al (2011) Quantifying air pollution attenuation within urban parks: an experimental approach in Shanghai, China. Environ Pollut 159:2155–2163. https://doi.org/10.1016/j.envpol.2011.03.009

Zohary D, Hopf M (2000) Domestication of plants in the Old World: the origin and spread of cultivated plants in West Asia, Europe, and the Nile Valley, 3rd edn. Oxford University Press, Oxford

Sivajanani Sivarajah (PhD) is an Assistant Professor and Research Chair of Urban Trees and its Environment at Université Laval, in Québec, Canada. Her research is transdisciplinary, applied, and collaborative, aimed at building sustainable and resilient communities, using nature-based solutions.

Md Abdul Halim (PhD) is a Schmidt AI in Science Postdoc Fellow at the University of Toronto. His research aims to design climate-smart management strategies for managed urban and forested ecosystems to combat the detrimental effects of climate change.

Sadia Butt (PhD) is an urban forest professional and ISA certified arborist with 20+ years of experience. Her research interests include urban forest ecology, community engagement, governance, and invasive species management.

Imrul Kayes (PhD) is working as an Assistant Professor at Noakhali Science and Technology University (NSTU), Bangladesh. He is currently doing his PhD in urban forestry and climatology at the University of Toronto.

Part IV

How Do Foragers Forage?

Learning to Find the "food beneath your feet": Urban Foraging, Social-Meet-ups, and Mobile Social-Ecological Memory in Philadelphia, USA

12

Patrick T. Hurley, Kristin McGillis, Marla R. Emery, and Brian J. Leffler

Abstract

Urban greenspaces contribute to the maintenance of social-ecological memory (SEM), or the knowledge, experiences, and practices a community develops about and within local ecosystems that support social-ecological resilience. Research on (SEM) associated with food production—or provisioning services—in urban greenspace demonstrates how knowledge about species, spaces, and practices is maintained or (re)produced. To date, however, SEM scholarship in cities has focused more narrowly on (1) urban allotment gardens and other agricultural sites explicitly designated and actively managed for food production and (2) a relatively limited set of associated species (i.e., agricultural cultivars). These spaces and species are generally characterized by clear spatial boundaries, governance dynamics, terms of access, and spatially fixed management, thereby leaving questions about how other urban green spaces and species in cities might support the development of SEM. To address this gap on how SEM is developed and maintained, where these activities take place, and which species are involved in its maintenance, we turn to research on urban foraging. We examine whether and how a wider set of species, spaces, and governance dynamics can foster SEM. Research from Philadelphia, Pennsylvania, USA documents how urban foraging supports engagement with species and spaces through social networks that arise using social media, and which lead to the maintenance *and* emergence

P. T. Hurley (✉) · K. McGillis
Department of Environmental and Sustainability, Ursinus College, Collegeville, PA, USA
e-mail: phurley@ursinus.edu

M. R. Emery
Norwegian Institute for Nature Research and Rubenstein School of Environment and Natural Resources, University of Vermont, Burlington, VT, USA

B. J. Leffler
National Park Service, Lewis and Clark National HIstoric Trail, Omaha, NE, USA

S. Dhyani, M. Sardeshpande (eds.), *Urban Foraging in the Changing World*,
https://doi.org/10.1007/978-981-97-0345-6_12

of *mobile* SEM practices. Our study of the Wild Foodies of Philly reveals how human–species interactions in diverse urban green spaces can foster SEM and the adaptation of SEM to new places and contexts throughout the city.

Keywords

Social-ecological memory · Urban foraging · Urban forests

12.1 Introduction: Social-Ecological Memory (SEM) and Urban Foraging

Urban greenspaces contribute to the maintenance and emergence of social-ecological memory (SEM), or the individual and collective memory associated with a community of practice in an ecosystem. In cities, studies of SEM often focus on self-provisioning and management activities in urban gardens (Barthel et al. 2004, 2013a, b; S. Shackleton et al. 2015). This work demonstrates how the tending of plants, harvesting of edible items from these species, and the management of the ecosystems in which these items are produced foster the development of knowledge, experiences, and distinctive practices by a community of practice within local ecosystem (Barthel et al. 2010a, b). As such SEM may reside in communities characterized by traditional ecological knowledge systems or become apparent in contemporary community-based resource management endeavors (Barthel et al. 2010a). SEM associated with food production in these gardens (re)produces knowledge about specific species and their care, the spaces and conditions in which these species flourish, and the practices necessary to support the resulting provisioning ecosystem services from these urban greenspaces (Barthel et al. 2010a, 2013a, b; Bendt et al. 2013). To date, however, urban SEM scholarship has focused narrowly on urban allotment gardens and other sites explicitly designated for food production and associated harvests (Barthel et al. 2015), with studies of SEM centering on a relatively narrow community of urban residents with access to these gardens and control over species selection in spaces that are characterized by clear spatial and governance boundaries.

A striking feature of urban SEM research is its focus on active use and consumption, in contrast to passive nature engagements more typically associated with urban settings. Besides the development of knowledge around food production and ecosystem services, SEM research describes the dynamics of knowledge maintenance, adaptation, degradation, and disappearance. The particular focus on allotment gardens in Europe reveals a distinct community of practice through which critical ecological knowledge is developed, based on both individual and supraindividual dynamics (see Barthel et al. 2015: 1325). Through use and management, participants carry forward knowledge about organisms and ecosystems. Barthel et al. suggest that the physical location of urban gardens provides vital "physical mnemonic cues... [without which SEM would] have nothing on which to work" (2015: 1328). Gardeners rely on "habits [of practice], oral traditions, written accounts, and self-organized systems of rules" (Barthel et al. 2010a, as cited in Barthel et al. 2015: 1326). Over time, gardeners learn from ecological feedbacks, enhancing understandings of and beneficial interventions in these social-ecological systems. Gardening serves as a living laboratory for transmitting information about locally adapted plants, soils, and micro-climates and the practices that facilitate consumptive use. While SEM scholars have begun considering the role additional urban greenspaces contribute to residents' resilience, using an "urban green commons" (UGS) lens (Colding et al. 2013), this approach still largely features UGS types that emphasize agricultural or horticultural management and species.

Meanwhile, urban foraging research raises important questions about how urban greenspaces might foster the maintenance and (re)production of SEM. Urban foraging consists of harvests of diverse species, for diverse uses (e.g., foods, medicines, fibers, and materials), and from species occurring in diverse urban greenspaces (Konijnendijk 2008). Foraging in cities supports access to foods and other resource uses for household needs (Emery et al. 2003; Wehi and Wehi 2009; Grabbatin et al. 2011; Hurley et al. 2013; Poe et al. 2013; Rebecca J. McLain et al. 2014; Kaoma and Shackleton 2015; S. Shackleton et al. 2015; Michelle Cocks et al. 2016; Synk et al. 2017; Landor-Yamagata et al. 2018; Johnson Gaither et al. 2020; Garekae and Shackleton 2020; Adeyemi and Shackleton 2023; Sardeshpande and Shackleton 2023) and through access to plant materials used to maintain cultural traditions (Grabbatin et al. 2011; Landor-Yamagata et al. 2018; Marquina et al. 2022; Poe et al. 2013). Through the harvesting and stewarding of these species, urban foragers connect with nature (Fischer and Kowarik 2020; Palliwoda et al. 2017; Schunko and Brandner 2022), find a deeper sense of belonging in the city (Marquina et al. 2022; Poe et al. 2014), and enact rules of use and stewardship (Charnley et al. 2018; R.J. McLain et al. 2017).

Foraging practices may be conspicuous or clandestine in nature (Brandner and Schunko 2022; Hurley et al. 2015; C. M. Shackleton et al. 2017), with harvests occurring in parks, remnant woodlands, cemeteries, institutional campuses, street trees, yards, and even in ornamental landscapes. Interstitial spaces and fringe ecologies (Brandner and Schunko 2022; Garekae et al. 2022; Hurley et al. 2008; Unnikrishnan and Nagendra 2015), such as unmanaged spaces characterized by spontaneous vegetation or species found between parcels with different owners, also provide access to species that support urban residents diverse needs and motivations (Svendsen et al. 2016; Grabbatin et al. 2011; Hurley et al. 2013; Emery and Hurley 2016). These fringe ecologies include sidewalk edges, cracks, and street verges and rights-of-ways, vacant lots, and in some cases, vegetation on yards not intended for harvest (Hurley et al. 2013, 2015; Rebecca J. McLain et al. 2014; Sardeshpande and Shackleton 2023).

Insights from urban foraging suggest that by focusing on urban gardens, scholars of SEM may risk overlooking a wider array of species, spaces, practices, and governance dynamics that contribute to social-ecological resilience of urban residents who are not gardeners and shape the maintenance or development of SEM among residents who are not gardeners. While gardens have relatively clear spatial and governance boundaries, urban foraging is a mobile practice dependent on diverse and deconcentrated spaces where spatial and governance boundaries are often unclear. Foraging locations may expand, contract, or shift over time in response to opportunistic discovery, changes in ownership, policy, or land use, or variations in species composition over time (Emery et al. 2003; Ginger et al. 2012; Hurley et al. 2013).

Urban forager knowledge appears to be acquired from previous experience elsewhere or through consultation of published sources, local experts, and online venues (Poe et al. 2014) in addition to social media (Hurley et al. 2015, 2022). By contrast, African American foraging practices in urbanizing areas of South Carolina, USA is transmitted between generations in kinship groups, but also to recent migrants to the area (Grabbatin et al. 2011). In New York City, USA, Mexican, and Chinese immigrants and their descendants apply their SEM to harvest culturally important items (Hurley et al. Forthcoming; Becker et al. this volume). Here knowledge transfer occurs through kinship groups, other personal relationships, and by conventional and social media. Likewise, foraging in both Berlin, Germany (Landor-Yamagata et al. 2018) and Vienna, Austria (Schunko and Brandner 2022) may be shaped by exposure to the practice as a child and by family members. These examples highlight some of the varied ways urban foragers learn about plants, their uses, the spaces where they are found, and associated ecologies, suggesting that in the twenty-first century, oral communication within kinship groups remains an important means of foraging SEM transmission, joined by new communication modes such as social media, which distribute knowledge across wider, mobile social networks.

Learning to find particular species for food or other uses in cities suggests foragers must pay attention to urban ecologies and dynamics in spaces other than allotment gardens. As such, urban foraging potentially challenges critical assumptions about where and how SEM develops in cities among urban residents. Thus, our research specifically asks: How can species other than cultivars related to self-provisioning contribute to SEM (re)production and, thus, the social-ecological resilience of urban residents? Which spaces not set aside for gardening and food production might facilitate this process (e.g., parks and other areas where rules, regulations, and management practices do not sanction this activity)? To what extent can SEM knowledge be transferred in and across these spaces through activities that are intended to share, or sometimes introduce, the experience urban foraging among interested residents?

This chapter explores how urban foraging contributes to the maintenance and development of mobile social-ecological memory in and across a diversity of urban spaces, while also pointing to some of the challenges to SEM development among urban foragers. Using qualitative research derived from examination of urban foraging tours in Philadelphia, Pennsylvania, USA, we specifically demonstrate how mobile SEM is maintained *but* also how urban foraging provides the opportunity for the development of SEM through an introduction to species across a diversity of urban spaces. We examine the Wild Foodies of Philly (hereafter referred to as Wild Foodies), a *Meet-up* group organized around foraging, to understand activities that contribute to the maintenance and emergence of SEM as part of the group's activities across the city. Meet-up groups are collections of individuals who use web-based utilities to organize events and online discussions on topics of interest (Hurley et al. 2015, 2022).

12.2 Case Study Context and Methods

12.2.1 Philadelphia, Pennsylvania, USA

The City of Philadelphia, Pennsylvania, USA is home to just under 1.6 million people (U.S. Census Bureau 2023), while the Philadelphia Metropolitan area is home to nearly 6.2 million people (U.S. Census Bureau 2021). The city and metro area are the sixth and seventh largest in the United States, respectively. While the city is 44.4% white alone and 43.6% African American alone (as compared to 15.9% Hispanic or Latino, 8% Asian alone, and 0.9% American Indian and Alaska Native alone and 0.2% Native Hawaiian and Other Pacific Islander alone), the metro area is overwhelming majority white (62% U.S. Census Bureau 2023). Social differences between the two geographies are also noticeable when considering rates of homeownership, household incomes, and levels of education.

The Philadelphia Metropolitan area extends from the Atlantic Coastal Plain in the east to the Piedmont Plateau in the west, with small portions of the city and large portions of the metropolitan area in New Jersey and Delaware occupying the coastal plain (Ricketts et al. 1999). Pine-dominated forests, tidal wetlands, and large suburban areas that were formerly agricultural fields characterize this part of the region. Meanwhile, parts of west and northern Philadelphia, including its larger parks, and much of the suburbs in the Pennsylvania portion of the metro area span the Piedmont Plateau (Ricketts et al. 1999). Here, forests are characterized by a rich mix of hardwood forests, freshwater wetlands, small creeks and streams, and agricultural areas transformed by urbanization over multiple centuries (Rhoads and Block 2007). Numerous native nut-, fruit-, and berry-producing species of trees and shrubs, and even some herbaceous plants, are common to forest types in both ecological provinces, with agricultural practices in the region frequently involving orchard species, such as the growing of apples, peaches, cherries, and other fruits (Hurley, personal observation).

Today, the city is home to dozens of parks and the wider region contains hundreds of public parks and game lands of diverse types that are owned and managed by municipal, county, and state governments as well as some lands owned by the national government (Hurley et al. 2015). These parks and open spaces include natural areas, sports fields, and greenways and recreational corridors that extend along area streams, rivers, and power corridors. Likewise, the region includes numerous nature reserves and open space areas owned and managed by private owners, including land trusts and conservancies. Further, large institutional campuses owned and managed by colleges, universities, and business parks are common in the region, with cemeteries and abundant street trees common in many municipalities. Hundreds of miles of transmission corridors associated with energy distribution also create vegetated greenspaces in the region. Foraging, or more specifically the removal of plants or plant materials, is generally a prohibited or restricted activity in most parks and preserves (though some exceptions apply, such as on state game lands (Hurley et al. 2015).

12.2.2 The "Wild Foodies of Philly" and Social Meet-ups in and Around the City

Founded in Philadelphia in September 2010, the Wild Foodies are a Meet-up group dedicated to "promot[ing] public - wild edibles - the only truly sustainable food source," doing so under the mantra "there's free food beneath your feet!" (Wild Foodies 2016). Wild Foodies maintains an extensive online library about useful species in the region (Wild Foodies 2016) that features myriad weblinks to additional sites that describe social and ecological aspects of species. During more than 10 years of existence, Wild Foodies' membership has grown steadily to just over 6700 individuals, with differing levels of engagement and activity among its members. Further, the group uses the website "meetup.com" to announce tour locations, or places where members learn "in the field" about foraging, and encounter forageable species and discussions of their uses. Events sometimes feature information about processing and use of specific plant materials (i.e., maple syrup). In addition to the meetup.com platform, the Wild Foodies also maintain a separate website featuring information about the utility of diverse plants found in the region. This site includes pages devoted to species identification and uses a discussion among members, and for sharing preparation and menu ideas as well as news stories about land-management issues in and stories about foraging in the region (Wild Foodies 2016).

Besides use of the social meet-up site and evolving online materials that focus on plant species and their uses for humans, a group organizer regularly sends out emails, announcing upcoming events or highlighting a "plant of the week." The latter includes additional information about a particular species and its uses, often with information on how to identify the species and, when appropriate, warnings about potential look-alikes that are dangerous Likewise, emails provide information on habitat preferences, interesting cultural facts, and specific locations where the species might be found, ecologically appropriate harvesting techniques, and ways to incorporate harvested materials into one's daily routine. Emails may also announce stewardship opportunities, including forageable species (e.g., invasives or aggressive native species) that land managers may wish to eradicate. Email exchanges by members often center on interest in a particular species, additional ways a species may be used, and timely information about where to harvest the species when it's in season.

12.2.3 Data Collection and Analysis

We draw on document analysis, participant observation, and in-depth interviews with Wild Foodies members to explore the dynamics and mechanisms urban foraging creates for the maintenance of social-ecological memory and associated learning about the utility of plants and associated urban ecologies. Our focus in this study was on documenting: (1) the diversity of species that the Wild Foodies engage with on their tours, (2) whether these engagements can provide a mechanism for maintaining knowledge of, or foster learning about (developing), the useful aspects of these plants and their local ecologies (or SEM associated with foraging), and (3) whether this knowledge was being maintained or being learned by participants at the "meet-ups."

We first reviewed online materials for past Wild Foodies tours and events to characterize locations, attendees, species encountered, and the use focus (i.e., food, medicines, other) of meet-ups, as well as information about any other aspects of the learning process. Our review of the Wild Foodies' Meet-up portal and website analyzed tour and event participants, the location types visited (e.g., parks, cemeteries, street verges or sidewalks, private backyards), the range of species encountered, and which human-plant uses associated with encountered species were the focus of the event (i.e., food, medicines, other). We reviewed event announcements and post-event materials for 2 years prior to the study summer. We also noted whether there were fees associated with events, what practical elements related to foraging were involved (e.g., general encounters with different species, focus on learning species identification, plant processing for use), and attendance levels. Where possible, we classified attendees into one of four age range categories: 0–20, 21–40, 40–60, and 60 or over.

Next we used participant observation to (A) determine the role different greenspace types play in either connecting again *or* introducing anew to participants the diverse plant species found in the region and (B) document the discussions participants with a variety of past foraging experiences had about how these species, including about the way materials harvested from certain species might be put to use in one's everyday routines. Participant observation also provided opportunities to better understand participants' interests in and prior experience with foraging, observe knowledge exchange dynamics, such as direct transmission through verbal discussion or use of online/hardcopy resource materials, and to recruit participants for follow-up interviews.

Members of our team attended four meet-up events during the summer of 2015 (mid-June and mid-July) in the metropolitan Philadelphia area. We kept field notes for each tour; we noted species encountered, discussion and types of questions associated with each species, and other aspects of the tours that related to knowledge transmission, including (a) how comfortable participants were generally with plant identification and use, (b) whether participants carried field guides (either printed or accessed online through their phones), (c) if they referenced these materials, and (d) how the participants went about teaching one another key points about the species, a species' uses, and the types of locations where found. We noted individual attendees' previous experience with foraging, their use of field guides or foraging-related apps, and ancillary activities by participants at these events (e.g., birdwatching). These observational details helped to demonstrate the ways that knowledge of and about species was being shared, including aspects of identifying particular species, species locational characteristics (e.g., along streams, in unmowed meadows), a species usefulness to humans (both contemporary and historical aspects), and considerations regarding a species function within the local ecosystem.

Finally, we conducted interviews with meet-up attendees to gather data on community characteristics of Philadelphia foragers, particularly types and degree of previous experience with foraging, and other information about the social dynamics at meet-ups. To follow up on our tour impressions, we conducted interviews with participants by phone. Interview questions focused on (a) characterizing individuals' reasons for attending the event, (b) their general interest in foraging, (c) extent of previous experience with foraging, (d) how they learned to forage and what sources informed their learning about the practice, (e) their ideas about the benefits from foraging and attending a foraging meet-up event, (d) the species these individuals had harvested, and (f) how they incorporate them into their daily lives. Interviews helped us better understand foragers' personal histories, extent of plant and ecological knowledge, and experience with the practice of harvesting plant materials for consumption or use. We gained insights into the ways foraging meet-ups support the dissemination of knowledge about individual species and local ecologies.

To put people–species interactions into their ecological and agricultural contexts, we further conducted a species-focused analysis to classify plants foraged during tours and discussed online as native, non-native, and/or invasive in the region (Table 12.2). We identified plants based on using common and scientific names online and with field guides in the field. Where common names were indicated, we matched these to likely species found in the region using county species inventories. Native/non-native determinations serve as a proxy for species importance within the regulating and supporting aspects of area ecosystem services, cultivar/non-cultivar designations serve as an indication of culturally important species in terms of aesthetic or other utilitarian values. Species nativeness designations were based on the U.S. Department of Agriculture's PLANTS Database (USDA 2016). We classify all species identified as introduced as non-natives. We classified species as invasive if designated as such by the Pennsylvania Department of Conservation and Natural Resources (PA DCNR 2016). The PA DCNR defines "invasive plants as those species that are not native to the state, grow aggressively, and spread and displace native vegetation." As such, this designation represents a regulatory designation. Some species not listed as invasive in this analysis may be considered invasive by land managers in our study area. Species cultivar designations were derived from the U.K. plant charity and database, *Plants for a Future* (2023).

In the following results, we first describe trends in the 2 years of meetup events prior to our participant observation in summer 2015. This information details the range of species and the frequency with which these species were encountered, the organizational themes of events, and the types of learning about these species that may have occurred. Drawing on this background, we next present insights from our participant observation at four social meet-up events. These data document the species encountered, frequency of encounters with species, and the organizational themes of the events. We describe the depth of discussions about particular species, event themes, and how these events foster the exchange of knowledge related to species of interest to foragers. We also offer observations about the types of individuals attending these events and what they seek from the experience. Next, we present findings from our in-depth interviews, in which we further probe species encounters and uses, but more importantly the range of experiences individuals have with foraging, the role foraging plays in their lives, and motivations for participating in Wild Foodie meet-ups. Taken together, these lines of analysis offer a window into the role foraging can play in maintaining and developing SEM.

12.3 Results

12.3.1 Assessing Meet-ups: Wild Foodies Excursions

Analysis of 2 years of Wild Foodies' social meet-ups reveals that between May 2013 and June 2015, 99 events were scheduled and held. Analyzing online RSVPs and event photos (with attendees of 3 or more) suggests 1385 event attendees (an average 14 people per event). Photo analysis suggests that a majority of attendees are between 20 and 60 years old (Table 12.1).

Thirty-seven events took place at nine city parks, an informal park, or a National Wildlife Refuge, the last two on the city's outskirts (Table 12.2). Outside the city, four events occurred at three different county parks, 13 events at four state parks in the metropolitan area, and seven at private nature preserves. Other urban and suburban greenspaces were frequent event sites, with seven events at cemeteries within the city and five at farms in the metropolitan area. Private businesses hosted 12 events, private museums hosted two, and private residents hosted 10, eight of these inside the city. It was not possible to identify location type for two events.

The majority of species were non-cultivars or cultivars not typically associated with culinary uses or food-oriented gardening activities. Events featured 128 taxa (Table 12.2), with 65 considered native, 64 non-native, and 29 invasive. Further, 46 of these 128 taxa are considered non-cultivars, 38 cultivars, and 40 have both cultivar and non-cultivar varieties in the USA. Of the 10 most frequently discussed taxa on tours, six are non-native or invasive and four include non-cultivars and cultivars. All were encountered at three or more events, and none are typical agricultural

Table 12.1 Study participant demographics

Gender		Age		Race		Length of time foraging (years)		
Female	Male	≤40	>40	White	Other	≤1	1–3	>3
10 (83%)	2 (17%)	7 (58%)	5 (42%)	11 (92%)	1 (8%)	1 (8%)	6 (50%)	5 (42%)

Table 12.2 Taxa encountered during meetups Species; native/non-native status; cultivar/non-cultivar species; number of past meetups and participant observation events at which encountered; interviewees mentioning a taxon; and locations where taxon was encountered

Species (likely Latin)	Status[a]	Cultivar[b]	Meetups	Partic Obs	Inter	Location types[c]
Past meetups						
1. Plantain (*Plantago major*)	NN	NC&C(r)	7	3	10	P,O,G,R,T
2. Chickweed (*Stellaria media*)	NN	NC	6	1	3	P,O,F,R
3. Garlic mustard (*Alliaria petiolata*)	I	NC	6	2	3	P,R,F,G
4. Lambs quarters (*Chenopodium album*)	NN	NC	6	2	3	P,O,R,G
5. Wineberries (*Rubus phoenicolasius*)	I	NC&C(r)	6	1	0	P,O
6. Wild violet (*Viola* spp.)	N&NN	NC&C	5	2	3	P,O,R,T
7. Nettles (*Urtica* spp.)	NN	NC&C	4	2	4	F,R
8. Spicebush (*Lindera benzoin*)	N	NC	4	1	2	P,R
9. Wood sorrel (*Oxalis* spp.)	N	NC	4	3	0	P,O
10. (Barren) strawberry (*Fragaria* spp.)	N	NC	3	2	1	P,O,R
11. Black walnut (*Juglans nigra*)	N	NC&C	3	0	0	O,R
12. Blackberries (*Rubus* spp.)	N&I	NC&C	3	0	1	O,C
13. Chicken of the woods (*Laetiporus* spp.)	N	NC	3	0	1	P,O
14. Daylilies (*Hemerocallis* spp.)	I	C	3	1	1	P,O,R
15. Fennel (*Foeniculum vulgare*)	NN	C	3	0	0	O
16. Hawthorne (*Crataegus* spp.)	N&I	NC&C	3	0	0	N,R
17. Honey locust (*Gleditsia triacanthos*)	N	NC&C	3	1	0	P,I
18. Mile-a-minute (*Persicaria perfoliata*)	I	NC	3	0	0	P,N,C
19. Mint/peppermint (*Mentha* spp.)	NN	C	3	0	1	O,R
20. Pine (*Pinus* spp.)	N&NN	NC&C	3	0	2	P,R
21. Sassafras (*Sassafras albidum*)	N	NC&C	3	0	3	P,C,I
22. Spiderwort (*Tradescantia* spp.)	I	NC	3	0	0	P,O
23. Amaranth (*Amaranthus* spp.)	NN	NC	2	1	2	P,R,G
24. Annie stalks (unknown)	UD	NA	2	0	0	P
25. Asiatic dayflower (*Commelina communis*)	I	NC	2	1	1	P,R
26. Autumn olive (*Elaeagnus umbellata*)	I	C	2	0	0	O
27. Birch (*Betula* spp.)	N&NN	NC&C	2	0	0	NH
28. Bittercress (*Barbarea vulgaris*)	I	NC	2	0	0	O,F
29. Burdock (*Arctium* spp.)	NN	NC	2	2	3	R,F
30. Cattails (*Typha* spp.)	N&I	NC	2	0	0	P,O
31. Cleavers (*Galium aparine*)	NN	NC&C	2	0	1	P,R
32. Echinacea (*Echinacea* spp.)	N	NC&C	2	0	0	R
33. Elderberry (*Sambucus* spp.)	N&NN	NC&C	2	0	1	O,NH
34. Evening primrose (*Oenothera* spp.)	NN	C	2	0	0	O,F
35. Feverfew (*Tanacetum parthenium*)	N&NN	NC	2	0	0	NH
36. Gingko (*Ginkgo biloba*)	NN	C	2	0	1	R,NH
37. (Wild) Grapes (*Vitis* spp.)	NN	NC	2	0	0	O
38. Holly (*Ilex* spp.)	N	NC&C	2	0	0	I,R
39. Honey mushrooms (*Armillaria* spp.)	N	NC	2	0	0	P,R
40. Hops (undetermined), N&I, C	N&I	C	2	0	0	NH
41. Hosta (*Hosta* spp.), NN	NN	C	2	1	2	R
42. Juneberries/serviceberries (*Amelanchier* spp.)	N&NN	NC&C	2	0	4	I

(continued)

Table 12.2 (continued)

Species (likely Latin)	Status[a]	Cultivar[b]	Meetups	Partic Obs	Inter	Location types[c]
43. Japanese Knotweed (*Fallopia japonica*)	I	NC	2	3	1	C
44. Kousa dogwood (*Cornus kousa*)	NN	C	2	1	0	P,I
45. Lemon balm (*Melissa officinalis*)	NN	C	2	0	0	O
46. Lemons (*Citrus limon*), NN, C	NN	C	2	0	0	P
47. Mallow (*Malva* spp.), I, NC	I	NC	2	1	0	P,O,C
48. Milkweed (*Asclepsias* spp.)	N	NC	2	0	1	C
49. Mimosa tree (*Albizia julibrissin*)	I	C	2	0	0	R
50. Motherwort (*Leonurus cardiaca*)	NN	C	2	0	0	R
51. Mugwort (*Artemisia vulgaris*)	NN	NC	2	3	3	P,C
52. Mulberries (*Morus* spp.)	N&I	C	2	1	7	I,R
53. Onions (*Allium* spp.)	NN	C	2	0	0	O,R
54. Paper mulberries (*Broussonetia papyrifera*)	I	C	2	1	0	P,R
55. Paw paws (*Asimina triloba*)	N	NC&C	2	0	0	P,O
56. Persimmon (*Diospyros virginiana*)	N	NC&C	2	0	1	O,I
57. Poke berry/weed (*Phytolacca americana*)	N	NC	2	1	1	P,C
58. Queen Anne's lace (*Daucus carota*)	N	NC	2	1	1	N,C
59. Ramps (*Allium tricoccum*)	NN	NC&C	2	0	1	P
60. Red clover (*Trifolium pratense*)	NN	NC&C	2	0	0	C
61. Sheep sorrel (*Rumex acetosella*)	NN	NC	2	0	0	P,N
62. Shiso (*Perilla frutescens*)	I	C	2	1	2	P,R
63. St. John's wort (*Hypericum perforatum*)	NN	NC	2	0	1	R
64. Thistle (*Cirsium* spp.), N&I, NC	N&I	NC	2	0	0	O
65. Wapato/arrowhead (*Sagittaria* spp.)	N&NN	NC	2	1	0	P
66. Yarrow (*Achillea millefolium*)	N&NN	NC	2	1	2	C
67. (Oak) Acorns (*Quercus* spp.)	N&NN	NC&C	1	1	1	R
68. Apple (*Malus* spp.), N, C	N	C	1	0	2	O
69. Apple mint (*Mentha suaveolens*)	NN	C	1	0	0	P
70. Bamboo (unknown)	I	UD	1	0	0	NH
71. Beefsteak mushrooms (*Fistulina hepatica*)	N	NC	1	0	0	P
72. Blewits (mushroom) (*Clitocybe* spp.)	N	NC	1	0	0	P
73. Brambles (1), UD (likely blackberry)	UD	NA	1	0	0	C
74. Brick top (mushroom) (*Hypholoma lateritium*)	N	NC	1	0	0	P
75. Carpetweed (*Mollugo verticillata*)	N&NN	NC	1	1	0	T
76. (Cat) greenbrier (*Smilax* spp.)	N	NC&C	1	0	1	O
77. Catnip (*Nepeta cataria*)	NN	C	1	0	0	R
78. Chamomile (*Matricaria chamomilla*)	NN	C	1	0	0	R
79. Chicory (*Cichorium intybus*)	NN	C	1	0	0	P
80. Chufa corms (1), N&NN, NC	N&NN	NC	1	0	0	P
81. Clover (*Trifolium repens*)	NN	C	1	1	2	O
82. Crab apples (*Malus* spp.)	NN	NC&C	1	1	2	P

(continued)

Table 12.2 (continued)

Species (likely Latin)	Status[a]	Cultivar[b]	Meetups	Partic Obs	Inter	Location types[c]
83. Creeping Charlie (*Glechoma hederacea*)	NN	NC	1	1	1	C
84. Curly/yellow dock (*Rumex crispus*)	NN	C	1	1	2	P
85. Dandelion (*Taraxacum officinale*)	NN	NC&C	1	1	7	G,F,R,T
86. Dock (*Rumex* spp.)	NN	C	1	0	0	NH
87. Fairy rings (mushrooms) (undetermined)	N	NC	1	0	0	O
88. Fleabane (Asteraceae)	N&NN	NC	1	0	0	C
89. Garden sorrel (*Rumex acetosa*)	N&NN	NC&C	1	0	4	R
90. Garlic (*Allium oleraceum*)	N&NN	NC&C	1	0	0	NH
91. Giant puffballs (*Calvatia gigantea*)	N	NC	1	0	0	P
92. (Wild) ginger (*Asarum* spp.)	N	NC	1	0	0	R
93. Hackberries (*Celtis* spp.)	N&NN	NC&C	1	0	0	I
94. Hen-of-the woods (*Grifola frondosa*)	N	NC	1	0	0	P
95. Honeysuckle (*Lonicera* spp.)	I	C	1	0	1	C
96. Horseradish (*Armoracia rusticana*)	NN	C	1	0	0	R
97. Hot peppers (*Capsicum* spp.)	NN	C	1	0	0	R
98. Inky caps (*Coprinopsis atramentaria*)	NC	NC	1	0	0	O
99. Lemongrass (*Cymbopogon* spp.)	NN	C	1	0	0	R
100. Kentucky coffeetree (*Gymnocladus dioicus*)	N	NC&C	1	0	0	P
101. Lesser Celandine (*Ficaria verna*)	I	NC	1	0	0	P
102. Linden (*Tilia* spp.)	N&NN	NC&C	1	0	0	R
103. Lotus (*Nelumbo* spp.)	N&NN	NC&C	1	0	0	NH
104. Lovage (*Levisticum officinale*)	NN	C	1	0	0	R
105. Marshmallow (*Althaea officinalis*)	NN	NC&C	1	0	0	C
106. Nodding star-of-Bethlehem (*Ornithogalum nutans*)	I	C	1	0	0	P
107. Partridge berry (*Mitchella repens*)	N	NC&C	1	0	0	O
108. Passionflower (*Passiflora* spp.)	N&NN	NC&C	1	0	0	NH
109. Patience dock (*Rumex patientia*)	NN	C	1	0	0	R
110. Pear-shaped puffballs (*Lycoperdon pyriforme*)	N	NC	1	0	0	P
111. Phragmites (*Phragmites australis*)	I	NC	1	0	0	P
112. Raspberry (*Rubus* spp.)	N&NN	NC&C	1	1	4	O
113. Red maple (*Acer rubrum*)	N	NC&C	1	0	0	P
114. Rose of Sharon (*Hibiscus syriacus*)	NN	C	1	1	0	R
115. Rose (*Rosa* spp.), N&I, NC&HC	N&I	NC&C	1	0	0	O
116. Rue (*Ruta graveolens*)	I	NC&C	1	0	0	R
117. Shagbark hickory (*Carya ovata*)	N	NC&C	1	0	0	O
118. Shaggy manes (mushroom) (*Coprinus comatus*)	N	NC	1	0	0	P
119. Stevia plants (*Stevia rebaudiana*)	NN	C	1	0	0	R
120. Sugar maple (*Acer saccharum*)	N	NC&C	1	0	0	P
121. Tree-of-heaven (*Ailanthus altissima*)	I	NC	1	0	0	C
122. Tulsi (*Ocimum tenuiflorum*)	NN	C	1	0	0	R
123. Watercress (*Ocimum tenuiflorum*)	NN	C	1	1	0	I

(continued)

Table 12.2 (continued)

Species (likely Latin)	Status[a]	Cultivar[b]	Meetups	Partic Obs	Inter	Location types[c]
124. White oak (*Quercus alba*)	N	NC&C	1	0	0	I
125. Wild cherries (*Prunus* spp.)	N&NN	C	1	0	1	O
126. Wild parsnips (*Pastinaca sativa*)	I	C	1	0	0	N
127. Wild raisins (*Viburnum* spp.)	N	NC	1	0	0	N
128. Willow (*Salix* spp.)	N&NN	NC&C	1	1	0	NH
New species—participant observation						
129. Jewelweed (*Impatiens* spp.)	N&NN	NC	0	2	3	P,R
130. Lady's thumb (*Polygonum persicaria*)	NN	NC	0	2	3	P,R P
131. Blue vervain (*Verbena hastata*)	N	NC	0	1	0	P
132. Bishop's weed (*Aegopodium podagraria*)	I	C	0	1	0	P
133. Crabgrass (*Digitaria* spp.)	NN	NC	0	1	0	P,R,T
134. Peach (*Prunus persica*)	NN	C	0	1	1	P
135. Pepper (grass)(*Lepidium* spp.)	I	NC	0	1	0	P
136. Purslane (*Portulaca oleracea*)	NN	NC&C	0	1	1	P
137. Quickweed (*Galinsoga* spp.)	N&NN	NC	0	1	0	T
138. Sage wort (*Artemisia* spp.)	N	NC	0	1	0	R,T
139. Sedum (*Sedum* spp.)	NN	C	0	1	0	R
140. Smartweed (*Polygonum* spp.)	NN	NC	0	2	0	P
141. Sow thistle (*Sonchus* spp.)	I	NC	0	1	0	T
142. Yucca flower (*Yucca* spp.)	N	NC&C	0	1	0	R,T
Additional species—interviews						
143. Morel (mushroom) (*Morchella* spp.)	N		0	0	2	O
144. Beech nuts (*Fagus* spp.)	N&NN	NC&C	0	0	1	P
145. Chanterelle (*Cantharellus cibarius*)	N	NC	0	0	1	P
146. Forsythia (*Forsythia* spp.)	NN	C	0	0	1	O
147. Gilampus (Unknown)	UD	NA	0	0	1	ND
148. Hazelnut (*Corylus* spp.)	N&NN	NC&C	0	0	1	P
149. Maple (*Acer* spp.)		NC&C	0	0	1	P,O
150. Prickly lettuce (*Lactuca serriola*)	NN	NC	0	0	1	P,R
151. Sunflowers (*Helianthus* spp.)	N	NC&C	0	0	1	P,R
152. Wild mustard (Unknown)	UD	NA	0	0	1	R
153. Wintergreen (*Gaultheria* spp.)	N	NC&C	0	0	1	P

[a] *N* native, *NN* nonnative, *I* invasive (PA DCNR 2016); *UD* undetermined species

[b] *NC* non-cultivar, *C* cultivar, (*r*) indicates species rarely appears as cultivar (Plants for a Future 2023)

[c] *P* park inside city, *O* park outside city, *N* nature reserve, *F* farm, *G* community garden (weeds), *C* cemetery, *R* private residence (i.e., yard), *I* institutional campus, *T* interstitial space (e.g., sidewalk), *NH* materials not harvested during an event (i.e., found at processing event only)

plants in the region. The frequency with which taxa were discussed at events varied from 10 times (plantain; *Plantago* spp.) to once (61 taxa), with another 44 discussed at two events (Fig. 12.1). Given that individual tours tend toward species of interest to both the leader and the participants, these numbers potentially overlook and undercount the total number of useful species present in an area.

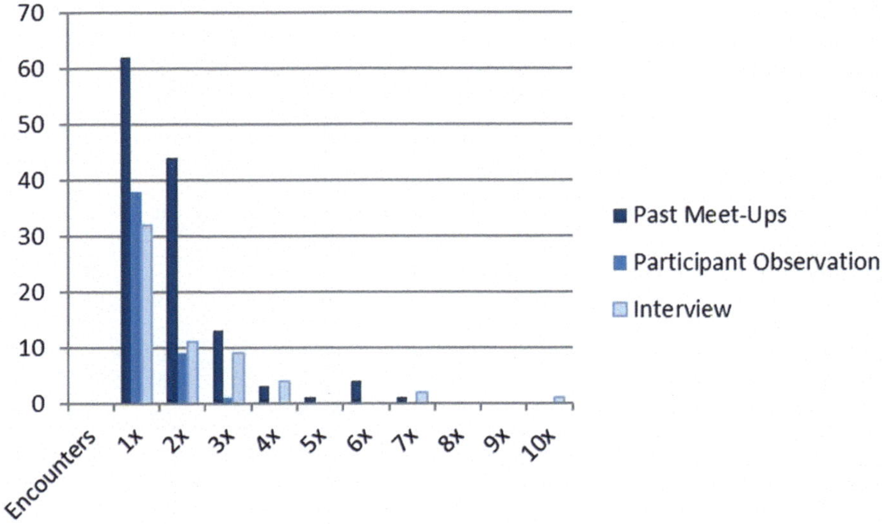

Fig. 12.1 Frequency of taxa encounters / mentions

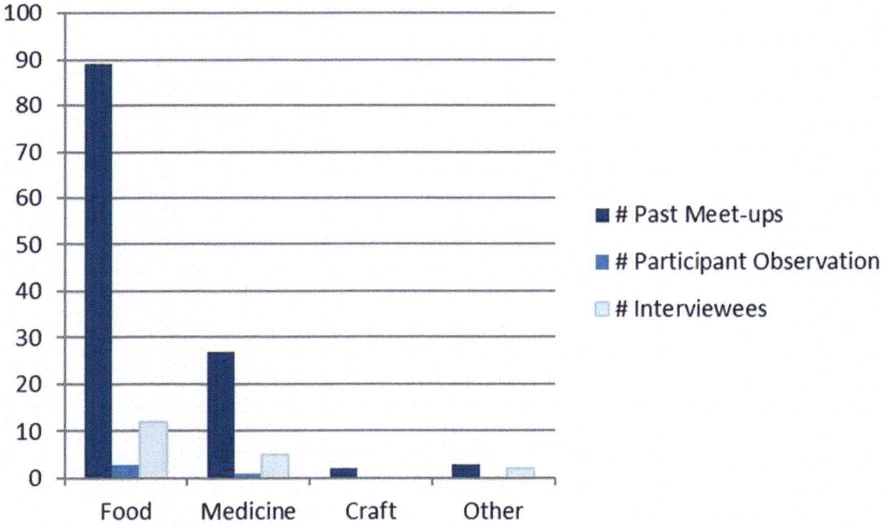

Fig. 12.2 Uses of plant taxa by frequency

Of the 99 events we analyzed, 89 focused on edible uses (Fig. 12.2). Another 27 events emphasized medicinal uses, two focused explicitly on crafts, and three on foraging education. Of 89 meet-ups, four focused on processing of materials and did not include any in-the-field identification or harvesting. Thirty-six meet-ups were intended as introductions to the practice of foraging and included a focus on learning species identification. An additional six focused on introduction, identification, and processing of species for specific uses. Three tours focused exclusively on the processing of harvested materials, and one emphasized the introduction of species and processing without any apparent attention to identification.

12.3.2 Participant Observation: Learning to Locate Useful Species

On the first meet-up at a public park just inside the City, we joined 10 people—4 men and 6 women—on a walking tour focused mainly on locating edible species along a forested streambank in a city park (Fig. 12.3). The organizer explained that she enjoyed bringing people together to share what they know about edible and medicinal plants, but emphasized how she herself was not an expert on the subject. Half of this group was in the 20–40 age range, while the other half was in the 40–60 category. At least seven individuals were new to foraging, while five had some experience. Participants discussed species identification and food-related uses, the need to avoid picking plants too close to the street, the advantages of young plants over mature plants, the importance of thoroughly washing plants, and how to use online sources for the identification of the plants they were encountering. There was no mention of concerns about medicinal plant uses. However, a brief discussion

Fig. 12.3 Collaborative identification of plants along a trailside in a city park

addressed plantain's use to relieve bug bite irritation. During this tour, only the organizer harvested any species for later use. Participants touched, smelled, and occasionally tasted discussed plants. Seventeen taxa were discussed, including non-cultivar species (e.g., wood sorrel) as well as non-native species (e.g., plantain) and invasive species (e.g., creeping charlie).

The second meet-up visited another forested stream trail in the same park (West Philadelphia). There were 10 attendees, of whom four were in the 20–40 years age range and seven in the 40–60 range. Six people were inexperienced foragers (having foraged for less than 6 months), while five were somewhat experienced (having foraged more than 6 months). Informal discussions during this event were almost exclusively about how to identify the plants encountered. There was no discussion about medicinal uses of any species encountered. The tour organizer (same woman as above previous tour) was again the only individual harvesting materials. One couple collected a few leaves from multiple plants for use in future identification. Tour participants encountered 14 taxa and, as with the previous park visit, these included non-cultivars, such as spicebush, and a number of non-native, invasive species (e.g., Asiatic dayflower). Food was the only species use discussed, including as trail snacks, ingredients in hearty salads or whole meals, or ingredients associated with historical beer-making techniques. Again, the organizer of this tour did not claim to be an expert on the topic of wild foraging, but rather emphasized her interest in fostering people's ability to gather and learn from one another through knowledge exchange.

The third event took place in Center City Philadelphia, where 13 people explored the interstitial spaces—street trees and tree pits, sidewalks, property edges, and alleys—of a historic neighborhood. Nearly all discussion about these plants focused on their edibility. Upon arriving at the organizer's home, participants discovered some previously harvested materials already prepared to sample (e.g., cooked amaranth and sow thistle). Six participants were in the 20–40 age range, while seven were in the 40–60 age range. Nine individuals were beginner foragers, while four were experienced foragers. An individual active in leading Wild Foodies' tours during the past 5 years led this tour. This organizer led the group from street to street, pointing out forageable plants, discussing how to identify the species, ways to prepare them, and other places where one might find them. No one on this tour harvested anything to take home. However, many people smelled and tasted plant materials along the way. During the walk, we encountered 26 species, many of which were invasive species (e.g., Asiatic dayflower, chickweed), but also ornamental species common to well-managed yards. Further discussion focused on properly cooking encountered species, with only minimal mention of using species as medicines (i.e., plantain for bug bites).

The fourth tour took place at a community-managed park in North-Central Philadelphia, characterized by well-maintained ornamental plantings, an open lawn area, and a fenced-in allotment garden. Participants stopped at landscaped

Fig. 12.4 Meet-up participants gather around a horticultural planting area in the community-managed Liberty Lands Park in Philadelphia

ornamental plantings and areas characterized by spontaneous vegetation in interstitial spaces, passing through lightly managed lawn (Fig. 12.4). Although the focus was on medicinal uses of species, participants also introduced, identified, and discussed species with edible parts. Twenty people attended the event: 13 in the 20–40 age range and seven in the 40–60 range. They included at least four beginning foragers and three more experienced people, with one individual proudly describing her more than 40 years of foraging to the group and another describing multiple decades as a master gardener.

Focusing on medicinal species, this tour was guided by a practicing herbal therapist, who wove in stories of treating clients with a variety of specific plants. During the hour-plus tour, participants saw and discussed 13 taxa (Fig. 12.5), including cultivars (e.g., peach tree), non-cultivars (e.g., yarrow), and non-native, invasive species (e.g., mugwort). Conversations engaged topics on how to find and use species believed to have medicinal benefits (Fig. 12.6), including how species relieve inflammation from bug bites and claims about potential cures for cancers. Here, the guide also discussed historical conceptions about locating medicinal plants. There was only minimal discussion about the use of field guides for plant identification, as the individual hosting the tour clearly stuck to plants she easily recognized.

Fig. 12.5 Participants on a medicinal species plant meet-up pass by a community garden (right), focusing their attention on the non-food cultivars and spontaneous species found outside this designated food production space

Fig. 12.6 Participants learn about the medicinal properties of bee balm (*Monarda* sp.), a native species planted here as an ornamental

12.3.3 Interviews with Meet-up Participants: Incorporating Useful Species into Daily Life

A majority of Wild Foodies tour participants interviewed were white (92%) and female (83%). Fifty-eight percent were 40 years of age or younger and 42% older than 40 years. Half reported that they had been foraging for 1–2 years. One individual had less than 1 year of experience; five had foraged for 3 or more years (Table 12.1). Reported uses for foraged plant materials and fungi included food (all interviewees), medicine (5 interviewees), and other purposes (2 interviewees). None described specific craft-related uses. Interviewees reported having foraged 63 taxa (Table 12.2). Frequency of mention provides an indication of the learning potential in a group context. Of 63 taxa, 32 were harvested by only one person, 11 by two people, 12 by three people, 3 by 4 people, 2 by 7 people, and 1 by 10 people (Fig. 12.1). No taxon was mentioned by everyone.

When considering their frequency and intentionality with foraging, respondents indicated their foraging tends to be impromptu rather than planned. That is, they foraged in the course of their daily lives, when they found plants and/or fungi they recognized and knew how to use, rather than planning outings specifically for this purpose. Interviewees rarely mentioned medicinal uses, although they were aware of this potential. Parks provide opportunities for these spontaneous foraging encounters. Interviewees most consistently described expanding their knowledge of locally occurring plants and their associated ecologies as an important benefit of the practice. This knowledge fosters a sense of independence and control over food sourcing, including in potential food shortages or emergencies.

Interviewees reported the biggest draw for meet-ups as the opportunity to gain new knowledge about species *and* ground this knowledge in new places (particular urban greenspaces and their local ecologies). Tours provided people with immediate answers to questions about identifying species safely (both species specific elements and ecosystem contexts) and the best practices for harvesting particular species or parts, instead of searching field guides or online references. Tours specifically allowed people to gain new knowledge and develop greater confidence about engaging with certain species in their surroundings. This includes among beginning foragers, who express worries about contradictory information found online. Indeed, tours served as a starting point for experimentation with foraging for many participants. These individuals may come to events with a list of questions to ask the person who is the leading the walk.

Learning about specific plants and their uses was not the only meet-up benefit reported by participants. Other benefits included meeting like-minded people, sharing knowledge with others, connecting to nature, gaining survival skills, and developing an enjoyable hobby focused on the environment. This interaction can allow individuals to connect to the land in meditative ways. Foraging can also provide access to other people with similar interests for newcomers to the city who do not know others. These engagements illustrate the multi-functionality of Wild Foodies events in meeting the differing needs of those attending, focusing on the aspects of resilience that extend beyond supporting, regulating, or even provisioning dynamics of these services to include aspects of well-being and cultural ecosystem services. In each case, these may reflect an individual's experience level and interests.

Interviewees report only partial understandings of dynamics associated with governance and local environments. When asked about foraging legality, most interviewees did not know rules for locations they foraged, nor how to find them. A few interviewees reported asking for permission, particularly when finding species on private property. Still, participants' most common concern was potential exposure to contaminants from animal wastes and/or pollution associated with motor vehicles. A smaller number of foragers worried about potential species misidentification. Regular communication on the Wild Foodies listserv reminds community members to respect local park rules, to take precautions in choosing where to harvest, and to follow best practices in washing harvested materials before consumption.

12.4 Discussion

Our results from 2 years of Wild Foodies' meet-ups reveal key dimensions about where events take place, which species are encountered and discussed, who participates in these events and why, which plant uses foragers prioritize, and what these suggest for the potential of urban foraging to support maintenance and emergence of SEM. First, parks and conserved lands were common locations for events inside and outside Philadelphia. These spaces provide access to a sufficiently diverse range of useful species that can support social-ecological learning. Second, meet-ups introduced individuals to various species during the growing season. Many species are native to regional forest ecosystems. Finally, meet-up events emphasized a range of uses and themes, of which food was the overwhelming organizing focus. However, unlike with the urban allotment gardening activities in urban SEM studies, the majority of species involved are non-cultivars or cultivars not typically associated with food-oriented gardening activities.

Our participant observation of tours revealed three key aspects of knowledge transfer and learning dynamics at Wild Foodies events: topical focus, location, and information resources. Meet-up events focus on single use for foraged materials, with species edibility predominating. Tour locations reflect diverse urban greenspace types where foraging occurs. Although parks are common locations, these spaces are far from uniform in terms of land-covers and management approaches. Tours also mobilize a suite of potential information resources in differing ways, with key variables being leader skill-sets, participant experience levels, leader-participant dynamics, and presence or absence of formal reference materials. These events allow individuals with greater degrees of existing SEM to share their knowledge with others, particularly newcomers. This sharing includes how to identify species in their local ecologies, appropriately harvest materials, use species in different ways (though generally in terms of food), and discuss a range of associated issues. Importantly, social meet-ups offer direct engagements with specific and diverse species across a range of urban greenspaces (and associate ecologies) that provide an introduction to ecological or landscape cues that foster the learning necessary to develop SEM.

Our interviews with tour attendees suggest a wide range of experience levels with foraging, both in terms of years engaged in the activity and the total number of species with which individuals participants are familiar. Thus, the degree of social-ecological learning at any given social meet-up can be highly variable. While some individuals present at meet-ups may have existing reservoirs of knowledge about some species, a limited number of individuals may have more extensive knowledge of, or experiences with, the useful species in the specific location being visited. Attendees may seek out these events specifically to gain knowledge about local ecosystems, the species they support, and develop confidence in identifying useful species of interest in new locations.

Collectively, these insights suggest that the pool of knowledge about Philadelphia area species and spaces that provide food, which is resident in Wild Foodies as a group, appears both to be uneven and dispersed among multiple individuals. Collective foraging knowledge exceeds that of any individual forager, with knowledge of and about forageable species collected in city-specific online material, shared via social media, and drawing on mobile foraging apps. Data from a review of the Wild Foodies website (2016) as well as our analysis of participant observations and interviews documented knowledge of 153 forageable taxa in the Philadelphia region. A majority (128, 84%) was encountered as part of tours during the 2013–2015 period. Participant observations documented 56 encountered taxa (36.6%), while interviewees mentioned 64 taxa (41.8%). On most tours, knowledge of the suite of species encountered depended on either different individual participants providing contributions to discussions that touched on encountered species or consultation of formal resources such as books, apps, other online materials to enhance knowledge development. Interestingly, interviewees discussed 11 taxa not encountered in earlier tours or participant observation.

Within the context of this uneven knowledgebase, foraging meet-ups operate as key experiences for and ad hoc sites where learning about edible species and their various social-ecological uses takes place. These moments of SEM development occur nearly entirely outside designated garden spaces. Indeed, diverse locations and landscape management regimes provide access to forageable taxa and opportunities for individuals to learn about species utility. Parks provide access to 88 (57.5%), private yards for 40 (26.1%) (cemeteries to 17 (11.1%), institutional campuses to 10 (6.5%), interstitial spaces to nine (5.8%), farms to seven (4.6%), and nature reserves to six (3.9%). We note that 21 species (13.7%) were found only in private yards, while 10 species (6.5%) were found only in cemeteries. This tally is largely based on analysis of 2013–2015 meet-ups, supplemented by information from participation in the fourth tour. Thus, these numbers likely underestimate forageable species distributions in and across urban greenspaces.

Tours regularly visit diverse urban greenspaces to introduce city residents primarily to food-producing, but also other myriad species, helping those individuals with established social-ecological memory and those seeking to acquire it to share information and learn about the habitats where these species can be found. Given these spaces are characterized by diverse land-use and land-management regimes, foraging appears to be critical for the learning that (re)produces social-ecological memory about provisioning and cultural ecosystem services from cultivar *and* non-cultivar species. Here, private yards, cemeteries, and even institutional campuses offer spaces where this knowledge can be transferred and is acquired.

Our data suggest, however, that it is necessary for newcomers to attend several meet-ups across multiple seasons to gain knowledge of a wide array of species, uses, and the locations where species are likely to be found. The importance of frequent encounters with a new species to solidify learning suggests patterns related to phenology and tour scheduling may be important to (re)production of social-ecological memory through foraging meet-ups. The prevalence of encounters with useful species in parks, thus, highlights their particular importance as places where access, landscape management regimes, and ecologies allow people to learn about and experiment with a range of species. Meet-up groups, such as Wild Foodies, have the potential to serve as a digital analog to allotment gardens, providing both virtual, and sometimes physical, sites for members to convene to gain, adapt, and transmit foraging SEM through highly mobile activities.

Learning to find food beneath one's feet includes valorizing species not typically viewed as desirable by managers and others, such as non-native, weedy, and invasive species frequently targeted for eradication. This recognition of the benefits of non-native species extends observations about their ecosystem service benefits (Tredici 2010) to dynamics associated with the maintenance and emergence of SEM. By incorporating these oft-maligned species into their lives, foragers often draw on social-ecological memory about the usefulness of plants developed *elsewhere* and share this knowledge through these learning networks. Field guides, social media, and social meet-ups enable the matching of plant mobilities with mobile ecological knowledge, suggesting that insofar as urban ecosystems feature novel ecologies, foraging, and other human interactions with these species also may lead to the emergence of novel ecological knowledges.

Unlike much urban gardening, the legal and regulatory contexts of foraging locations range from ambiguous to prohibited, rendering these practices precarious. While harvesting plant materials and fungi is prohibited in most parks in U.S. cities, learning about usefulness species is not. Indeed, some Wild Foodie events have enjoyed the co-sponsorship and participation of local park and nature reserve staff, even though harvesting in these spaces is prohibited. Foraging appears to be tacitly accepted in some parks and informally permitted in other urban greenspaces. Our results suggest benefits from conventional environmental education and collaborative learning focused on plant and fungi identification, uses, and stewardship for the physical and emotional well-being of urban residents. Parks may wish to consider whether their regulations can be adapted to foster provisioning ecosystem services and resilience that result from social-ecological memory of foraging without compromising their other missions. Outside parks, findings suggest opportunities to foster new land-management norms for everyday landscapes that support the (re) production of SEM derived through foraging.

12.5 Conclusions

By documenting how Wild Foodies activities bring together individuals in, and to learn about, the urban ecologies of diverse greenspaces, our research reveals that encounters with species focused on foraging fosters learning about plant species that contributes to the development of mobile SEM. Foraging meet-ups introduce metropolitan residents to a diverse array of forageable plants and fungi found in urban settings. SEM is maintained by and developed among people interested in free and local food, supported by mechanisms that foster social interactions around this knowledge sharing, outdoor recreation, and enhanced connections to nature. Social meet-ups focused on urban foraging bring people together with different degrees of knowledge to openly engage in an activity that might otherwise be undertaken as a surreptitious solitary pursuit (Brandner and Schunko 2022). Although individual foragers are the actors who hold memories and learn, they are embedded within social communication networks that facilitate SEM sharing (Barthel et al. 2015; Davidson-Hunt 2006). The research demonstrates that socially organized meet-ups facilitate information sharing between (1) some urbanites with longstanding knowledge of plant and fungi uses they may be adapting to new locations and (2) a greater number of newcomers seeking to learn about these engagements with local ecologies and their species as a way of enriching their daily lives. Wild Foody tours function as outdoor laboratories for learning both harvesting techniques and experiencing flavors and other sensations associated with foraged foods. Frequently, they also allow foragers to introduce non-foraging friends or family to an activity they enjoy and the knowledges upon which this practice depends.

The spatial development and maintenance of urban SEM are potentially more variable and clearly more mobile within cities than is currently recognized in the urban SEM literature focusing on urban gardens. Findings highlight the importance of parks and non-cultivar species for SEM retention by knowledgeable urbanites and SEM emergence among neophytes. Forager-species interactions include engagements with urban greenspaces of all kinds. Thus, SEM may depend on well-tended sites not specifically intended for natural resource use, such as parks and legacy agricultural ecosystems that persist in expanding urban areas, or in the novel ecosystems found in existing urban areas. In this sense, SEM is bound up in regional and global social-ecological processes that preserve native ecosystems in some urban spaces while creating emergent ecologies in other urban spaces. Foraging may foster resilience through maintenance of bioregionally specific SEM, but also through adaptation of mobile SEM to new places and ecological configurations.

Social-ecological knowledge, experiences, and practices are bundled together as SEM that is retained, reified, and transmitted through oral communication, written accounts, group events, learning and imitating practices and rules-of-use exemplified by other foragers (Barthel et al. 2010a, 2015; Charnley et al. 2018), and increasingly through social media and app-based platforms. Critical social interactions that foster these engagements rely on social media to both organize events at known times as well as to celebrate and foster knowledge gained through these interactions.

Urban foraging, and the mobile harvesting practices that constitute this practice, connects diverse people to myriad species, in diverse urban ecologies, and thereby supports SEM maintenance and development among foragers whose experience differs. Learning to find food beneath one's feet in cities is also learning how to adapt SEM to new situations and contexts. Here, internet resources and social media networks are critical to SEM development in spaces outside of gardens, where it is mobilized and enacted through human-species engagements with, and that take advantage of, urban ecologies of myriad kinds.

A focus on urban foraging can reveal the potential of new urban spaces, species, and practices to engender SEM in previously unrecognized spaces and among previously unrecognized residents. Considerations of mobile SEM in urban social-ecological systems suggest that a wider diversity of spaces and species interaction are at play. While we recognize the research presented here focuses extensively on the benefits from these ecosystems for human material and non-material resilience, this preliminary work in connection with other emerging insights on urban foraging suggests that future work should consider the multiple ways that urban foraging helps to constitute the ecosystems in which it is embedded and how these may reinforce or influence the provision of ecosystem services for nonhuman aspects of social-ecological resilience.

Urban foraging's diversity of species, practices, and greenspaces may be critical to the maintenance *and* development of SEM that link ecosystem services to residents. This insight is particularly true for those species not associated with urban gardening, and for which management may be largely beyond control of the forager. Moreover, as a community of practice, whether culturally differentiated or not, SEM appears to be maintained by and emerge unevenly within urban foragers. Social meet-ups by residents in the city who do not know one another function to transmit this knowledge and support the development of critical new social-ecological knowledge, and thus SEM, by relative newcomers to a city. This SEM is focused overwhelmingly on the foods and material benefits associated with species to which these urban foraging tours—and the range of species a diversity of urban greenspaces—provide access. Greater attention to the dynamics of urban foraging and the role this community of practice plays in producing SEM in diverse cities will provide scholars of social-ecological resilience with deeper insights into the myriad pathways, interactions, species, and challenges critical to developing and maintaining SEM.AcknowledgmentsWe thank the organizers of the Wild Foodies social meet-ups we attended, attendees at these events who welcomed our presence, and the interview participants who shared additional information about their foraging practices and experiences.

Funding This research was funded by the Office of the Vice President for Academic Affairs at Ursinus College.

References

Adeyemi O, Shackleton CM (2023) Understanding foraging practices in Lagos Metropolis to redesign urban greenspaces in support of human-nature interactions. Urban For Urban Green 79:127805. https://doi.org/10.1016/j.ufug.2022.127805

Barthel S, Colding J, Elmqvist T, Folke C (2004) History and local management of a biodiversity-rich, urban cultural landscape. Ecol Soc 10. https://doi.org/10.5751/ES-01568-100210

Barthel S, Folke C, Colding J (2010a) Social–ecological memory in urban gardens—retaining the capacity for management of ecosystem services. Glob Environ Chang 20(2):255–265. https://doi.org/10.1016/j.gloenvcha.2010.01.001

Barthel S, Sörlin S, Ljungkvist J (2010b) Innovative memory and resilient cities: echoes from ancient constantinople. In: The urban mind

Barthel S, Parker J, Ernstson H (2013a) Food and green space in cities: a resilience lens on gardens and urban environmental movement. Urban Stud 52:1321. https://doi.org/10.1177/0042098012472744

Barthel S, Parker J, Folke C, Colding J (2013b) Urban gardens-pockets of social-ecological memory. In: Greening in the red zone, pp 145–158. https://doi.org/10.1007/978-90-481-9947-1_11

Barthel S, Parker J, Ernstson H (2015) Food and green space in cities: a resilience lens on gardens and urban environmental movements. Urban Stud 52(7):1321–1338. https://doi.org/10.1177/0042098012472744

Bendt P, Barthel S, Colding J (2013) Civic greening and environmental learning in public-access community gardens in Berlin. Special Issue Urban Ecosyst Serv 109(1):18–30. https://doi.org/10.1016/j.landurbplan.2012.10.003

Brandner A, Schunko C (2022) Urban wild food foraging locations: understanding selection criteria to inform green space planning and management. Urban For Urban Green 73:127596. https://doi.org/10.1016/j.ufug.2022.127596

Charnley S, Mclain R, Poe M (2018) Natural resource access rights and wrongs: nontimber forest products gathering in urban environments. Soc Nat Resour 31:1–17. https://doi.org/10.1080/08941920.2017.1413696

Cocks M, Alexander J, Mogano L, Vetter S (2016) Ways of belonging: meanings of 'Nature' AMong Xhosa-speaking township residents in South Africa. J Ethnobiol 36(4):820–841. https://doi.org/10.2993/0278-0771-36.4.820

Colding J, Barthel S, Bendt P, Snep R, Van der Knaap W, Ernstson H (2013) Urban green commons: Insights on urban common property systems. Global Environmental Change 23(5):1039–1051

Davidson-Hunt I (2006) Adaptive learning networks: developing resource management knowledge through social learning forums. Hum Ecol 34:593–614. https://doi.org/10.1007/s10745-006-9009-1

Del Tredici P (2010) Spontaneous urban vegetation: reflections of change in a globalized world. Nat Cult 5(3):299–315. https://doi.org/10.3167/nc.2010.050305

Emery MR, Hurley PT (2016) Ethnobiology in the City: embracing the urban ecological moment. J Ethnobiol 36(4):807–819. https://doi.org/10.2993/0278-0771-36.4.807

Emery MR, Ginger C, Newman S, Giammusso MRB (2003) Special Forest products in context: gatherers and gathering in the eastern United States. U.S. Department of Agriculture, Forest Service, Northeastern Research Station. https://doi.org/10.2737/ne-gtr-306

Fischer LK, Kowarik I (2020) Connecting people to biodiversity in cities of tomorrow: is urban foraging a powerful tool? Ecol Indic 112:106087. https://doi.org/10.1016/j.ecolind.2020.106087

Gaither J, Cassandra AA, Madden M, Alford S, Wynn A, Emery M (2020) 'Black Folks Do Forage': examining wild food gathering in Southeast Atlanta Communities. Urban For Urban Green 56:126860. https://doi.org/10.1016/j.ufug.2020.126860

Garekae H, Shackleton CM (2020) Urban foraging of wild plants in two medium-sized south African towns: people, perceptions and practices. Urban For Urban Green 49:126581. https://doi.org/10.1016/j.ufug.2020.126581

Garekae H, Shackleton CM, Tsheboeng G (2022) The prevalence, composition and distribution of forageable plant species in different urban spaces in two medium-sized towns in South Africa. Global Ecol Conserv 33:e01972. https://doi.org/10.1016/j.gecco.2021.e01972

Ginger C, Emery MR, Baumflek MJ, Putnam DE (2012) Access to natural resources on private property: factors beyond right of entry. Soc Nat Resour 25(7):700–715. https://doi.org/10.1080/08941920.2011.633596

Grabbatin B, Hurley PT, Halfacre A (2011) 'I still have the old tradition': the co-production of Sweetgrass basketry and coastal development. Geoforum 42(6):638–649. https://doi.org/10.1016/j.geoforum.2011.06.007

Hurley PT, Halfacre AC, Levine NS, Burke MK (2008) Finding a 'Disappearing' nontimber forest resource: using grounded visualization to explore urbanization impacts on Sweetgrass Basketmaking in greater Mt. Pleasant, South Carolina. Prof Geogr 60(4):556–578. https://doi.org/10.1080/00330120802288941

Hurley PT, Grabbatin B, Goetcheus C, Halfacre A (2013) Gathering, buying, and growing Sweetgrass (Muhlenbergia Sericea): urbanization and social networking in the Sweetgrass Basket-making industry of Lowcountry South Carolina. In: Voeks R, Rashford J (eds) African ethnobotany in the Americas. Springer, New York, NY, pp 153–173. https://doi.org/10.1007/978-1-4614-0836-9_6

Hurley PT, Emery MR, McLain R, Poe M, Grabbatin B, Goetcheus CL (2015) Whose urban Forest? The political ecology of foraging urban nontimber forest products. In: Isenhour C, McDonogh G, Checker M (eds) Sustainability in the Global City: myth and practice, New directions in sustainability and society. Cambridge University Press, Cambridge, pp 187–212. https://doi.org/10.1017/CBO9781139923316.014

Hurley PT, Becker S, Emery MR, Detweiler J (2022) Estimating the alignment of tree species composition with foraging practice in Philadelphia's urban forest: toward a rapid assessment of provisioning services. Urban For Urban Green 68:127456. https://doi.org/10.1016/j.ufug.2021.127456

Kaoma H, Shackleton C (2015) The direct-use value of urban tree non-timber forest products to household income in poorer suburbs of south African towns. Forest Policy Econ 61:104–112. https://doi.org/10.1016/j.forpol.2015.08.005

Konijnendijk C (2008) The forest and the city—the cultural landscape of urban woodland. The forest and the city: the cultural landscape of urban woodland. https://doi.org/10.1007/978-1-4020-8371-6

Landor-Yamagata JL, Kowarik I, Fischer LK (2018) Urban foraging in Berlin: people, plants and practices within the metropolitan green infrastructure. Sustainability 10(6). https://doi.org/10.3390/su10061873

Marquina T, Emery M, Hurley P, Gould R (2022) The 'Quiet Hunt': the significance of mushroom foraging among Russian-speaking immigrants in New York City. Ecosyst People 18:226–240. https://doi.org/10.1080/26395916.2022.2055148

McLain RJ, Hurley PT, Emery MR, Poe MR (2014) Gathering 'Wild' food in the city: rethinking the role of foraging in urban ecosystem planning and management. Local Environ 19(2):220–240. https://doi.org/10.1080/13549839.2013.841659

McLain RJ, Poe MR, Urgenson LS, Blahna DJ, Buttolph LP (2017) Urban non-timber forest products stewardship practices among foragers in Seattle, Washington (USA). Urban For Urban Green 28:36–42. https://doi.org/10.1016/j.ufug.2017.10.005

PA DCNR (2016) DCNR invasive plants. http://wwwdcnrstatepaus/cs/groups/public/documents/document/dcnr_20026634pdf. 28 July 2017

Palliwoda J, Kowarik I, Lippe M (2017) Human-biodiversity interactions in urban parks: the species level matters. Landsc Urban Plan 157:394–406. https://doi.org/10.1016/j.landurbplan.2016.09.003

Poe MR, McLain RJ, Emery M, Hurley PT (2013) Urban forest justice and the rights to wild foods, medicines, and materials in the city. Hum Ecol 41(3):409–422. https://doi.org/10.1007/s10745-013-9572-1

Poe MR, LeCompte J, McLain R, Hurley P (2014) Urban foraging and the relational ecologies of belonging. Soc Cult Geogr 15(8):901–919. https://doi.org/10.1080/14649365.2014.908232

Rhoads A, Block T (2007) Natural areas inventory update. Submitted to the Montgomery County Planning Commission, June 30

Ricketts TH, Dinerstein E, Olson DM, Loucks CJ, Eichbaum W, DellaSala DA, Kavanagh K, Hedao P, Hurley P, Carney K (1999) Terrestrial ecoregions of North America: a conservation assessment, vol 1. Island Press

Sardeshpande M, Shackleton C (2023) Fruits of the city: the nature, nurture and future of urban foraging. People and Nature 5(1):213–227. https://doi.org/10.1002/pan3.10428

Schunko C, Brandner A (2022) Urban nature at the fingertips: investigating wild food foraging to enable nature interactions of urban dwellers. Ambio 51(5):1168–1178. https://doi.org/10.1007/s13280-021-01648-1

Shackleton S, Chinyimba A, Hebinck P, Shackleton C, Kaoma H (2015) Multiple benefits and values of trees in urban landscapes in two towns in northern South Africa. Landsc Urban Plan 136:76–86. https://doi.org/10.1016/j.landurbplan.2014.12.004

Shackleton CM, Hurley PT, Dahlberg AC, Emery MR, Nagendra H (2017) Urban foraging: a ubiquitous human practice overlooked by urban planners, policy, and research. Sustainability 9(10). https://doi.org/10.3390/su9101884

Svendsen ES, Campbell LK, McMillen HL (2016) Stories, shrines, and symbols: recognizing psycho-social-spiritual benefits of urban parks and natural areas. J Ethnobiol 36(4):881–907. https://doi.org/10.2993/0278-0771-36.4.881

Synk CM, Kim BF, Davis CA, Harding J, Rogers V, Hurley PT, Emery MR, Nachman KE (2017) Gathering Baltimore's bounty: characterizing behaviors, motivations, and barriers of foragers in an urban ecosystem. Urban For Urban Green 28:97–102. https://doi.org/10.1016/j.ufug.2017.10.007

U.S. Census Bureau (2023) QuickFacts: Philadelphia County, Pennsylvania. https://www.census.gov/quickfacts/philadelphiacountypennsylvania

U.S. Census Reporter (2021) Philadelphia-Reading-Camden. https://censusreporter.org/profiles/33000US428-philadelphia-reading-camden-pa-nj-de-md-csa/

Unnikrishnan H, Nagendra H (2015) Privatizing the commons: impact on ecosystem Services in Bangalore's lakes. Urban Ecosyst 18(2):613–632. https://doi.org/10.1007/s11252-014-0401-0

USDA (2016). http://plants.usda.gov/java/nameSearch

Wehi P, Wehi W (2009) Traditional plant harvesting in contemporary fragmented and urban landscapes. Conserv Biol 24:594–604. https://doi.org/10.1111/j.1523-1739.2009.01376.x

Wild Foodies (2016). http://www.meetup.com/Wild-Foodies-of-Philly

Patrick T. Hurley (PhD) is Professor and Chair of Environmental Studies at Ursinus College. When not researching urban foraging and its relation to urban forests in the USA, Dr. Hurley works with students to steward the College's food forest.

Kristin McGillis is an alumnus of the Ursinus Environmental Studies department and now works as a quantitative analyst in Philadelphia, Pennsylvania.

Marla R. Emery (PhD) is a Scientific Advisor with the Norwegian Institute for Nature Research (NINA). Dr. Emery is a globally recognized scholar on indigenous and local knowledge and served as the Co-Chair of the Assessment of Sustainable Use of Wild Species, a report of the Intergovernmental Science-Policy Platform on Biodiversity and Ecosystem Services (IPBES).

Brian J. Leffler (PhD) is an ethnobiologist who centers his work around Indigenous peoples, protected areas, and traditional ecological knowledge. Dr. Lefler currently serves as the Cultural Anthropologist and Tribal Partnerships Specialist for the Lewis and Clark National Historic Trail, administered by the National Park Service.

Urban Foraging for Social-Ecological Resilience in the Global South

13

Dhanya Bhaskar, Sharmistha Rej,
and Samudyatha Ramananda

Abstract

Urban foraging, a traditional human–nature interaction, has been fairly well documented globally. However, its relevance is being realized in terms of livelihood support, culture, healthy lifestyle, closeness to nature, and recreation. In this chapter, we present a review of urban foraging practices—both traditional and emerging ones in cities of the Global South with a view to highlight the diversity of foraging practices in various social-ecological contexts and the host of factors that enable or hinder urban foraging. Further, we use a case study of changing foraging practices around an urban protected forest in Bengaluru, India's fastest-growing metropolis to bring out the impacts of urbanization on wild food foraging.

The review illustrates that foragers use a variety of indigenous species with high economic potential and cultural preference. Generally, urban foragers belong to lower income groups, and foraging is important for them in securing livelihood and nutritional security. The chapter also documents how urban foraging has been recently gaining traction as a new-age sustainable dietary practice, preferred by urban elites. The case study on Bannerghatta National Park in Bengaluru revealed the declining importance of foraging from urban forests in the context of rapidly changing urban socioeconomic milieu and cultural preferences. The study further highlights the role of urban planning and policies in reviving and sustaining this low ecological footprint consumption practice for synergizing various sustainable development goals.

D. Bhaskar (✉) · S. Rej · S. Ramananda
Indian Institute of Forest Management, Bhopal, India
e-mail: dhanyab@iifmbhopal.edu.in

S. Dhyani, M. Sardeshpande (eds.), *Urban Foraging in the Changing World*,
https://doi.org/10.1007/978-981-97-0345-6_13

Keywords

Urbanization · Wild food · Nutrition · Diet · Policies · Traditional knowledge · Protected area · Provisioning service

13.1 Introduction

Countries of the Global South, viz. India, China, and Nigeria will account for more than a third of the projected growth in the world's urban population by 2050 (UN-Habitat 2020). Fast-growing cities in the Global South are marked by unplanned urban sprawl, high social inequality, environmental pollution, and degradation (Arku and Marais 2021). Urban foraging and use of wild edible species are being increasingly recognized as important adaptation strategies for supporting food and livelihood security of marginalized communities (Somesh et al. 2021) as well as cultural and recreational means for enhancing human well-being (Shackleton et al. 2017a) in these cities.

Urban foraging is the practice of identifying and collecting wild food such as flora, fungi, and other natural materials from formal and informal urban green spaces (Shackleton et al. 2017a). Gatherers gain significant benefits from foraged materials for livelihood uses (Triman 2013). Foraging can facilitate the easy availability of micronutrients (Ray et al. 2020), contributing to the nutritional security of foragers (Gopal and Nagendra 2014). It might also enhance the cultural adequacy of food systems, community participation (Bunge et al. 2019), and human–nature interactions (Fischer and Kowarik 2020) and could promote the integration of wild foods in urban green spaces to develop "edible green spaces" (Russo et al. 2017). Foraging practices might enable consumers to start engaging in larger discussions of land use, food justice, and the effects of climate change (Hurley et al. 2015). Promoting urban foraging could be a potential strategy for synergistically achieving SDGs—especially the goals related to poverty, hunger, gender equality, good health and well-being, sustainable cities, responsible production and consumption, climate action, and biodiversity in a decentralized, cost-effective, and equitable manner (Fig. 13.1).

Urban foraging could include gathered biological resources which are not planted or cultivated within urban and peri-urban settings without economic transaction (Bunge et al. 2019). Studies indicate that while foragers generally do not follow any specific criteria for selecting foraging locations, public urban green spaces with high accessibility, low contamination, and infrequent visitors are preferred (Branders and Schunko 2022). The species used for urban foraging and local practices are in constant flux as urban ecological and social settings change. This necessitates that the urban foragers have a broad understanding of various species, harvest places, and rights of access, as well as the ability to adapt their activities to changing circumstances (Shackleton et al. 2017a). Although the trend of wild food consumption is declining in many cities, wild edibles with cultural values and the recreational function of gathering remain popular (Reyes-García et al. 2015). Foraging could be

Fig. 13.1 Potential role of urban foraging in achieving SDGs (Image credit: Sarang KT)

a potential coping strategy for short-term relief, mid-term recovery, and long-term resilience against disruptive changes such as the COVID-19 pandemic and gradual ones such as climate change and other environmental vulnerabilities (Sardeshpande et al. 2021).

Urban foraging has the potential for enhancing citizen stewardship by fostering informal green space co-management and use (Sardeshpande and Shackleton 2020a). In developed countries, policies have been framed around urban green spaces and foraging with the specific aim of promoting public edible landscapes (McLain et al. 2012). "Forestry Management Plan" for Seattle (Hurley and Emery 2018), the "edible city" of Andernach, Germany (Kosack 2016), and food forests and edible parks of Boston (Spiller and Kashwan 2023), etc. are noteworthy examples of this regard. On the contrary, policies in Global South often don't recognize or even hinder urban foraging, indicating the need for better appreciation of the multiple social-ecological benefits of this practice.

In this chapter, we review urban foraging practices—both traditional and emerging ones in cities of the Global South, to understand their patterns, current status, and the various challenges faced by foragers. We further use a case study of an urban protected forest in Bengaluru, India's fastest-growing metropolis to illustrate changing preferences for foraging in response to urbanization.

13.2 Review of Urban Foraging in the Global South

13.2.1 Methodology

A comprehensive search of peer-reviewed literature was carried out during March and April 2023, using keywords such as urban foraging, foraging practices, wild food, wild edibles, wild fruits, food forests, green spaces, etc. Databases including Scopus, Science Direct, Google Scholar, PubMed, and Research Gate were searched. From the articles that appeared in the search, relevant literature from the cities of Asia, Africa, and Latin America was filtered out for the review. The reference section of each article was further explored to find additional articles. We also searched Indian newspapers including The Hindu, and the Times of India, and magazines such as Down to Earth to garner reports on the new and emerging practices of urban foraging in the country. Reports such as Mexico Hungry Cities Report, World Cities Report (UN-Habitat), Bangalore Development Authority Master Plan 2031 (Draft), etc. were examined to collect pertinent information. Authors' personal observations on cultural practices in which wild edibles are used also supported this review.

Papers that were finally filtered in for the review were thoroughly examined to identify the major recurring themes and categories. Data from each paper were tabulated under the themes and simple statistics including percentages were computed to represent the frequency of the themes.

13.2.2 Results and Discussion

Our literature search yielded 66 papers, out of which we selected 33 papers that discuss foraging in urban and peri-urban contexts in the Global South countries. Out of these, six were from Asia, 20 were from the African continent, six were from Latin America, and one covered both Latin American and African cities. The papers belonged to the period 2011–2023. Very few research papers on urban foraging could be found from Latin American and Asian contexts, except the Indian city of Bengaluru, while African countries are the most studied in these aspects. From these 33 papers, we further screened in 28 papers (six from Asia, seventeen from the African continent, four from Latin America, and one on both Latin American and African cities) that contained enough details for systematic analysis of information. These papers described themes including plant species foraged, patterns of foraging, the socioeconomic profile of foragers, and the changing status of foraging (Table 13.1) as detailed in Sects. 13.2.2.1 to 13.2.2.4 below. Data from the rest of the papers were used to supplement the results of the systematic review.

13.2.2.1 Urban Foraging: What and Why?

A variety of species ranges from completely wild to fully domesticated by foragers (Pisto and Lebese 2014). In the papers reviewed, the number of foraged species ranges from 35 to 448, mostly plants and fungi, and fish from lakes. In addition to these, foragers also collected produce such as honey—both for consumption and

Table 13.1 Characteristics of foraging and foragers in different urban areas of the Global South (blanks indicate no data available on the specific feature in the paper)

Continent	Country	City/town	Material foraged	Use
Africa	Ethiopia	Berehet	Vegetables, fruit, gum, roots, leaves, and flower nectar	Food, medicine
	South Africa	Potchefstroom and Thabazimbi	Wild vegetables & fruits mostly, fruits used as snacks	Food
	South Africa	Potchefstroom and Thabazimbi	Wild vegetables and fruits, branches	Food, fuelwood
	South Africa	Queenstown, Phalabrowa, Grahamstown, Fort Beaufort, Bela-Bela, Bathurst, Zeerust, Tzaneen	Edible insect,mushroom, leafy vegetables, bushmeat	Food, medicinal
	South Africa	Cape Town	Flower, fruit, leaf, bark, rhizome, bulb, stem, tuber, animal	Food, medicine, fiber, cut flower trade, fuelwood
	South Africa	Mtubatuba, Kwambonambi, Richards Bay, Empangeni, Eshowe, Stanger, Ballito, Durban, Scottburgh, Park Rynie, Port Shepstone, Port St Johns	Fruits, bark, vegetables, leaves	Food, medicine
	South Africa	Mtubatuba, Carnarvon,Witrivier, Jan Kempdorp, Lutzville, Port Edward, Grahamstown, Villiersdorp	Wild edible fruits	Food
	South Africa	Durban Mtubatuba, Empangeni Stanger, Ballito	Fruit	Food, medicine
	South Africa	Limpopo and North West provinces	Edible Fruits	Food
	South Africa	Richards Bay, Dundee, and Harrismith	Fruits, vegetables, edible insects, bushmeat, fish, mushrooms	Food, medicine
	South Africa	Batswana	Fruit, flower, bark, gum, bulb, edible oil are use for soap making	Food, beverage (tea, beer), medicine
	Saharan Africa	Tiassale, Bamenda, Moshi, Palapye, Phalaborwa, Queenstown	Vegetables, Fruits, Honey, wood	Food, medicinal, Fuelwood
	Uganda	Kampala	Fruits, leafy vegetables, condiments	Food, Medicine, and Combined
	Zimbabwe	Masvingo	Fruits, vegetables, tuber, bark, root, honey, branches, termites, mice, rabbits, caterpillars, birds	Food, medicine, fuelwood
	Rwanda	Kigali	Only fruit	Food
	Nigeria	Lagos	Fruit, vegetables, flower, seed, root, tuber, bark, branches, fishes	Food, medicine, decoration, fuelwood

(continued)

Table 13.1 (continued)

Continent	Country	City/town	Material foraged	Use
	Democratic Republic of Congo	Municipalities (Makiso, Tshopo, Lubunga, Kisangani, Mangobo, Kabondo) of Kisangani Capital, Oriental Province	Food (vegetables, nuts, pulses, roots, and tubers, mushroom), fish, insect, caterpillar	Food
South Africa & Latin America	–	–	Fruits, vegetables, leaves, mushrooms, insects, beverage materials	Foods, beverage, medicine, fodders
Latin America	Brazil	Recife	Fruits, vegetables, leaves	Food
	Brazil	Lepa City & Panc na city	Fruits, edible fowers, leaves, vegetables	Food
	Brazil	São Paulo	–	Food
	Mexico	Roma-Condesa, Mexico City	Fruits, flowers, leaves, beans, stems, tuber	Food, drink, medicine
Asia	India	Bengaluru	Fruits, oil, coirs, fibers, flowers, leaves, Whole plants	Food, shades, medicinal, sacred purposes, cooking fuel
	India	Bengaluru	Fruits, flowers, vegetables, wood, timbers, roots, shoots	Food, fodder, medicine, fuel wood, furniture (temple renovation, door and window for houses)
	India	Bengaluru	Leaves, fruits, flowers, stem, roots, whole plant parts	Food, Medicine, Cultural purpose
	China	Dulong Jiang, Yunnan	Wild fruits, spices, nuts, tea leaves, liquor-making materials, oils	Food
	China	Shaanxi (Zhen town) in Dali valley	Mushrooms, fruits, vegetables,	Food, medicine
	Vietnam	Cham Island	Vegetables, leaves, top of the stem, flowers, fruits	Food

No. of species foraged	Major species	Details of foragers				
		Age	Gender	Education/employment	Socio-economic status/category	Location
146	*Ziziphus spina christi, Balanites aegyptiaca, Rosa abyssinica, Carissa spinarium, Ficus sur, Asparagus africanus, Carissa spinarium, Cordia africana and Ximenia americana*	More than 30 years	Males forage more as compared to females	–	Low-income people	–
–	–	22 to 81 years with an average of 45.5 (middle-aged)	Generally males	Nearly half of the foragers are unemployed, and the majority of the people having secondary education	Low-income households	Most people grew up in rural settings. Residents in the outskirts of town were more likely to forage than those in the core/ inner part of town
–	–	Middle-aged	Generally males	Unemployed with the intermediated level of education	Poor people	Local people who grew up in rural areas,91.8% of people foraged during their childhood
–	–	–	–	–	–	Urban Dwellers
250 plants species 198 animals	*Carpobrotus spp., Poaceae spp., Proteaceae spp., Fabaceae spp., Ocotea bullata*	–	–	Low education level. small-scale, informal, commercial operators who fulfill fluctuating demands for wild-harvested products. Ethnic origin population(Rastafarians) and longstanding residents with strong traditional links to Cape biodiversity	Low-income	Urban low-income residential area

(continued)

Table 13.1 (continued)

No. of species foraged	Major species	Details of foragers				
		Age	Gender	Education/employment	Socio-economic status/category	Location
–	*Trichelia dredgeana, Harpephyllum caffrum, Diospyros whyteana, Ficus sur*	–	–	–	–	–
51	*Opuntia spp., Rubus spp., Citrullus lanatus, Physalis peruviana, Harpephyllum caffrum, Carissa bispinosa, Syzygium cumini*	Slightly older people of average age 44-45 years	–	Unskilled and unemployed	Low-income households	Urban and rural (Rural households use more wild foods and NTFPs than urban households)
51	*Syzygium cordatum, Tabernaemontana elegans, Phoenix reclinata, vernacular isundu*	–	–	Unskilled and unemployed	–	Predominantly urban
–	*Mangifera indica, Sclerocarya birrea subsp. Caffra*	–	–	Employed and unemployed	Urban poor and migrants	Informal areas, low-cost housing areas (under Reconstruction and Development Programme), townships
–	–	15–49 years	Female	–	Generally, below the food poverty line	Living in rural, peri-urban, urban
–	*Acacia karoo, Celtis africana, Pappea capensis, Rhus pyroides, Ziziphus mucronata, Rhus lancea, Acacia erioloba, Acacia tortilis, Mimusopszeyheri, Grewia flava, Ensete ventrico-sum, Diospyros lycioides, Dovyalis caffra, Grewia occidentalis, Citrillus lanatus, Diospyros lycioides, Vangueria infausta, Mimusops zeyheri, Celtis africana, Kedrostis hirtella, Chenopodium album, Amaranthus, Carissa edulis, Boscia albitrunca*	–	Generally females, few males	–	Predominantly poor, Local communities and migrants	–

–	–	–	–	–	–	People from urban, peri-urban, and rural areas
–	*Momordica foetida, Vernonia amygdalina, Hoslundia opposita, Bidens pilosa, Aloe vera, Cyanthillium cinereum, Amaranthus dubius, Artocarpus heterophylus, Psidium guajava, Mangifera indica, and Persea Americana*	Mostly young people (Less than 51 years)	–	–	Low-income households	Predominantly living in peri-urban
–	*Azanza garkeana, Strychnos*	–	–	–	Low-income residents	Peri-urban area
–	*Psidium guajava, Persea americana, Musa spp.*	–	–	–	Wealthy and poor both	People living in informal and scattered settlements
35	*Magnifera Indica, Carica papaya, Terminalia catappa, Vernonia amygdalina, Spinacia oleracea, Ocimum gratissimum, Talinum fruticosum, Azadirachta indica, Citrus sinensis, Psidium guajava, Telfairia occidentalis, Musa acuminata, Corchorus olitorius, Cymbopogon citratus, Solanum lycopersicum, Moringa oleifera, Musa paradisiaca, Colocasia esculenta, Lawsonia inermis, Citrus aurantiifolia, Saccharum officinarum, Aloe barbadensis miller, Manihot esculenta, Malus domestica, Elaeis guineensis, Iresine herbstii, Morinda lucida, Anacardium occidentale, Cocos nucifera*	45 to 65	Generally, females	Uneducated, unemployed, and retried	A high-income group who settled in Lagos for last 4years	People living in the core of the city
–	*Manihot esculenta, Amaranthus, Arachis hypogaea, Musa spp., Persea americana, Allium fistulosum, Solanum melongena*	Generally more than 30–35	Women	–	–	Urban Areas

(continued)

Table 13.1 (continued)

No. of species foraged	Major species	Details of foragers				
		Age	Gender	Education/employment	Socio-economic status/category	Location
35	*Adansonia digitate, Berchemia discolor, Diospyros mespiliformis, Sclerocarya birrea , Physalis peruviana, Dipteryx alata, Mauritia flexulosa, Lophira lanceolata*	–	–	–	–	–
93	*Syzygium malaccense, Eugenia uniflora, Mangifera indica, Handroanthus impetiginosus*	Middle-aged people	Generally, males	Gardeners are more involved in the foraging practices	High-income people	Urban areas
–	*Ocimum gratissimum, Ocimum carnosum, Alternanthera brasiliana, Cymbopogon citratus, Pelargonium graveolens, Cnidoscolus aconitifolius, Moringa oleifera, Cinnamomum verum, Laurus nobilis, Murraya koenigii, Sambucus nigra, Malvaviscus arboreus, Pereskia aculeata, Costus spicatus, Plectranthus grandis, Plectranthus ornatus,Maranta arundinacea,Curcuma longa, Ipomoea batatas,Musa sp.*	–	–	Experts and educated people	Both wealthy and poor people	Urbanites and outsiders
–	*Amaranthus spp. (deflexus and viridis), Plantago tomentosa, and Taraxacum officinale*	–	–	–	Some groups of city	Urban

36	*Musa spp, Coffea arabica, Ficus carica, Psidium guajava, Citrus reticulata, Citrus aurantifolia, Eriobotrya japonica, Citrus aurantium, Carica papaya, Prunus persica, Punica granatum, Agave spp, Pimpinella anisum, Ocimum spp., Tagetes spp., Erythrina americana, Dysphania ambrosioides, Foeniculum vulgare, Piper auritum, Lavandula dentata, Oreganum marjorana Mentha spp., Rosmarinus officinalis, Salvia officinalis, Thymus vulgaris, Persea americana, Capsicum annum, Apium graveolens, Sechium edule, Capsicum annuum, Cucurbita pepo, Phaseolus vulgaris, Zeamays, Solanum tuberosum, Opuntia spp., Solanum lycopersicum*	–	–	–	Poor, landless people, Business groups, and affluent groups who live and work in that area	–
46 tree sp. & 95 sp. of herbs, shrubs, creeper	*Moringa oleifera, Cocos nucifera, Azadirachta indica, Ficus religiosa, Muntingia calabura, Muntingia calabura, Mangifera indica, Psidium guajava*	–	Females	–	Low-income people	People living in informal settlements
–	*Mangifera indica,Syzygium cumini,Madhuca longifolia, Artocarpus heterophyllus,Tamarindus indica ,Ficus racemosa, Ficus mysorensis, Ficus benghalensis*	–	–	–	Poor people, Local communities, and migrants	

(continued)

Table 13.1 (continued)

No. of species foraged	Major species	Details of foragers				
		Age	Gender	Education/employment	Socio-economic status/category	Location
76	*Acalypha indica, Alternanthera pungens, Alternanthera sessilis, Alternanthera trigyna, Amaranthus spinosus, Amaranthus tricolor, Amaranthus viridis, Andrographis paniculata, Boerhavia diffusa,Celosia argentea, Centella asiatica, Chenopodium giganteum, Cleome monophylla, Commelina benghalensis, Cyanotis cristata, Digera muricata, Eclipta prostrata, Euphorbia heterophylla, Euphorbia hirta, Glossocardia bosvallia, Leucas aspera, Lycopersicon esculentum, Mimosa pudica, Ocimum tenuiflorum, Oxalis corniculata, Oxalis latifolia, Phyllanthus amarus, Phyllanthus fraternus, Plectranthus amboinicus, Portulaca oleracea, Senna tora, Solanum americanum, Solanum viarum, Sonchus oleraceus, Trianthema portulacastrum, Tribulus terrestris , Argemone mexicana, Bambusa bambos , Barleria mysorensis, Calotropis gigantea, Chenopodium giganteum, Dodonaea viscosa, Helicteres isora, Lantana camara, Ocimum basilicum, Senna occidentalis, Sida acuta, Solanum erianthum, Solanum rudepannum, Basella alba, Coccinia grandis, Cardiospermum halicacabum, Cocculus hirsutus, Cucurbita maxima, Dioscorea esculenta, Ipomoea batatas, Ipomoea staphylina, Lagenaria siceraria, Luffa spp., omordica charantia, Tylophora indica, Passiflora foetida, Ziziphus oenopolia, Acacia caesia, Langium salviifolium, Azadirachta indica, Euphorbia tirucalli, Imonia acidissima , Moringa oleifera, Phyllanthus acidus , Phyllanthus emblica, Pithecellobium dulce, Psidium guajava, Senna surattensis, Sesbania grandiflora, Tamarindus indica*	–	Generally females	–	Socially and economically deprived groups	Higher percentage of foraging in the peri urban areas or in the least urbanized sites

148	*Maianthemum atropurpureum, Cardamine tangutorum, Aralia elata,* Cornus *capitate, Saurauia napaulensis, Angiopteris esculenta, Cardiocrinum giganteum, Dioscorea polystachya, Caryota obtus, Gnetum montanum,*	People are mostly young (most of the people were in the age group of 20 to 39)	Males & females	Farmers, wage workers, traders with low education levels, and youth with higher education level	Low-income groups	–
113 plant species and 12 fungi	*Pimpinella spp., Pueraria lobata, Pteridium aquilinum, Polygonatum spp., Sinacalia tangutica.* Mushroom like *Cantherellus cibarius, Agaricales spp., Ramaria spp., Grifola umbellata, Abelia engleriana* treated by locals as one folk taxon and *Grifola umbellata*	All age categories from children to grown-ups	Both males and females	Un-employed	–	–
43	*Strophioblachia fimbricalyx, Zanthoxylum nitidum, Glochidion zeylanicum, Pierre cambodiana, Cratoxylum maingayi, Barringtonia macrostachya, Diplazium esculentum, Nephrolepis falcata, Pasiflora foetida*	–	–	–	Local people, tourists	

(continued)

Table 13.1 (continued)

Source of collection	Pattern of foraging		Current status	Reasons for current status		Reference
	Season	Frequency		Enabling factors	Constraining factors	
Source of collection	Pattern of foraging		Current status	Reasons for current status		Reference
	Season	Frequency		Enabling factors	Constraining factors	
Fallow land and forest patches	–	–	Declining	–	Agriculture expansion	Alemayehu et al. (2015)
Domestic garden, vacant space, riparian zone, peri-urban farms	–	–	–	–	–	Garekae and Shackleton (2020a)
Vacant spaces, riparian zones, domestic gardens, parks, Institutional spaces, cemeteries, farms, and dumping sites	Fruits and vegetables were foraged more during the rainy season while for firewood, the peak was during the dry and cold winter season	Few times per week or per month for most resources, except medicinal plants that are sourced only a few times per year	–	Culture, health, livelihood needs, and leisure motivate people for foraging	Lack of time, inadequate foraging spaces, and limited knowledge of practices	Garekae and Shackleton (2020b)
Undeveloped municipal land in peri-urban area, private homesteads	–	Species declining	–	Inappropriately harvesting, killed or reproductively harmed through the harvesting processes and incur bark stripping damage for medicinal use	–	Shackleton et al. (2017b)
Conservation area	–	–	Increasing	Indiscriminate extraction to meet the demand	–	Petersen et al. (2012)

Green spaces	–	–	Declining	–	Lack of information about wild indigenous edible species and their sustainable harvest is the main barrier to planting of foraging-friendly species. Illegal dumping is the largest challenge in open space management, also other illegal activities such as drug use, sand mining, theft, and vandalism were also cited in open spaces. lacking suitable policy, also normative views of pristine nature may hinder foraging	Sardeshpande and Shackleton (2020a)
Gardens, yard, Wilderness area, vacant plots, roadside	–	Wild edibles foraged throughout the year	–	–	–	Sardeshpande and Shackleton (2020b)
Wild forested areas, on the streets in their neighborhoods, vacant lots, abandoned farmland, rivers, valleys, the beach, and hills	–	Most respondents forage 1–3 days a week and few forages daily in week and some only in fruiting season (3–30 days annually)	Declining	–	loss of urban green infrastructure by road and housing development and restrictions by land managers	Sardeshpande and Shackleton (2023)

(continued)

Table 13.1 (continued)

Source of collection	Pattern of foraging		Current status	Reasons for current status		Reference
	Season	Frequency		Enabling factors	Constraining factors	
Homesteads, municipal land or underdeveloped private land, and other informal and formal places in urban areas and peripheries of the town	–	–	–	Cash Income for households	–	Kaoma and Shackleton (2015)
Forest, homestead, proximity of area of peri urban and rural area	Decrease in self-collection with a decrease in agro-ecological gradient	Throughout the year	–	–	–	Chakona and Shackleton (2019)
Cultivated land	Summer season, especially during summer rains	–	Declining	–	Cultivation and convenient processing of exotic and cultivated plants, loss of traditional knowledge about wild edibles	Pisto and Lebese (2014)
Pockets of nature in the peri-urban area	–	–	–	Cash or non-cash Income for households	–	Schlesinger et al. (2015)

Public spaces (e.g., roadsides and wetlands) in peri-urban areas and private spaces in urban areas (e.g., neighbor's gardens and vacant plots)	Wild plant collection is lower during the rainy season and higher during the dry season. Seasonal variation is lower for medicinal plant collection than for food plant collection	Mostly wild plants were collected less than once per week during the reported months of collection. However, 8% of the food plants were collected on daily basis	–	Livelihood Needs and traditions of native people	–	Mollee et al. (2017)
–	Dry (mice, termites and rabbit) wet (fruit, medicine, fishing) and throughout the year(hunting birds, animals)	–	Declining	–	Disappearing forest, overexploitation of forest resources	Murwendo (2011)
Cultivated forests, Urban woodlots, Street trees, Institutional gardens, Domestic gardens	–	–	Declining	–	Informal house construction, concretization without planning	Seburanga et al. (2014)
Domestic gardens, streets, unused lands, institutional grounds, lakes, riparian fringes	–	Food is gathered a couple of times a week and medicine a few times a month. Fish and water leaves were collected daily basis	Declining	–	Unavailability of foraging sites and species	Adeyemi and Shackleton (2022)

(continued)

Table 13.1 (continued)

Source of collection	Pattern of foraging		Current status	Reasons for current status		Reference
	Season	Frequency		Enabling factors	Constraining factors	
Forest	Wild edible plants are more foraged during July to September due to availability. July to Oct hunting catterpillars	–	Declining	–	Less knowledge about the nutritional value of wild food, rarely consumed due to the distance to be walked to collect certain species in primary forest or the workload to collect and prepare the wild edible plant, high workload involved in cracking the nuts	Termote et al. (2012)
Forests, agroforestry and urban landscapes, deserts, fallows, natural lands, and plantations	–	–	–	–	Lack of motivation	Sardeshpande and Shackleton (2019)
Squares & parks, sidewalks, forests, empty lots, private lot edges, road edges, beaches	–	–	Declining	–	Concern about pollution, lack of motivation and legislation for biodiversity conservation	Guenat et al. (2023)
Community gardens, streets, kitchen gardens	Summer rains were beneficial to kitchen garden	–	–	Foraging popularized by foraging tours and cultivation of wild edibles in neighborhood waste lands	–	Rigo (2021)

Freeways, local roads, Municipalities parks	–	–	–	Consumption	Different gradients of elemental concentrations were obtained according to the traffic categories. Freeways presented higher concentrations of several elements than local roads or parks. For the WEPs collected along freeways and some plants along arterial roads, the concentrations of Pb exceeded safety levels for human consumption. Foraging in large urban centers should be performed preferentially in low-traffic areas	Amato-Lourenco et al. (2020)
Unmaintained concrete raised bed, pavement bed, wasteland	–	–	Improving	Policies implemented by the city government for the resilience of a city		Hare and Isla (2021)
Potted plant, sack gardening, scattered trees	–	Use all species for a different purpose throughout the year	Declining	–	Lack of free spaces	Gopal and Nagendra (2014)
Lake and wooded groves	–	–	Declining	–	Change in status, management, use & perception of thopes	Mundoli et al. (2017)

(continued)

Table 13.1 (continued)

Source of collection	Pattern of foraging		Current status	Reasons for current status		Reference
	Season	Frequency		Enabling factors	Constraining factors	
Unused land, Parks, farmlands	Most of the species were collected from June to September (monsoon) and then October to December (post-monsoon & winter)	–	Declining	–	Urbanization	Somesh et al. (2021)
Barren lands, agricultural land, home gardens	Throughout the year		Declining	–	Loss of traditional knowledge and changes in land use	Cheng et al. (2022)
Forest	–	Wild fruits and fungi are occasionally gathered while wild vegetables have high frequency to use	Improving (wild vegetables) and Declining(mushroom)	High commodification of wild vegetables, cultural preferences and knowledge sharing	Foraging for mushrooms is declining because cultivated mushrooms are easily available	Kang et al. (2013)
Forest fringes, roadside, wetlands, paddy field, open spaces, etc.	Wild edibles vegetables are abundant in winter while in the summer season main customers are restaurant and homestay and tourists	Several times in years (average 3 times in a week)	Improving	Cultivation of wild species	–	Thoa et al. (2013)

sale (Murwendo 2011) and edible insects such as termites, caterpillars, and meat (Termote et al. 2012) and flavoring material such as *Rhus pyroides, Acacia erioloba* for beverages (Pisto and Lebese 2014). Edible fruit-bearing species such as *Mangifera indica, Carica papaya, Terminalia catappa, Psidium guajava, Persea Americana,* and *Sclerocarya birrea subsp. Caffra,* and leafy vegetables are found to be commonly used in most foraging studies. The foraged material is primarily used for edible purposes, followed by medicinal uses. Fuelwood and decorative uses are also mentioned in some cases.

Studies in urban areas of Africa reveal wild edible herbs as important components of the diets of poor communities (Shackleton et al. 2013). In South Africa, 96% and 79% of the households surveyed were found to consume wild vegetables and fruits, respectively (Garekae and Shackleton 2020a), and up to 73% of wild edible species documented in various regions of Ethiopia were used as supplementary foods (Alemayehu et al. 2015). Twenty-two out of the 70 woody species and nearly 50% of all tree individuals documented in Accra, Ghana have potential to provide food (fruits, seeds, nuts, leaves, oils, etc.) and more than 70% of the food-tree species could supply critical nutrients such as Fe, Zn, Mo, I, and Ca (Nero et al. 2018).

In the Dali Valley of Shaanxi City, central China, though 12 fungi and 113 wild plants with high cultural preference were listed in folk taxonomy, there was no interest in collecting wild mushrooms as edibles (Kang et al. 2013). Local food systems integrating wild edibles have emerged as an important option to combat urban food and nutritional deficiency in Latin America (Capron et al. 2017) where 81% of the total population lives in urban areas (World Population Prospects 2022). Wild edibles and unconventional food plants including medicinal and aromatic species foraged from greenspaces were consumed as nutritional substitutes by all age groups and strengthened human–nature relationships in City Lapa in Brazil during the COVID pandemic (Rigo 2021). In Uchoa municipality of Sao Paulo in Brazil, yield potential models estimated that the edible street tree species have the capacity to cater to 63.7% of the total annual caloric deficit in city dwellers and meet the minimum daily fruit intake requirements for 3% of its total population (Brito and Borelli 2020).

The review shows the importance of traditional and indigenous knowledge and cultural practices in sustaining urban foraging. In India, both in rural and urban settings, wild plants are consumed during traditional festivals, for instance-—in Bengal on the day before Diwali called "*14 saag parbo*" people consume 14 types of locally available leaves that are of cultural importance. Those are *Marsilea quadrifolia, Clerodendrum infortunatum, Tinospora cordifolia, Cordia dichotoma, Trichosanthes dioica, Enhydra fluctuans, Alternanthera sessilis, Sesbania sesban, Azadirachta indica, Cassia sophera, Brassica campestris, Chenopodium album, Brassica oleracea, and Costus speciosus.* These leaves are easy to digest, vitamin-rich, and provide health benefits such as boosting immunity during autumn. During the "*Ganesh Chaturthi*" festival in Maharashtra, women collect leaves from home gardens or surrounding areas including riverbanks and roadsides in the ritual of *21 patri*. At present, some of these plants are *Curcuma aromatica, Argyreia cuneata, Pandanus* especially *Pandanus odoratissimus* and *Pandanus tectorius* have become rare in urban areas (Banerjee 2014) and are sourced from peri-urban spaces. In the spring festival of Bihu in Assam, 10 dishes are prepared with leafy vegetables mostly collected from the wild or from local markets (pers. obs). In Dulong Jiang township in southwest China's Yunnan province, traditional knowledge about local flora has supported the nutritional adequacy of local diets, and around 148 species of wild edible plants are consumed and protected by associated traditional knowledge (Cheng et al. 2022).

13.2.2.2 Urban Foraging: Who, When, and Where?

In the studies reviewed, foraging rates varied from 16% of the people surveyed in India (Somesh et al. 2021) to 33% in Brazil (Guenat et al. 2023) to more than 50% in Africa (Mollee et al. 2017). All age-groups—from children foraging for fruits to young males practicing gardening to elderly people have been found foraging, though middle-aged (average 40 ± 11 years) people emerged as the most common category. Both men and women practice foraging in different urban contexts—studies reviewed from Latin America reported men as the main foragers, in Asian countries like India female foragers are more active while in Africa foraging by both men and women are reported in almost equal numbers. In the studies reviewed, we also found that foragers are mostly unemployed (Garekae and Shackleton 2020a; Adeyemi and Shackleton 2022). Foragers belong to low-income groups in 80% of the reviewed papers. Wild edibles contribute not only to nutritional security but also to the household income of the socio-economically deprived groups, for instance, in small towns in the Limpopo and North West provinces of South Africa, fruits were found to contribute to 1.3% to 4.3% of the total income of poor households (Kaoma and Shackleton 2015). In Accra, the diversity of food-bearing tree species was higher in the poorer neighborhoods compared to the wealthier neighborhoods (Nero et al. 2018). Foraging by higher income groups and both by higher and lower income groups were reported in 10% of the studies each.

Spatial locations of foragers vary from peri-urban to urban, often in informal settlements (Seburanga et al. 2014; Gopal and Nagendra 2014); however, people who grew up in rural settings tend to prefer foraging even in urban locations (Garekae and Shackleton 2020a). Seventy percent of the papers that provided data about locations of foragers and foraging sites mentioned peri-urban and rural sites as more active concerning foraging as compared to core urban areas. Schlesinger et al. (2015) in a study of 6 medium-sized cities from 5 African countries demonstrated a clear increase in the percentage of households engaged in foraging at increasing distance from the city center; 50% of households interviewed in the city foraged, while the proportion was 59% in the peri-urban and 76% in the rural.

Seasonal variations are conspicuous in foraging, e.g., dry and cold season for hunting, caterpillar collection and firewood gathering, wet/rainy season for fruits and vegetables, etc. The frequency of collection varies from a few times per month to a daily basis, however, not much seasonal variation is observed for medicinal herbs that are collected only once or a few times a year. Foraging from a variety of green spaces including urban and peri-urban forests and wooded groves, farms, home gardens, scattered trees and potted plants, vacant/undeveloped municipal lands, sidewalks and pavements, public parks, wetlands and riparian zones, institutional spaces, and cemeteries has been recorded in urban settings. Community kitchen gardens are also gaining popularity in recent times, in apartment complexes and institutional spaces. Records for foraging even from dumping sites in the literature reviewed indicate that risk of contamination is not a major criterion for deciding on the foraging locations in the Global South, unlike in the Global North where foragers prefer traffic-free foraging sites to avoid contamination of food produce (Brandner and Schunko 2022). Only one of the 28 papers reviewed cited concerns about pollution as a constraining factor for foraging, despite studies (e.g., analysis of lead contamination in edible weeds such as *Amranthus* spp. in São Paulo by Amato-Lourenco et al. 2020) indicating the presence of pollutants beyond permissible limits in edible plants in areas of heavy traffic including freeways and arterial roads. This apparent lack of concern could also be related to the low-income status of foragers in the Global South whose lack of purchasing power restricts access to expensive and nutritious food and force them to rely on freely available wild edible materials even from polluted sources.

13.2.2.3 Status of and Threats to Urban Foraging

Around 80% of the reviewed studies reported a decline in the practice of urban foraging; land use changes induced by urbanization, concretization, and agricultural expansion and accompanying loss of green spaces and access to them are the major reasons. Loss of traditional knowledge, lack of awareness, and lack of motivation are the next important reasons. Some people even considered gathering practice as a stressful activity due to insufficient foraging spaces and long distances to be traveled for forgeable resources (Adeyemi and Shackleton 2022) and high workload involved in collection and processing of wild produce (Termote et al. 2012). The absence of a conducive policy environment was an important reason in two cases. In some studies, interest in foraging for certain wild species was found to have declined after their commercial cultivation started, as in the case of wild mushrooms in Shaanxi, central China (Kang et al. 2013).

Preferences for culture, health and leisure needs, traditional attachment, recent promotion of cultivation and commercialization in the form of foraging tours, and favorable policies are reported as major enabling factors for urban foraging in the papers reviewed. Studies that reported flourishing urban foraging practices (only 4 out of 28 studies reviewed) reveal such intensified interest when business groups with commercial interests are involved, resulting in the commodification of wild species. Wherever subsistence users remain as the major foragers, a decline in the practice is conspicuous. In Shangri-La, under increasing external influences from the high tourist influx, foragers are now selling their finds at local markets, and the resultant overharvesting and commercialization threaten the sustainability of a variety of wild species (Walsh 2018). In Cape Town, South Africa, high levels of poverty and population growth have led to a booming informal industry based on the harvesting and consumption of wild products mostly from nature conservation areas leading to unsustainable harvesting—nearly 70% of harvested flora and 100% of all collected fauna, including red-listed species are either killed or reproductively harmed through the harvesting processes (Petersen et al. 2012).

13.2.2.4 Policy and Research Needs for the Promotion of Urban Foraging

Urban green space governance and policies at various scales, especially at the city level have a direct influence on the practice of urban foraging. In South African cities, in addition to a lack of knowledge about wild indigenous species and sustainable harvesting practices, ambiguity in policies and "normative views of pristine nature" may hinder foraging practices (Sardeshpande and Shackleton 2020a). Free access to foraging spaces and the development of potential components of foraging in urban green spaces make people more secure about foraging (Adeyemi and Shackleton 2022), but the absence or inadequacy of policies in many cities often restricts access of marginalized groups to green spaces or promotes undesirable exotic and non-fruit-bearing species in community parks and open spaces (Sardeshpande and Shackleton 2020a).

A review of Urban Forestry Management Plans of 58 Brazilian cities (Brito and Borelli 2020) showed that 84% of them mentioned "fruit" or "food," but only five considered the positive aspects of edible species while others mentioned negative aspects of food species prohibiting their planting. Reasons for prohibiting certain edible species vary from being considered exotic invasive species [e.g., *Hovenia dulcis* (Japanese raisin tree) and *Eisenia japonica* (loquat)], large fruits that might cause accidents when falling from the trees [*Mangifera indica* (mango) and *Persea americana* (avocado)], presence of thorns [*Citrus limon* (lemon)], toxic components [*Schinus terebinthifolius* (Brazilian peppertree)] or aggressive tree roots [*Bombacopsis glabra* (castanha-do-Maranhao)], potential to attract bats [*Terminalia catappa* (tropical almond) and *Myrciaria cauliflora* (jaboticaba)], to being not amenable to pruning [palms such as *Bactris gasipaes* (pupunha)] and production of "many fruits" [Surinam cherry (*Eugenia uniflora*)]. Also, the practice of foraging itself was discouraged by the requirement of permits for harvesting the products, under policies for biodiversity conservation (Guenat et al. 2023).

In India, traditional knowledge is rarely incorporated into urban planning and policies (Somesh et al. 2021). Some state-level legislations, e.g., Karnataka village offices abolition act, 1961 resulted in the abolition of traditional management institutions that had historically ensured sustainable extraction of common resources in city peripheries and replaced them with multiple government departments that often lack coordination and work at cross purposes (Mundoli et al. 2017). In recent urban development projects in India such as AMRUT and Smart City Mission, there is also no focus on urban resilience planning with adequate recognition of multiple uses of urban commons (Mundoli et al. 2017).

Mexico City is a noteworthy case in the Global South, due to its innovative efforts to achieve urban resilience that are also fostering urban foraging. With

23% of its population in the "food insecure" category, the city has launched the "sustainable local food system for Mexico City" initiative and has enacted the "Urban Orchards Act" that promotes urban foraging practices in the interstitial green spaces to mitigate the sudden, complete loss of access to food from the long-distance market (Hare and del Valle Isla 2021). A review of land management policies in 11 urban centers including Durban in the Indian Ocean Coastal Belt area of South Africa, revealed that the existing policies are either silent on wild resource extraction from public green spaces or allow harvest in a sustainable manner—for example the Durban Open Space Management system, while not explicitly mentioning the removal of wild fruits or animals as food, allows removal of barks for medicinal use as long as the extraction do not damage the plant (Sardeshpande and Shackleton 2020a). Except for a few such conducive initiatives, gaps are evident in the policy environment for promoting urban foraging in the Global South.

In addition to addressing these policy lacunae, it is also important to promote research on wild flora to increase the production of seeds and seedlings for distribution to traditional farmers, as done with conventional vegetables (Thoa et al. 2013). Wholesale nurseries and cultivation efforts driven by conservation agencies could help in meeting the local demand for wild produce from informal sector (Petersen et al. 2012). Collaboration of civil society with the government to promote knowledge and awareness is important in the utilization of indigenous plant resources for food security, nutrition, and income generation (Pisto and Lebese 2014). An analysis of financial potential of establishing food trees in urban public green areas of Villa El Salvador in Peru revealed that the involvement of local producers in the maintenance of public green could provide financial benefits for the municipality as well as for the producers (from edible products that can be harvested) (Lafontaine-Messier et al. 2016).

13.2.3 New-Age Practices of Urban Foraging

While the literature from Global South mostly emphasizes the role of wild edibles in supporting nutritional, income, and livelihood needs, a recent interesting trend is a shift in the perception of wild edibles from "famine food" to "delicacies" (Reyes-García et al. 2015) or "sustainable superfoods" in both Global North and South. Urban foraging is gaining popularity in the post-COVID world as people observed the benefits of locally sourcing their own food during the pandemic. Urban foraging has been taking a new course where bakers, food artists, organic farmers, etc. take a keen interest in sourcing materials including edible weeds, flowers, and decorative material locally. In this subsection, we document some emerging practices and movements around urban foraging in different cities, mostly collated from Indian dailies (Table 13.2).

Table 13.2 New practices of urban foraging

Location	Organization/initiative	Description	Pathway	References
Delhi National Capital Region	"Garden To Table" and "Delhi I Love You" movements, Foraging walk in Lodhi Gardens	Awareness creation and traditional knowledge sharing of foraging	Tourism	Matange (2016)
Goa	"Growing and eating collectives" in food courts	Creating innovative items using fresh seasonals and wild edibles	Consumer offering	Bhuyan (2022)
Hyderabad	'Festival of uncultivated food' hosted by Deccan Development Society	Showcasing various delicacies with edible weeds	Consumer offering	Avadhani (2019)
Bengaluru	Sarjapura Curries	Sharing traditional knowledge of uses of wild edible and supports farmers in planting and marketing traditional wild herbs	Farmer support using technology platforms	Roopa (2022)
Mumbai	Bombay canteen, professional chefs of Mumbai	Promoting traditional recipes with wild edibles and seasonal ingredients, creating apps that guide users to edible plants in nearby parks	Consumer offerings and use of technology/social media	Ved (2022)
China	Local eateries in National Parks	Serving regional cuisines using wild vegetables	Tourism and consumer offerings	Kang et al. (2013)

13.3 Foraging in Urban Protected Areas: Case Study from Bengaluru, South India

Bengaluru, India's third largest metropolis is well studied with respect to urban foraging practices. A total of 76 species were found to be foraged (Somesh et al. 2021). Foraging for wild greens from green spaces including parks and unused lands and dry lake beds during the summer season and for fruits such as mangoes (*Mangifera indica*) and jamun (*Syzygium cumini*) from wooded groves in urban and peri-urban regions have been reported (Somesh et al. 2021; Mundoli et al. 2017). Native tree species including drumstick (*Moringa oleifera*) and mango provide supplementary dietary security in slum settlements in the city (Gopal and Nagendra 2014). Urban foragers in Bengaluru are mostly women (81%) belonging to the socially deprived classes (Somesh et al. 2021).

While substantial literature on foraging in a variety of green spaces is available from the city, urban forests, and protected areas (PAs), sources of wild food in Bengaluru are not well documented. Urban PAs act as ecological shields lessening the negative impacts of urban expansion on rural peripheries. PAs have also traditionally been nurturing people-nature linkages and ecosystem services such as provisioning of food, fodder, and fuelwood; climate, soil, and water regulation; recreation and cultural services for urban residents. Despite undergoing significant land use changes on account of urbanization, many tangible and intangible benefits are still derived from urban PAs. In this section, we use a case study of a protected forest in Bengaluru, to showcase the changing practices of foraging in response to the drivers of urbanization and biodiversity conservation.

13.3.1 Study Site

Bannerghatta National Park (BNP), the largest urban green space in Bengaluru spans an area of 260 km^2 with 12 reserve forests and 4 wildlife ranges (Location map in Fig. 13.2). Comprising predominantly deciduous and scrub forests, BNP serves as a critical habitat for Asian elephants and other fauna and acts as an important ecological corridor connected to other PAs. There are 16 enclosure (EL) villages located within the park boundary and 77 villages in the eco-sensitive zone (ESZ) demarcated at 100 m to 1 km radius of the park boundary.[1]

As home to traditional forest-dependent communities such as *Iruligas* and *Hakki Pikkis,* the BNP landscape has a long history of pastoralism, agriculture, and forest produce collection (Kamat 1980). Collection of honey, wild fruits, diverse varieties of greens, lac, bamboo for basket weaving, and firewood from BNP forests has been documented by many studies (Varma et al. 2009; Gopalakrishna et al. 2015). However, significant changes occurred in the use and protection status of the forests

[1] (MoEFCC), Ministry of Environment, Forests and Climate Change. 2020. No. 25/12/16-ESZ-RE. Gazette of India: Extraordinary, Part II, Section 3 (ii), Annexure IV. 1–42. https://kspcb.karnataka.gov.in/sites/default/files/inline-files/16bannerghatta%20MOEF.pdf.

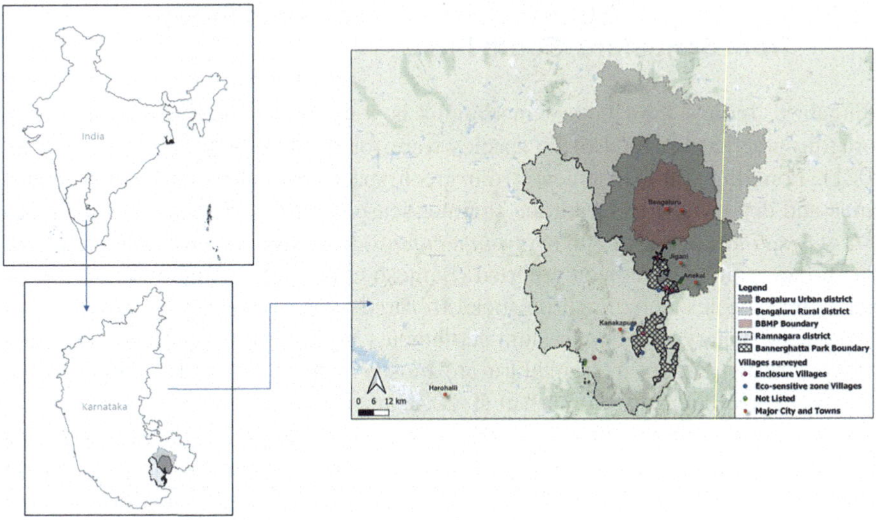

Fig. 13.2 Location map of BNP indicating the study villages

after the declaration of the area as a "national park" in 1974 under the Wildlife Protection Act, 1972. The National Park status prohibits livestock grazing in forests and collection of forest products for commercial purposes, while forest produce removal for personal *bonafide* needs of people living in and around the forests is allowed.[2]

13.3.2 Methodology

To understand the trends in forest use for provisioning uses of forests such as livestock grazing, collection of fuel wood and other non-timber forest produce (NTFPs), and foraging for wild fruits and greens, we surveyed 102 respondents from 6 EL villages (dominated by *Iruliga* and *Hakki Pikki* communities) and 6 ESZ villages in BNP. Semi-structured interviews were conducted, randomly sampling 8–10 households in each village based on the household list obtained from local *anganwadis (childcare centers/schools).* We used photo-elicitation methods wherein images of provisioning services including livestock grazing, collection of fuel wood, wild fruits and greens, medicinal herbs, and other NTFPS such as honey and broom grass (*Thysanolaena maxima*) captured from the BNP landscape were shown to the respondents in order to visually elicit the relevance of each use. Respondents were then asked to score the importance of each service for them on a scale of 1 (not important) to 5 (very important), both for the present and past (20 years ago when

[2]The Wildlife Protection (Amendment) Act, (WPA). 2002. Ministry of Environment, Forest, and Climate Change. Government of India. Section 35- (6), (7). Pg-20. https://moef.gov.in/wpcontent/uploads/2018/03/MINISTRY_OF_LAW_AND_JUSTICE.pdf.

urbanization pressure started mounting around Bengaluru) time and explain the reasons for a change in preference, if any, over time.

Mean scores were compared between past and present times and between EL and ESZ villages to examine the temporal and spatial variations in the perception of these uses by local communities. We used Wilcoxon's sign rank test to assess the variation of importance scores between past and present times and between the EL and ESZ villages.

13.3.3 Results

Our study showed that respondents ranked the collection of fuel wood as the most important provisioning use of BNP forests in the past in both EL (mean score of 4.92) and ESZ (mean score of 4.64) zones as the dependence on fuel wood for cooking and other household purposes was higher earlier (Fig. 13.3). This was followed by grazing as the next valued use from BNP in both EL (4.76) and ESZ (4.59) villages in the past, given the traditional livelihood dependence on rearing cattle, goats, and sheep. Collection of wild fruits and greens was preferred more in EL villages (4.39) than in the ESZ villages (4.18) in the past, owing to their proximity to the forest, easy accessibility of the produce, and the presence of traditionally forest-dependent communities. The use of medicinal herbs from forests was also valued in the past in both EL (3.06) and ESZ (3.08) zones. This is due to people's dependence on indigenous health systems that was sustained by the rich knowledge of traditional healers about wild medicinal resources. This has eventually changed due to improved medical facilities facilitated by growing urban proximity that has affected both the collection of these medicinal plants and also the knowledge about their uses.

At present, the collection of fruits and greens (2.02) and other non-timber forest products (2.01) has been ranked the highest in terms of importance in the EL zone while the collection of fuel wood (1.31) and fruits and greens (1.29) are ranked higher in the ESZ zone. Interestingly, the higher importance for the collection of local fruits, greens, and NTFPs compared to other uses in EL villages at present shows a lack of alternatives for these products and the value attached to their nutrition, medicinal properties, and

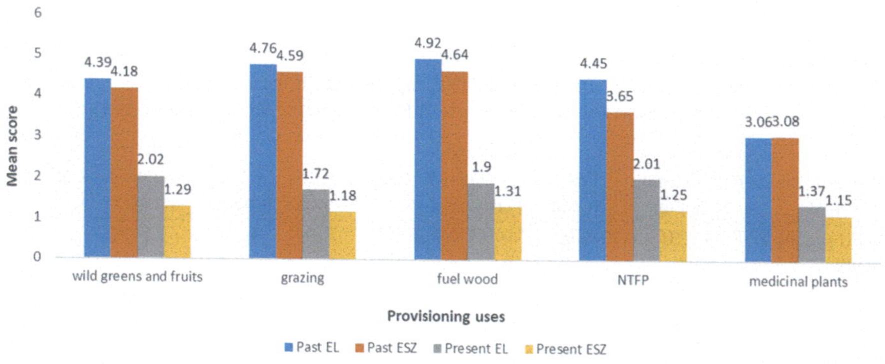

Fig. 13.3 Importance of provisioning uses of BNP forests

traditional or cultural use by indigenous communities inhabiting the EL zone. For most other provisioning uses, better substitutes from non-forest sources are available at present. Fuelwood use has been substituted by the availability of LPG stoves for cooking, grazing in forests has been restricted by the PA regulations, and also shift in the type of livestock raised from sheep and indigenous cattle to hybrid dairy cows that are usually stall-fed for meeting the huge demand for milk from Bengaluru city.

Respondents revealed that they have traditionally collected many fruits like stone fruit (*Aegle marmelos*), jackfruit (*Artocarpus heterophyllus*), mango (*Mangifera indica*), jamun (*Syzygium cumini*), tamarind (*Tamarindus indica*), and a variety of wild greens like *seege soppu* (*Acacia sinuata*), *kaachi soppu* (*Solanum nigrum*) and bamboo shoots from forests. The nomadic tribal community—*Hakki pikkis* used to hunt small birds and animals like rabbits, wild boars, etc., but hunting is now banned in the PA. People reported the collection of other forest produce like broomstick grass (*Thysanolaena maxima (Roxb.)*), bamboo (*Bambusa* sp.) leaves for basket weaving, *arali* flowers (*Nerium oleander*), and neem (*Azadirachta indica*) leaves, for wedding functions and other festivals. A general perception across zones was that the diversity of flowering and fruiting trees and produce such as fruits, greens, and fish has reduced over time and the area under forest plantations has increased. Spatial analysis of BNP's land cover shows the expansion of Eucalyptus and teak plantations between 1973 and 2007, changing the ecology of the park (Adhikari et al. 2017).

Iruligas, a forest-dwelling community that was traditional foragers, used to earn their livelihood by selling forest produce to other communities who were not engaged in forest produce collection. Earlier, community grazing practices enabled people, especially women, to go to the forest in groups which allowed them to collect wild foods such as seasonal greens. But as grazing is banned now, such group visits are not practiced, and women do not prefer to venture out alone in forests just for foraging.

The Wilcoxon signed-rank test statistics were significant at $p = 0.001$ for all the provisioning uses assessed, indicating significant variation in perceptions between past and present. A significant difference in perceptions also exists for all uses except medicinal plant collection between communities in EL and ESZ villages (Table 13.3). All provisioning uses from forests have significantly declined from the past to the present. Also at present, people from the EL zone had a significantly

Table 13.3 Respondents' perception of provisioning uses of forests in past and present time periods and in the EL and ESZ zones around BNP. Significant differences in the perception were tested using Wilcoxon signed-rank test (*** $p < 0.001$, ** $p < 0.01$, NS, non-significant)

Provisioning uses of forests	Present mean	Past mean	Wilcoxon signed-rank test statistic (Z)	EL mean	ESZ mean	Wilcoxon signed-rank test statistic (Z)
Wild greens and fruits	1.50	4.08	−9.55***	2.02	1.29	−2.72**
Grazing	1.37	4.35	−10.27***	1.72	1.18	−2.64**
Fuelwood	1.46	4.52	−10.33***	1.9	1.31	−2.76**
Other NTFPs	1.51	3.82	−8.98***	2.01	1.25	−2.80**
Medicinal plants	1.26	2.86	−7.82***	1.37	1.15	−1.43 (NS)

higher preference for all uses of forests except for medicinal use which showed no spatial variation.

13.3.4 Discussion

The case study shows that the level of use and access to BNP forests have drastically changed over the years, especially in the last 10 years. The changes observed arise out of the interplay of two drivers—urbanization and conservation. While the forest department cites the need for strict regulations for forest protection due to extensive livestock grazing, illegal logging, forest degradation, and land diversions in and around BNP, the traditional community uses of the forest have been compromised. Tightening forest protection and consequent increase in wildlife population have led to increased human-wildlife conflict, livestock loss, crop damage, and human life loss, especially from elephants (Prasad et al. 2016; Bantalapad et al. 2017). Damage to useful fruits and greens even within the forests by elephants and other animals like monkeys was mentioned by people. At the same time, urbanization facilitates better access to urban centers and markets making many products easily available. With urban expansion and growing community preference for other jobs, better lifestyles, and diverse types of food, the need to forage has come down, leading to a waning appreciation for wild food from forests. In our discussions with local communities, people associated forests with qualities like "pleasant environment," "clean air," etc. reflecting a view of forests as spaces for biodiversity conservation, and aesthetic and recreational benefits concomitant with the reduced appreciation of provisioning uses.

The case study of BNP confirms the trends in most Global South cities on two counts: a) socioeconomic changes unleashed by urbanization result in decreased interest in wild food collection and consumption and b) biodiversity conservation policies restrict wild food foraging. In the case of BNP, the declaration of the PA has hindered access to forests for wild food collection. The conflicts and synergies between urban biodiversity conservation/land management policies and foraging practices have been extensively discussed in studies from both Global North and South. In Brazilian cities, biodiversity conservation policies often do not allow the harvest of tree produce in public land (Guenat et al. 2023), in parks of New York city removal of any form of vegetation is unlawful, and in the city of Philadelphia, NTFP harvesting is prohibited outright in some county parks (McLain et al. 2013). On the contrary, South Africa policies in some cities are conducive to the sustainable extraction of wild materials from public spaces (Sardeshpande and Shackleton 2020a) while collecting plants for personal, non-commercial use is allowed by the German Nature Conservation Act (Fischer and Kowarik 2020). Thus the impact of conservation policies on urban foraging is quite varied globally. In BNP, though local people are allowed to collect wild plants for personal uses, the restricted access to the protected area and the prohibition of livestock grazing has curtailed such wild food gathering now. Much of the foraging had traditionally happened as a collective activity during livetsock grazing.

As revealed in the review, conservation policies enshrining a pristine view of nature' tends to exclude local uses of forests and biodiversity and with urbanization and decreasing livelihood dependence on forests, even local communities have started viewing BNP as a space of biodiversity conservation, aesthetic, and recreational benefits while ignoring its provisioning uses. But it is important to note that collection of wild food is the highest ranked service from forests in villages closer to the forests at present, though its importance has temporally declined, indicating the cultural attachment of indigenous communities to this practice.

13.4 Conclusions and Recommendations

Learnings from the literature review along with the case study present interesting insights into various aspects of urban foraging. The review shows a wide diversity of urban foraging practices in the Global South in terms of the profile of foragers, species, patterns and locations of foraging, etc.; however, a few common threads that run across the studies are the importance of this practice for low-income urban residents and the declining trend in the practice mostly triggered by urbanization driven changes and lack of policies that facilitate urban foraging. As unregulated land use changes associated with urban sprawl are projected to further increase in the Global South, it is important that supportive policies for urban foraging are formulated and implemented for the benefit of urban poor and marginalized communities.

Given the multiple socioeconomic benefits of urban foraging as a low ecological footprint and inclusive food system, mainstreaming foraging practices could offer low-cost solutions for sustainable urban development. In BNP with a large tourist footfall, government-run ecotourism programmes can integrate foraging tours, leveraging the traditional knowledge of forest-dwelling communities, and providing them with income as well. Commercial ventures and business interests valorizing urban foraging have an important role to play, but adequate safeguards should be installed to ensure the sustainable use of wild species. Fixing regulations/quotas to prevent unsustainable extraction and developing foraging guidelines specifying seasons of foraging, methods of foraging, etc. based on the rarity/abundance of species could be effective measures to prevent overexploitation. Developing such guidelines and regulations would also require rigorous scientific research and inventorying of less known wild species in cities and monitoring their population while building awareness on their nutritional, medical, or cultural significance. Moreover, enforcing quotas and permits might require strict policing and specialized biodiversity knowledge and could often result in perverse outcome of indiscriminate resource extraction, in the absence of checks (Petersen et al. 2012). "Social fencing" of resources wherein adequate opportunities are provided for local stakeholders to sustainably use the resources for own consumption and for other livelihood benefits (income from sales, promotion of foraging-based tourism, etc.) could be a more inclusive strategy, especially when traditional ecological knowledge of indigenous communities that is cognizant of natural resource limits is leveraged in such initiatives.

Urban planning has a crucial role in protecting green spaces for foraging through devising zoning regulations and averting harmful urban land use changes. In Bengaluru, the "cluster mode" of urban expansion proposed in Revised Master Plan–2031 (BDA 2017) encompassing multiple growth nodes and a green network could potentially sustain food production landscapes in the city's peripheries; however, the RMP's imagination of green spaces as recreational parks and lake networks seems to prioritize recreational uses of these spaces by the urban elite (Patil et al. 2018). A pro-poor imagination recognizing green networks as "edible spaces" should therefore be incorporated into urban planning visions. In an effort to meet the climate and biodiversity targets of the country, urban afforestation schemes are receiving increased attention in India, both from the government (e.g.. *Nagarvan or* urban forest scheme of the Ministry of Forests, Environment, and Climate Change)[3] and from private players including corporates funding plantation projects under Corporate Social Responsibility Schemes (Lewis 2023). There is immense opportunity to realign such afforestation programmes from "recreational" forests to "edible" forests to augment their ecological benefits as well as social outcomes. Restoration and beautification projects for urban lakes and development of lakefronts more often than not result in the clearing of wild vegetation around lakes, thus reducing the availability of greens for foragers. To avoid such negative consequences, lakes and other urban blue-green spaces should be managed in a participatory mode with citizen involvement, as "multi-use commons" that serve diverse provisioning, and cultural uses of different stakeholder groups. It is important to graduate lake and green space restoration projects to "rewilding" projects to encourage wild flora conservation and use. Recognizing the value of such wild spaces for foraging in urban development policies and projects will go a long way in promoting resilient cities and accomplishing SDGs in an inclusive manner in the Global South.

Acknowledgments The case study on BNP was supported by a research grant from the Department of Biotechnology, Government of India under the Indo-German collaborative project "The rural-urban interface of Bengaluru: A space of transitions in agriculture, economics and society." We wish to thank the respondents around BNP for participation in our survey. We also thank the reviewers for their valuable feedback. Ms. Rakhi Yadav and Mr. Sarang KT are acknowledged for their assistance in data handling and analysis.

Ethics Approval The primary survey for the case study was conducted with approval from the Institutional Review Board of Azim Premji University, Bengaluru, India vide document no. IRB/Ext-DBT/02/April 14, 2020.

[3]Ministry of Environment, Forests and Climate Change (MoEFCC). 2020. NagarVan Yogana (Revised guidelines version), National Afforestation and Eco-Development Board. https://cdn.climatepolicyradar.org/navigator/IND/2020/nagar-van-yojana-nvy-forestation-scheme_075ec1d35f35bba60ec055ad9efbac50.pdf.

References

Adeyemi O, Shackleton CM (2022) Understanding the foraging practices in Lagos metropolis to redesign urban greenspaces in support of human-nature interactions. Urban For Urban Green 79. https://doi.org/10.1016/j.ufug.2022.127805

Adhikari S, Fik T, Dwivedi P (2017) Proximate causes of land-use and land-cover change in Bannerghatta National Park: a spatial statistical model. Forests 8(9):342. https://doi.org/10.3390/f8090342

Alemayehu G, Asfew Z, Kelbessa E (2015) Plants diversity and ethnobotany in Berehat district, North Shawa zone of Amhara region (Ethiopia) with emphasis on wild edible plants. J Med Plant Stud 3(6):93–105

Amato-Lourenco LF, Ranieri GR, Souza VCO, Barbosa F Jr, Saldiva PHN, Mauad T (2020) Edible weeds: are urban environments fit for foraging? Sci Total Environ 698:133967. https://doi.org/10.1016/j.scitotenv.2019.133967

Arku G, Marais L (2021) Global south urbanisms and urban sustainability—challenges and the way forward. Frontiers in Sustainable Cities 3:692799. https://doi.org/10.3389/frsc.2021.692799

Avadhani R (2019) Exploring uncultivated greens: about 80 people participate in the festival of uncultivated foods. Retrieved December 17, 2022, from https://thehindu.com/news/cities/Hyderabad/exploring-uncultivated-greens/article29127441.ece

Banerjee S (2014) Ganesh festivities take a toll on rare plants. The Hindu. Retrieved July 8, 2023, from https://www.thehindu.com/news/national/other-states/ganesh-festivities-exacting-heavy-toll-on-endangered-plant-species-in-western-ghats/article6356918.ece

Bangalore Development Authority (BDA) (2017) "Revised Master Plan for Bengaluru – 2031 (Draft), Volume 1-Vision Document," 34–40. Retrieved December 16, 2022, from https://bda-bangalore.org/eng/bdaqlink

Bantalapad M, Gayathri A, Krishnan A (2017) Seasonal crop raiding of fruit trees by Asian elephants: an insight into foraging preferences from croplands abutting Bannerghatta National Park, Bengaluru, Karnataka, India. Asian J Environ Ecol 2(2):1–12

Bhuyan A (2022) On a journey to find where the wild things grow. Retrieved December 17, 2022, from https://vikalpsangam.org/article/on-a-journey-to-find-where-the-wild-things-grow/

Brandner A, Schunko C (2022) Urban wild food foraging locations: understanding selection criteria to inform green space planning and management. Urban For Urban Greening 73:127596. https://doi.org/10.1016/J.UFUG.2022.127596

Brito VV, Borelli S (2020) Urban food forestry and its role to increase food security: a Brazilian overview and its potentialities. Urban For Urban Greening 56:126835. https://doi.org/10.1016/j.ufug.2020.126835

Bunge A, Diemont SAW, Bunge JA, Harris S (2019) Urban foraging for food security and sovereignty: quantifying edible forest yield in Syracuse, New York using four common fruit- and nut-producing street tree species. J Urban Ecol:1–14

Capron G, Arellano SG, Wigle J, Diez AL, Monterrubio A, Hidalgo H, Morales J (2017) The urban food system of Mexico City. México. Hungry cities report no. 7. Hungry cities partnership

Chakona G, Shackleton CM (2019) Food insecurity in South Africa: to what extent can social grants and consumption of wild foods eradicate hunger? World Dev Perspect 13:87–94

Cheng Z, Lu X, Lin F, Naeem A, Long C (2022) Ethnobotanical study on wild edible plants used by Dulong people in northwestern Yunnan, China. J Ethnobiol Ethnomed 18(1):3. https://doi.org/10.1186/s13002-022-00501-3. PMID: 35062974; PMCID: PMC8781162

Fischer LK, Kowarik I (2020) Connecting people to biodiversity in cities of tomorrow: is urban foraging a powerful tool? Ecol Indic 112. https://doi.org/10.1016/j.ecolind.2020.106087

Garekae H, Shackleton CM (2020a) Foraging wild food in urban spaces: the contribution of wild foods to urban dietary diversity in South Africa. Sustainability 12

Garekae H, Shackleton CM (2020b) Urban foraging of wild plants in two medium-sized south African towns: people, perceptions and practices. Urban For Urban Greening 49:1–10

Gopal D, Nagendra H (2014) Vegetation in Bangalore's slums: boosting livelihoods, well-being and social capital. Sustainability 6:2459–2473

Gopalakrishna SP, Kaonga ML, Somashekar RK, Suresh HS, Suresh R (2015) Tree diversity in the tropical dry forest of Bannerghatta National Park in eastern Ghats, southern India. Eur J Ecol 1(2):12–27

Guenat S, Bailey-Athias JP, Fischer LK (2023) Urban foraging in Brazilian public greenspaces. Ambio 52(7):1248–1261

Hare M, del Valle Isla AP (2021) Urban foraging, resilience and food provisioning services provided by edible plants in interstitial urban spaces in Mexico City. Local Environ 26(7):825–846

Hurley PT, Emery MR (2018) Locating provisioning ecosystem services in urban forests: forageable woody species in New York City, USA. Retrieved December 23, 2022, from https://digitalcommons.ursinus.edu/environment_fac/13

Hurley PT, Emery MR, Mclain R, Poe M, Grabbatin B, Goetcheus CL (2015) Whose urban forest? The political ecology of foraging urban non timber forest products. In: Isenhour C, McDonagh G, Checker M (eds) Sustainability in the Global City: myth and practice. Cambridge University, New York, NY

Kamat SU (1980) A concise history of Karnataka: from pre-historic times to the present. Archana Prakashana, Bangalore

Kang Y, Łuczaj Ł, Kang J (2013) Wild food plants and wild edible fungi in two valleys of the Qinling Mountains (Shaanxi, Central China). J Ethnobiol Ethnomed 9(26). https://doi.org/10.1186/1746-4269-9-26

Kaoma H, Shackleton CM (2015) The direct-use value of urban tree non-timber forest products to household income in poorer suburbs in south African towns. Forest Policy Econ 61:104–112. https://doi.org/10.1016/j.forpol.2015.08.005

Kosack L (2016) Urban gardening—edible cities (case study Andernach). University of Bonn (sustainability). Retrieved January 23, 2023, from https://bonnsustainabilityportal.de/2016/06/09-06-2016-university-of-bonn-urban-gardening-edible-cities-case-study-andernach

Lafontaine-Messier M, Gélinas N, Olivier A (2016) Profitability of food trees planted in urban public green areas. Urban For Urban Greening 16:197–207. https://doi.org/10.1016/j.ufug.2016.02.013

Lewis C (2023) Now corporates splurge on Miyawaki urban forests, no longer civic preserve. Times of India. Retrieved July 11, 2023, from http://timesofindia.indiatimes.com/articleshow/99394126.cms?from=mdr&utm_source=contentofinterest&utm_medium=text&utm_campaign=cppst

Matange AT (2016) Go foraging for edible plants: one person's weed can be another person's green. DownToEarth. Retrieved December 17, 2022, from https://www.downtoearth.org.in/news/food/go-forage-56034

McLain R, Poe M, Hurley PT, Lecompte-Mastenbrook J, Emery MR (2012) Producing edible landscapes in Seattle's urban forest. Urban Forest and Urban Greening 11:187–194

McLain R, Hurley PT, Emery MR, Poe M (2013) Gathering "wild" food in the city: rethinking the role of foraging in urban ecosystem planning and management. Local Environ. https://doi.org/10.1080/13549839.2013.841659

Mollee E, Pouliot M, McDonald MA (2017) Into the urban wild: collection of wild urban plants for food and medicine in Kampala, Uganda. Land Use Policy 63:67–77

Mundoli S, Manjunatha B, Nagendra H (2017) Commons that provide: the importance of Bengaluru's wooded groves for urban resilience. Int J Urban Sustain Dev 9(2):184–206

Murwendo T (2011) Improving urban livelihoods at household level through sustainable utilisation of peri-urban forests in Masvingo City. J Sustain Dev Afr 13:299–313

Nero BF, Kwapong NA, Jatta R, Fatunbi O (2018) Tree species diversity and socioeconomic perspectives of the urban (food) Forest of Accra, Ghana. Sustainability 10:3417. https://doi.org/10.3390/su10103417

Patil S, Bhaskar D, Vanjari RS, Purushothaman S (2018) Urbanisation and new agroecologies: the story of Bengaluru's peripheries. Rev Environ Dev Econ Polit Wkly 53(41):71–77

Petersen LM, Moll EJ, Collins R, Hockings MT (2012) Development of a compendium of local, wild-harvested species used in the informal economy trade, Cape Town, South Africa. Ecol Soc 17(2):26. https://doi.org/10.5751/ES-04537-170226

Pisto FS, Lebese MR (2014) Traditional uses of wild edible plants in arid areas of South Africa. J Hum Ecol 48(1):23–31

Prasad A, Jayashankar M, Avinash K (2016) Perspective survey on the impact of vertebrates in agri-horticultural ecosystems in and around selected villages adjoining the western margins of Bannerghatta National Park. The Bioscan 1(2):1295–1298

Ray A, Ray R, Sreevidya EA (2020) How many wild edible plants do we eat—their diversity, use, and implications for a sustainable food system: an exploratory analysis in India. Front Sustain Food Syst 4. https://doi.org/10.3389/fsufs.2020.00056

Reyes-García V, Menendez-Baceta G, Aceituno-Mata L, Acosta-Naranjo R, Calvet-Mir L, Domínguez P, Garnatje T, Gómez-Baggethun E, Molina-Bustamante M, Molina M, Rodríguez-Franco R, Serrasolses G, Vallès J, Pardo-de-Santayana M (2015) From famine foods to delicatessen: interpreting trends in the use of wild edible plants through cultural ecosystem services. Ecol Econ 120:303–311

Rigo N (2021) Teaching and learning about unconventional food plants in an edible urban landscape: a brief report. In: Jacob MCM, Albuquerque UP (eds) Local food plants of Brazil. Ethnobiology. Springer, Cham. https://doi.org/10.1007/978-3-030-69139-4_21

Roopa N (2022) India's wild greens make a comeback. *Aljazeera*. Retrieved December 17, 2022, from https://www.aljazeera.com/features/2022/6/13/indias-modern-foragers

Russo A, Escobedo FJ, Cirella GT, Zerbe S (2017) Edible green infrastructure: an approach and review of provisioning ecosystem services and disservices in urban environments. Agric Ecosyst Environ 242:53–66

Sardeshpande M, Shackleton C (2019) Wild edible fruits: a systematic review of an under-researched multifunctional NTFP. Forests 10(6):467

Sardeshpande M, Shackleton C (2020a) Urban foraging: land management policy, perspectives, and potential. PloS One 15(4):e0230693. https://doi.org/10.1371/journal.pone.0230693

Sardeshpande M, Shackleton C (2020b) Fruits of the veld: ecological and socioeconomic patterns of natural resource use across South Africa. Hum Ecol 48(6):665–677. https://doi.org/10.1007/s10745-020-00185-x

Sardeshpande M, Shackleton C (2023) Fruits of the city: the nature, nurture and future of urban foraging. People Nat 5:213–227. https://doi.org/10.1002/pan3.10428

Sardeshpande M, Hurley PT, Mollee E, Garekae H, Dahlberg AC, Emery MR, Shackleton C (2021) How people foraging in urban greenspace can mobilize social–ecological resilience during Covid-19 and beyond. Front Sustain Cities 3:686254. https://doi.org/10.3389/frsc.2021.686254

Schlesinger J, Drescher A, Shackleton CM (2015) Socio-spatial dynamics in the use of wild natural resources: evidence from six rapidly growing medium-sized cities in Africa. Appl Geogr 56:107e115

Seburanga JL, Kaplin BA, Zhang QX, Gatesire T (2014) Amenity trees and green space structure in urban settlements of Kigali, Rwanda. Urban For Urban Greening 13(1):84–93

Shackleton R, Shackleton C, Shackleton S, Gambiza J (2013) Deagrarianisation and forest revegetation in a biodiversity hotspot on the wild coast, South Africa. PloS One 8(10):e76939

Shackleton CM, Hurley PT, Dahlberg AC, Emery MR, Nagendra H (2017a) Urban foraging: a ubiquitous human practice overlooked by urban planners, policy and research. Sustainability 9:1884

Shackleton CM, Blair A, Lacy PD, Kaoma H, Mugwagwa N, Dalu MT, Walton W (2017b) How important is green infrastructure in small and medium-sized towns? Lessons from South Africa. Landsc Urban Plan 3050:1–9. https://doi.org/10.1016/j.landurbplan.2016.12.007

Somesh D, Rao R, Murali R, Nagendra H (2021) Patterns of urban foraging in Bengaluru city. Urban For Urban Green 57:126940. https://doi.org/10.1016/j.ufug.2020.126940

Spiller KA, Kashwan P (2023) Food forests are bringing shade and sustenance to US cities, one parcel of land at a time theconversation.com/food-forests-are-bringing-shade-and-sustenance-to-us-cities-one-parcel-of-land-at-a-time-197388

Termote C, Meyi MB, Dhed'a Djailo B, Huybregts L, Lachat C, Kolsteren P, Damme P, P.V. (2012) A biodiverse rich environment does not contribute to a better diet: a case study from DR Congo. PloS One 7(1):e30533. https://doi.org/10.1371/journal.pone.0030533

Thoa PTK, Cuong LP, Yen NTK (2013) Biodiversity indices and utilization of edible wild plants: a case study of the Cham Island in Quang Nam Province, Vietnam. J Res Environ Sci Toxicol 2(9):167–174

Triman J (2013) Urban foraging: an "Intimate Connection" with nature in the city. Biophilic cities. Retrieved December 17, 2022, from https://www.biophiliccities.org/urban-foraging

UN-Habitat (2020) World cities report 2020: the value of sustainable urbanization. Nairobi: United Nations Human Settlements Programme (UN-Habitat). Retrieved December 20, 2022, from https://unhabitat.org/sites/default/files/2020/11/world_cities_report_2020_abridged_version.pdf

United Nations Department of Economic and Social Affairs, Population Division (2022) World Population Prospects 2022: Summary of Results. UN DESA/POP/2022/TR/NO.3. Retrieved December 20, 2022, from https://www.un.org/development/desa/pd/sites/www.un.org.development.desa.pd/files/wpp2022_summary_of_results.pdf

Varma S, Anand VD, Gopalakrishna SP, Avinash KG, Nishant MS (2009) Ecology, conservation and management of the Asian elephant in Bannerghatta National Park, southern India. A Rocha India/ANCF. Asian Elephant Ecology and Conservation Reference Series No. 1, pp 13–52

Ved S (2022, April 22) India's urban foragers are demonstrating the benefits of sourcing your own food. Vogue. Retrieved December 17, 2022, from https://www.vogue.in/culture-and-living/content/urban-foragers-demonstrate-the-benefits-of-sourcing-your-own-food

Walsh M (2018, October 19) Uprooting paradise: how foraging is transforming Shangri-La. Sixth tone. Retrieved December 17, 2022, from https://www.sixthtone.com/news/1003081

Dhanya Bhaskar (PhD) is Chairperson, Center for Policy Studies, and Associate Professor of Ecosystem and Environment Management at the Indian Institute of Forest Management, Bhopal. Her work is located at the interface of ecological and livelihood sustainability in forest, rural, and urban landscapes.

Sharmistha Rej has completed her Master's in Science from Calcutta University. She is currently pursuing a PhD in Forestry and has a keen interest in urban ecology.

Samudyatha Ramananda has completed her Master's from The University of Trans-Disciplinary Health Sciences and Technology (TDU), Bengaluru. She is interested in understanding people–nature interactions and policy analysis.

Understanding Urban Foraging Concepts in Fast Expanding Urban Areas in Indian Himalayan Region: Lessons from Himachal Pradesh, India

14

Radhika Sood and Shalini Dhyani

Abstract

The rapid expansion of urban areas in the Indian Himalayan Region presents a complex interplay between urban development and the preservation of natural ecosystems. As concrete landscapes encroach upon once-pristine habitats, a subtle yet profound connection between humans and their environment comes to light in the form of urban foraging. However, as the paradigm of the urban city evolved throughout the twentieth century, foraging lost its visibility to urban planners and researchers, concealing its social, cultural, ecological, and economic significance predominantly because of loss of greenspaces. Mountain areas are no exception to this. Urban areas are fast growing because of the increasing tourism pressure and to fulfill their recreational demands. This becomes crucial as increasing pressure in Himachal Pradesh in India over the last few decades has affected the urban green spaces. Despite that, a local connection from nature still persists and continues through urban foraging. Present chapter is an attempt to understand the concept of urban foraging from fast growing urban areas of Himachal Pradesh, India through key personal interviews, telephonic method, and secondary literature review. Total 18 species were identified in Kangra city followed by 17 species in Kullu and Manali city depicting interlinkage between urban foraging and daily consumption pattern. The local people still follow foraging for medicinal plants, fruits, and others as per their traditional linkages and knowledge systems. The chapter also brings insights from the interlinkages between bio-cultural aspects and urban foraging.

R. Sood · S. Dhyani (✉)
CSIR-National Environmental Engineering Research Institute, Nagpur, Maharashtra, India

Academy of Scientific and Innovative Research (AcSIR), Ghaziabad, Uttar Pradesh, India

S. Dhyani, M. Sardeshpande (eds.), *Urban Foraging in the Changing World*,
https://doi.org/10.1007/978-981-97-0345-6_14

Keywords

Urbanization · Tourism · Mountains · Wild edibles · Medicinal plants · Bio-culture

14.1 Introduction

Edible urbanism integrates three key sustainability themes: food security, resilience, and social inclusion and boosts metabolism, energy, and nutrition of the citizens (Russo and Cirella 2019). Urban foraging is a large dimension that requires identification of foraging dynamics at spatial and temporal scale, potential site identifications, importance of foragers and their purpose of foraging and its types (Fig. 14.1). In recent years, the practice of urban foraging has gained popularity due to its potential to provide fresh, nutritious, and affordable food to urban dwellers. It is influenced by various factors such as socio-economic status, availability of land, cultural and religious practices, environmental factors, and accessibility to markets (Rombach and Dean 2023). Foraging, or gathering wild food from natural spaces, is common in metropolitan settings throughout the world, but the literature primarily concentrates on the global north (Sardeshpande and Shackleton 2020). According to a study conducted in Vienna, large, biodiverse Public Urban Green Spaces (forests, school grounds, cemeteries, parks, courtyards, allotments, school grounds, road and wastelands, and margins of sport fields) and its easy access, little contamination, and few visitors are particularly suitable for urban wild food foraging (Brandner and

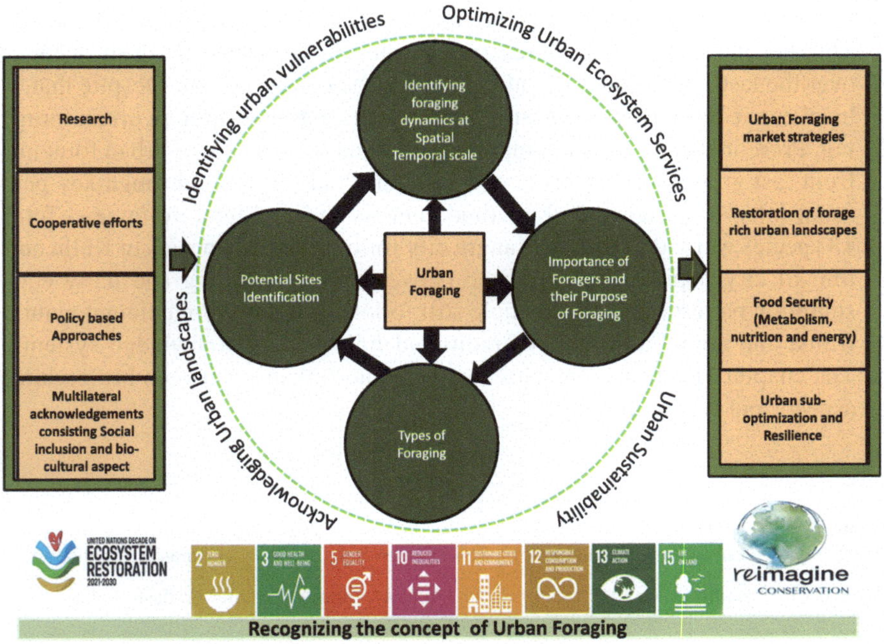

Fig. 14.1 Understanding the concept of urban foraging for long-term conservation strategies in urban landscapes

Schunko 2022). The extensive ethnography-based interviews in Berlin, Germany concluded 125 taxa from a variety of official and informal green space types reflecting the foragers' cultural backgrounds, motives, and the pattern of utilizing flora and its specific parts (Landor-Yamagata et al. 2018). During COVID-19 crisis, the US had a surge in local food provisioning interest, underscoring the importance of productive agriculture in urban settings to fill the demand of people (Clouse 2022). Yet, urban planners, policymakers as well as authorities are yet to acknowledge the practise of urban foraging since it has not been effectively recorded and reported (Dhyani and Kadaverugu 2020). Therefore, this study is an attempt to understand urban foraging concept in urban sprawls and acknowledging urban cities in Himachal Pradesh, India.

14.2 Foraging in Himachal Pradesh

Himachal Pradesh is one of the mountainous states of India and is located in the Western Himalayas. It is distinguished by an extreme landscape with numerous peaks and huge river systems, and altitudinal gradients are characterized by different vegetation types, soil types, and climatic conditions (Chawla et al. 2012). More than 95 tree species are native to Himachal Pradesh and are representative of the Western Himalayan flora, whereas 150 species of trees are exotics and the state is divided into Shivaliks (1500 m), Middle Himalayan region (1500–3000 m), and Himadri (>3000 m) (IFSR 2019). Foraging is common in rural areas of the Western Himalayan region and deeply ingrained in their culture (Bargali et al. 2021). Foraging in Himachal Pradesh has been widely practiced for centuries, and it is an important source of food and medicine for the local people especially in Himachal Pradesh. Roadside *Rhododendron* sp. and *Cannabis sativa* are collected to make chutneys (Dip) by the people even in fast growing urban sprawls of Shimla, Mandi, Kullu, Kangra, and Chamba districts in Himachal Pradesh. Another example includes the consumption of tender shoots of *Urtica dioica* near roads in the state. Foraging of medicinal plants for their therapeutic properties has been used for centuries by various cultures around the world (Haq et al. 2023). There are 643 species of medicinal plants that have been documented from the state, and their numerous plant parts are utilized to cure a wide range of illnesses even in urban set ups (Samant et al. 2007). However, medicinal plant foraging has been typically reported from rural areas, forests, or fields, where these plants grow naturally. Urbanization in the Western Himalayas has accelerated due to tourism, hydroelectric power development, and technological advancement, but foraging is still evident in the cities of the region as study on the impact of Himalayan cities (Kashmir, Dehradun, and Shimla) on[1]UGI quality (UGIQ) pinpointed that in larger cities, locations with more complex sections, interspersed with built-up areas, tend to have superior UGIQ in Shimla; however, areas with complete patches of forest tend to have the lowest population pressure (Anees et al. 2023). For instance, in the high urban parts of Himachal Pradesh, where there are less public parks, children frequently enjoy the

[1] UGI quality (UGIQ) is a metric or indicator used to evaluate the condition, performance, or effectiveness of urban green infrastructure in meeting its intended goals.

presence of scattered *Prunus salicina, Pyrus communis, Juglans regia*, and *Pinus gerardiana* while playing without taking into account the species present. Therefore, acknowledging urban foraging becomes crucial in the Himalayan states.

14.3 Interlinking Bio-cultural Aspect and Urban Foraging

The practice of urban foraging has gained popularity due to its potential benefits, such as promoting sustainable food systems, reducing food waste, and fostering a connection with nature (Glaros 2023). At the same time, urban foraging has also been linked to bio-cultural aspects, such as the preservation of traditional knowledge and cultural practices related to food and nature (Fontefrancesco et al. 2023). Interlinking bio-cultural aspects and urban foraging can be a new area of research that focuses on the intersection between culture formation, practices and its impact on urban foraging in cities. Himachal Pradesh, a state in northern India, is a particularly interesting area of study due to its rich biodiversity and unique cultural traditions. For instance, Lahaul and Spiti district is home to 98 medicinal plants, 14 species of them have been used for bioprospecting into, 20 traditional foods and drinks, unique traditional architecture, patterns of resource use, agricultural practices, and their means of subsistence (Rathore and Shashni 2023). Bio-cultural aspects emphasize the interdependence between biological and cultural diversity that influence behavior, and the practice of urban foraging involves the collection of food and other resources in urban setups (Khongsdier 2007). Understanding urban foraging in a state with a unique cultural background is important because urban foraging provides opportunities for nurturing and increasing bio-cultural learning. For instance, *Morchella esculenta* which naturally grows in the school premises of Kullu district of Himachal Pradesh is collected by the students to prepare food items by women of the house. Therefore, it becomes crucial to acknowledge the importance of urban foraging and urban foragers and their role in passing on information about species and locations for foraging. The interlinkage of bio-cultural aspects and urban foraging, being a new concept should be acknowledged considering the importance of edible and medicinal plant foraging.

14.4 Insights from Kangra

Kangra is located at the foothills of the Himalayas, making it a major transit point of trade between Punjab and Himachal Pradesh. Among the 12 districts in the state, Kangra, being an urban city has huge potential to be recognized as edible district of Himachal Pradesh. The information was derived from key informant interviews and telephonic method. Women were preferred for interviews due to their role in maintaining household chores and activities. Total 18 species were identified, and the pattern of consumption was also observed. Table 14.1 represents foraging species having an average elevation of 733 m amsl of Kangra city situated in Himachal Pradesh.

Table 14.1 Preferred foraged species in Kangra city having average elevation of 733 m amsl in Himachal Pradesh

S.No.	Scientific name	Local name	Family	Preparation name (P)/raw consumption (R)	Parts foraged	Season of foraging	Place of foraging Abandon lots, parks, riverside walks, and roadside
1.	*Aegle marmelos*	Bel/ Belpatri	Rutaceae	Murabba and Juice(P)	Fruit	May–June	Roadside
2.	*Bombax ceiba*	Semal	Bombacaceae	Semal ki Sabzi (P)	Buds and calyx	May–October	Parks, roadside
3.	*Colocasia* sp.	Kachalu	Arecaceae	Patrodu (P), Khatti bhujji (P), Khandole (P), and Ghandyali ki sabzi(P)	Leaves, stem, and corm	July to September	Riverside walks
4.	*Cordia dichotoma*	Lasiade	Boraginaceae	Lasiade ki sabzi(P)	Fruit	May–June	Abandon lots, roadside
5.	*Diplazium esculentum*	Lungru/ Lingadi	Athyriaceae	Lungru ki sabzi(P)	Fronds	May–June	Riverside walks
6.	*Ficus benjamina*	Tremble	Moraceae	Tremble ki Sabzi(P)	Fruit	May–June	Abandon lots, parks, and roadside
7.	*Ficus carica*	Dhura	Moraceae	Bhruni ki sabzi(P)	Tender leaves	Feb-March	Abandon lots, parks, and roadside
8.	*Mangifera indica*	Aam	Anacardiaceae	Ambua ki sabzi (P)	Raw or ripened fruits	May to July	Abandon lots, parks, and roadside
9.	*Moringa oleifera*	Sainjan	Moringaceae	Sainjan ke phulan ki bhurji (P)	Flowers	February	Riverside walks and road
10.	*Morus alba*	Shehtoot	Moraceae	Jam and Jellies (P)	Fruit	May–June	Parks and roadside
11.	*Murraya koenigii*	Kadi patta	Rutaceae	Kadi patta (R)	Leaves	January–December	Abandon lots and roadside
12	*Nasturtium officinale*	Chhunali	Brassicaceae	Chhunali ki Bhuji (P)	Whole plant	November to February	Riverside walks
13.	*Phyllanthus emblica*	Amla	Phyllanthaceae	Pickel and Murabba (P)	Fruit	December–March	Parks and roadside

(continued)

Table 14.1 (continued)

S.No.	Scientific name	Local name	Family	Preparation name (P)/raw consumption (R)	Parts foraged	Season of foraging	Place of foraging Abandon lots, parks, riverside walks, and roadside
14.	*Prunus persica*	Addu	Rosaceae	Addu ki Sabji(P)	Fruit	July–August	Abandon lots, parks, and roadside
15.	*Pyrus* sp.	Nakh	Rosaceae	Nashpati ki Sabzi(P)	Fruit	July–August	Abandon lots, parks, and roadside
16.	*Rhododendron arboreum*	Barah	Ericaceae	Barah ki Chutney(P)/Juice(P)	Flowers	March–May	Roadside
17.	*Syzygium cumini*	Jamun	Myrtaceae	Jamun (R)	Fruit	June–July	Abandon lots, riverside, roadside, and parks
18.	*Ziziphus mauritiana*	Ber	Rhamnaceae	Ber (R)	Fruit	February–April	Abandon lots, riverside, and roadside

14.5 Edible Cities of Kullu and Manali

In the state of Himachal Pradesh lies Kullu district, located south of the main Himalayan divide, and at elevations above 3500 amsl, the region can sustain a temperate Himalayan forest cover of coniferous as well as deciduous broad-leaved species, including historically significant oak and birch species and commercially valuable deodar (Himalayan cedar) and fir species (Gardner 2002). There are certain other species which are distributed across 1500 m to 2000 m altitudinal gradient like *Pyrus* sp., *Prunus* sp., *Diplazium esculentum, Rhododendron arboreum,* etc. These species are also present in Kullu and Manali and other cities of Himachal Pradesh. Total 17 species were identified from the cities through secondary literature and detailed personal interviews. Telephonic method was also carried out for cross-verification. Table 14.2 represents foraging species, parts foraged, season and place of foraging of Kullu and Manali cities in Himachal Pradesh.

Table 14.2 Foraging species of Kullu and Manali cities in Himachal Pradesh

S.No.	Scientific name	Local name	Family	Preparation name (P)/raw consumption (R)	Parts foraged	Season of foraging	Place of foraging
1.	*Aesculus indica*	Khanor	*Sapindaceae*	*Seed starch extract (Seekh) and Flour (P)*	*Seeds*	October	Abandon lots, riverside walks
2.	*Cannabis sativa*	Bhang	*Cannabaceae*	*Bhang ki Chutney (P)*	*Leaves*	April–October	Abandon lot, riverside, and roadside
3.	*Chenopodium album*	Bathua	Chenopodiaceae	Bathua ka Saag (P)	Leaves	April–October	Abandon lots, riverside walk, and roadside
4.	*Corylus jacquemontii*	Thangi	*Betulaceae*	*Raw consumption (R)*	*Seeds*	July–September	Roadside
5.	*Hippophae* sp.	Drilbu/ Chharma	Elaeagnaceae	*Raw consumption (R)*	*Berries*	October–November	Parks roadside
6.	*Juglans regia*	Khod/Akhrot	Juglandaceae	Raw consumption as dry fruit and used in filling Siddu (local dish) (P,R)	Kernels	July–September	Abandon lots, parks, and roadside
7.	*Malus domestica*	Seo, Seb	*Rosaceae*	*Scory (Dried Apple Chips), Juice, Wine, Jam, Chutney, Jelly (P)*	*Fruit*	September–October	Abandon lots, parks, and roadside
8.	*Morchella esculenta*	Guchhi	Morchellaceae	Gucchi ka Madra and Gucchi ki Sabzi (P)	Fruiting body	March–July	Abandon lots
9.	*Myrica esculenta*	Kafal	*Myricaceae*	*Raw consumption (R)*	*Fruit*	May–June	Parks and roadside
10.	*Pleurotus ostreatus*	Keun	*Pleurotaceae*	*Keun ki Sabzi (P)*	*Fruiting body*	April–October	Abandon lots
11.	*Prunus armeniaca*	Chulli	*Rosaceae*	*Chulli (P,R)*	*Fruits, seed*	May–August	Roadside
12.	*Prunus avium*	Cherry	Rosaceae	Raw fruit eaten (P)	Fruit	May–July	Abandon lots and roadside
13.	*Prunus domestica*	Palam	Rosaceae	Chutney, Jam and Jelly (P,R)	Fruit	September–October	Roadside
14.	*Prunus dulcis*	Badam	*Rosaceae*	*Raw consumption (R)*	*Fruit*	June–September	Abandon lots, parks, and roadside
15.	*Punica granatum*	Dahru	*Punicaceae*	*Anardana (dried fruit) for chutney preparation (P)*	*Fruit*	*August–September*	Parks and roadside
16.	*Rubus ellipticus*	Akhe	Rosaceae	*Raw consumption (R)*	*Fruits*	May–June	Riverside walk and roadside
17.	*Taxus wallichiana*	Rakhal	*Taxaceae*	*Rakhal ki Chai (P)*	*Bark*	December–February	Roadside

14.6 Way Forward

Populations in Himalayan regions diversify their food sources as represented in Tables 14.1 and 14.2, therefore educating them through capacity building and awareness programs especially women about urban foraging in mountain terrain can aid in the development of resiliency and food security. Apart from tourism and other source of income of livelihood in Himachal Pradesh, urban foraging can provide people the chance to increase their household income. For food foraging sustenance, governance needs to be more effective, institutions and tools need to work together more closely to improve the quality standard of foraging (Movilla-Pateiro et al. 2021). Democratic procedures and social inclusion can serve as important building blocks for the transdisciplinary urban foraging knowledge production and the formulation of sustainable decision-making processes (Fig. 14.1). Strict regulations and quality control measures in place to ensure the safety of wild food resource to inform the public about the benefits and risks of urban foraging become vital (Mclain et al. 2014). Efforts should be made to respect cultural values and social norms related to food consumption, and different user groups should work collaboratively to identify and address conflicts related to access to edible resources in the state. Perceived barrier for foraging in the city includes social acceptance and access to the green spaces (Schunko and Brandner 2022). In order to bridge the perception gaps between supporters of urban foraging and public authorities and to foster cooperative efforts for sustainable urban foraging, multilateral acknowledgment and research in implementing preventative measures by foragers and the challenges of urban foraging are required for promising urban foraging practices (Schunko et al. 2021). To reduce current threats to biodiversity and to foster urban foraging as a sustainable and biodiversity-friendly practice, human–nature interaction coordinated with environmental policy-based approaches, environmental education, and green space management becomes important (Fischer and Kowarik 2020). The idea of incorporating edible plants serves as a good justification for promoting and protecting native edible plants as well as for reconsidering the layout and use of urban open areas (Hajzeri and Kwadwo 2019). Therefore, understanding the concept of urban foraging in urban landscapes has potential for long-term conservation strategies like community-based conservation initiatives and fulfillment of sustainable development goals (Plate 14.1).

Plate 14.1 (**a**) *Pleurotus ostreatus* on sale, (**b**) *Nasturtium officinale foraging* (**c**) Urban landscape of Lahaul District and (**d**) *Hippophae* sp.

Conflict of Interest Authors declare no conflict of interest to draft and publish the chapter.

References

Anees MM, Banzhaf E, Wang J, Joshi PK (2023) Quality index approach for analysis of urban green infrastructure in Himalayan cities. Land 12(2):1–21. https://doi.org/10.3390/land12020279

Bargali H, Mathela M, Sharma R, Sharma M, Yaming D, Kumar A (2021) Plant studies in Himachal Pradesh, Western Himalaya: a systematic review. J Mt Sci 18(7):1856–1873. https://doi.org/10.1007/s11629-020-6401-z

Brandner A, Schunko C (2022) Urban wild food foraging locations: understanding selection criteria to inform green space planning and management. Urban Forestry and Urban Greening 73:127596. https://doi.org/10.1016/j.ufug.2022.127596

Chawla A, Parkash O, Sharma V, Rajkumar S, Lal B, Gopichand, Singh RD, Thukral AK (2012) Vascular plants, Kinnaur, Himachal Pradesh, India. Check List 8(3):321–348. https://doi.org/10.15560/8.3.321

Clouse C (2022) The resurgence of urban foraging under COVID-19. Landsc Res 47(3):285–299. https://doi.org/10.1080/01426397.2022.2047911

Dhyani S, Kadaverugu R (2020) Food security and cultural benefits from urban green spaces: exploring urban foraging as a silently growing global movement. Clim Change Environ Sustain 8(2):219–225. https://doi.org/10.5958/2320-642X.2020.00022.8

Fischer LK, Kowarik I (2020) Connecting people to biodiversity in cities of tomorrow: is urban foraging a powerful tool? Ecol Indic 112:106087. https://doi.org/10.1016/j.ecolind.2020.106087

Fontefrancesco MF, Zocchi DM, Pieroni A (2023) The intersections between food and cultural landscape: insights from three mountain case studies. Land 12(3):676. https://doi.org/10.3390/land12030676

Forest Survey of India (2019) State of forest report, 2019. Available at: https://www.fsi.nic.in/forest-report-2019

Gardner JS (2002) Natural hazards risk in the Kulu district, Himachal Pradesh, India. Geogr Rev 92(2):282–306. https://doi.org/10.1111/j.1931-0846.2002.tb00008.x

Glaros A (2023) Sustainable food systems transformation, from critique to practice: describing and designing food futures (Doctoral dissertation, University of Guelph)

Hajzeri A, Kwadwo VO (2019) Investigating integration of edible plants in urban open spaces: evaluation of policy challenges and successes of implementation. Land Use Policy 84:43–48. https://doi.org/10.1016/j.landusepol.2019.02.029

Haq SM, Khoja AA, Lone FA, Waheed M, Bussmann RW, Casini R, Mahmoud EA, Elansary HO (2023) Keeping healthy in your skin—plants and fungi used by indigenous Himalayan communities to treat dermatological ailments. Plan Theory 12(7). https://doi.org/10.3390/plants12071575

Khongsdier R (2007) Bio-cultural approach: the essence of anthropological study in the 21st century. Anthropologist 3:39–50

Landor-Yamagata JL, Kowarik I, Fischer LK (2018) Urban foraging in Berlin: people, plants and practices within the metropolitan green infrastructure. Sustainability 10(6). https://doi.org/10.3390/su10061873

McLain RJ, Hurley PT, Emery MR, Poe MR (2014) Gathering "wild" food in the city: rethinking the role of foraging in urban ecosystem planning and management. Local Environ 19(2):220–240. https://doi.org/10.1080/13549839.2013.841659

Movilla-Pateiro L, Mahou-Lago XM, Doval MI, Simal-Gandara J (2021) Toward a sustainable metric and indicators for the goal of sustainability in agricultural and food production. Crit Rev Food Sci Nutr 61(7):1108–1129. https://doi.org/10.1080/10408398.2020.1754161

Rathore S, Shashni S (2023) Indigenous knowledge system and livelihood option of natives of Lahaul and Spiti District, Himachal Pradesh. In: Climate change adaptation, risk management and sustainable practices in the Himalaya. Springer, Cham, pp 213–235

Rombach M, Dean DL (2023) Exploring key factors driving urban foraging behavior in garden and non-garden locations. Foods 12(5):1–18. https://doi.org/10.3390/foods12051032

Russo A, Cirella GT (2019) Edible urbanism 5.0. Palgrave Commun 5(1):1–9. https://doi.org/10.1057/s41599-019-0377-8

Samant SS, Pant S, Singh M, Lal M, Singh A, Sharma A, Bhandari S (2007) Medicinal plants in Himachal Pradesh, north western Himalaya, India. Int J Biodivers Sci Manag 3(4):234–251. https://doi.org/10.1080/17451590709618177

Sardeshpande M, Shackleton C (2020) Urban foraging: land management policy, perspectives, and potential. PloS One 15(4):1–26. https://doi.org/10.1371/journal.pone.0230693

Schunko C, Brandner A (2022) Urban nature at the fingertips: investigating wild food foraging to enable nature interactions of urban dwellers. Ambio 51(5):1168–1178. https://doi.org/10.1007/s13280-021-01648-1

Schunko C, Wild AS, Brandner A (2021) Exploring and limiting the ecological impacts of urban wild food foraging in Vienna, Austria. Urban For Urban Green 62:127164. https://doi.org/10.1016/j.ufug.2021.127164

Radhika Sood is an AcSIR PhD scholar at CSIR-NEERI, Nagpur working on assessing ecosystem health and mapping cultural ecosystem services in high conservation value forest ecosystems in the Indian Himalayan Region.

Shalini Dhyani (PhD) is a Principal Scientist at the CSIR-National Environment Engineering Research Institute, India. She is Asia Vice Chair and Steering Committee member in the Commission on Ecosystems Management of IUCN and Visiting Scientist at CIFOR-ICRAF.

Part V

Practitioner Perspectives

Foraging as Sustenance and Dissent in the Villages Bordering Bengaluru

15

Dhanya Rajaram

Abstract

We walk with the village women of Sarjapura as they forage wild weeds, as one of the last generations to hold this knowledge. The article also focuses on the efforts made by a few individuals to pass on this knowledge to wider audiences as new threats like unsustainable development and changing lifestyles come as barriers to foraging. We also take a look at the edible weeds and their local uses in culinary and medicinal preparations.

Keywords

Ethnocultural ritual related to women · Edible weeds · Ancient knowledge · Changing landscapes

15.1 Sumptuous Weeds that Feed and Fortify

Nagarathnamma's yellow saree flutters in the wind, as she bends over and plucks out greens after cursorily giving them a once over (Fig. 15.6). The earth is wet from the rains from the night before. Her saree bulges like a kangaroo's pouch, the loose end or pallu holding the greens she has already amassed. A whole assortment of edible plants is nestled in there, each with a distinct taste and quality. In her 60s, she is one of the few women who can identify the wild edible greens growing around Volagerekallahalli, her village (Fig. 15.1).

There is Kakke soppu (*Solanum nigrum*) (Fig. 15.2a) or black nightshade, a perennial green with sweet black berries that children fight over to eat, so delicious it is. The leaves are egg shaped and grow alternate to one another along the stem.

D. Rajaram (✉)
Bangalore, India

S. Dhyani, M. Sardeshpande (eds.), *Urban Foraging in the Changing World*, https://doi.org/10.1007/978-981-97-0345-6_15

Fig. 15.1 No need for bags: Foraged weeds nestled in the folds of a saree

a)Anne soppu (*Celosia argentea*)

b)Kakkesoppu (*Solanum nigrum*)

c) Attika Mamidakku (*Trianthema portulacastrum*)

Fig. 15.2 (**a**) Kakke soppu (*Solanum nigrum*). (**b**) Anne soppu (*Celosia argentea*). (**c**) Atika mamidaku (*Trianthema portulacastrum*)

The leaves and tender parts of the stem possess a subtle taste when cooked. They are much sought after. Then there are the arrow-shaped bright green leaves of Anne soppu (*Celosia argentea*) (Fig. 15.2b) or plumed cockscomb with gorgeous spear-heads of pink flowers. Its leaves are eaten by itself or mixed with other greens. The flowers and seeds are used to cure ailments of the bowel system by alternate medical practitioners but the women unwittingly use the greens simply as a freely available source of nutrition.

Attika mamidaku (*Trianthema portulacastrum*) (Fig. 15.2) or Horse purslane is your classic definition of a weed with a creeping stem and thick spade-shaped leaves. Its pink flowers tend to be solitary and stalkless. The leaves are boiled and mashed into curries with onion and tomatoes and a tempering of spices. Wild Amaranthus grow in several different forms and have local names. *Cenchilakku* (*Digera muricata*) (Fig. 15.3) is a false Amaranth, and the entire plant is boiled and made into curries. It is often used with Anne soppu to make *Masoppu*, a mashed green curry that includes the creaminess of added pulses, like cooked split pigeon pea.

Fig. 15.3 Preferred mixed green leafy edibles foraged for various traditional food preparations. (**a**) Cenchillaku (*Digera muricata*). (**b**) Kommeakku with Flower(*Boerhavia diffusa*). (**c**) Kommeakku (*Boerhavia diffusa*). (**d**) Joyweed, Ponnanganni, Hongane(*Alternanthera sessilis*)

Amaranthus can be purple or green and carry quite some weight, being a stocky and heavy plant with open clusters of seeds that look like poppy. Of the several sisters in Amaranthus, Joyweed (*Alternanthera sessilis*) (Fig. 15.3) or *Hongane* is the most beloved as it grows on lake beds and is tender and toothsome. Sometimes the women in the village choose to cook it alone as a showstopper. Then there is *Pulsakku* (*Oxalis*) or Sorrel, the telltale sour, clover shaped leaves of three that never grow more than a foot in length, always choosing to be ground cover. It is seldom used alone and is always used sparingly due to its sour taste.

Looking for and identifying the greens is part of Nagarathnamma's makeup. She has done it since childhood with her mother Ammayakka and the countenance of each edible plant is second nature to her. After collecting many a handful of tender, wild and edible greens, she sometimes adds her haul to her daughters or neighbors and it becomes a veritable mountain of greens that gets divided and cooked into a curry ready to satiate a large family's hunger.

A wide number of recipes and techniques abound for the cooking of the leaves depending on the factors of whether there is more of a single green or a mixture of many greens. Kakke soppu (*Solanum nigrum*) and Hongane (*Alternanthera sessilis*) or any of the other greens (except *Oxalis*) are often cooked by themselves if found in copious amounts, but can be used in a pinch combined with other greens.

They are often combined with pulses and boiled to a soft mush and then added to a masala prepared with onion, tomatoes, green chillies, ginger, garlic, cooked in oil and tempered with mustard, cumin, and turmeric. Sometimes grated or ground fresh coconut is added to it for a bit of sweetness and sometimes it is cooked like a green soup with no pulses but just the boiled and ground greens with onion, tomatoes, and spices. Following is a recipe where the greens shine without any pulses or coconut. It's called Miram Saaru, "miram" meaning pepper.

15.2 Recipe for Miram Saaru

Start by washing the greens with liberal amounts of cold water. Then boil or pressure cook with two onions, a few tomatoes, several peeled garlic, chilies, and a few peppercorns. Boil until the leaves are tender and mushy. Once cooked, cool the mixture and grind it into a pulp using a blender but try to keep it coarse. The women in the village usually use a wooden masher to achieve this. Pour sesame

Fig. 15.4 Key green leafy ingredient for Miram Saaru

or coconut oil into a hot frying pan and add mustard seeds and cumin. Allow these to sputter then add an onion diced into fine pieces and sauté. Add crushed ginger garlic paste. Cut a few sour tomatoes finely and sauté. The masala is ready. Add the masala to the green mixture and boil for 10 min or so till the flavors meld. Miram Saaru is ready (Fig. 15.4).

15.3 A Variety of Mixed Greens

Usually the mixed green curry is accompanied by *Ragi Mudde*. *Ragi Mudde* is a traditional staple of finger millet flour cooked in copious amounts of water till it turns solid then shaped into hefty balls and cooled till they hold shape easily. These brown orbs are pinched into small mouthfuls and eaten with the green curry. Rice is also another accompaniment to curry. Table 15.1 appended to this article lists ethnobotanical information on these greens.

Table 15.1 Names, life cycle, and uses of select foraged plants

Local and scientific name	Family	Common English name	Life cycle	Culinary and medicinal uses
Kakke soppu (*Solanum nigrum*)	Solanaceae	Black nightshade	Common, short lived perennial herb	Commonly used in curries or sauteed. Medicinally known to cure coughs, indigestion, and bacterial infections
Anne soppu (*Celosia argentea*)	Amaranthaceae	Red spinach/ plumed cockscomb	Annual (mid spring to summer)/can also be perennial	Commonly used in curries. Medicinally used to cure osteoporosis and jaundice among many other ailments. Flowers used to treat hemorrhoids and piles
Attika Mamidaku (*Trianthema portulacastrum*)	Aizoaceae	Horse purslane	Annual (April to June) can also be perennial. Red variety abundant, green variety appears earlier in the season	Commonly used for culinary purposes. Medicinally used as an antidote for alcohol poisoning
Cenchilakku (*Digera muricata*)	Amaranthaceae	False Amaranth	Annual (flowering August to September)	Commonly used for culinary purposes. Flowers and seeds are medicinally used to cure urinary disorders
Hongane (*Alternanthera sessilis*)	Amaranthaceae	Joyweed	Perennial	The leaves and young shoots are cooked and eaten. Medicinally used to treat bronchitis and asthma. New breastfeeding mothers eat it for increased lactation
Kommeakku / Punarvana (*Boerhavia diffusa*)	Nyctaginaceae	Hogweed	Annual (germinates during the first rains of May and regerminates throughout the monsoon). Can be perennial in wet conditions	Commonly eaten as food. Medicinally used to cure obesity and is considered a powerful antioxidant

(continued)

Table 15.1 (continued)

Local and scientific name	Family	Common English name	Life cycle	Culinary and medicinal uses
KundliKommu (*Boucerosia umbellata*)	Apocynaceae	Umbelled Caralluma	Perennial succulent	Medicinally used to treat stomach disorders
Kuppe Akku (*Acalypha indica*)	Euphorbiaceae	Indian Nettle	Annual (flowering and fruiting between July–December)	Commonly cooked. Highly prized in Tamil Sidda medicine as a rejuvenating herb. Also known to have an effect on domestic cats, similar to catnip

15.4 Golden Hands that Forage

It costs nothing to forage except time and the women who forage believe that their families are energetic and strong the whole year, due to the wild and hardy nature of these plants. As they collect the weeds, the women are unknowingly performing an act that is tens of thousands of years old. Their proto-ancestors did the very same to feed their clans. It is more than just a handful of greens, and it is an ancient and ethnocultural ritual.

Most of the young women in the same village today do not possess this skill. Buying from the local grocer has become so normative that even the farmers depend on the grocer for their daily kitchen needs. They cultivate large acres of mono crops like white pumpkin or chilly or worse still, flowers, which are expressly grown for the market to be sold wholesale to vendors who will take it to the far reaches of the city. The farmers hop on a two-wheeler as easily as anyone else to go buy vegetables and greens.

Edible weeds are also in the wane due to loss of habitat. Hapless and widespread construction by builders, land use change, and loss of topsoil are all contributing factors to the widespread loss of weeds.

In Hosahalli where Suresh Kumar's farm (Fig. 15.6), **Sarjapura curries** (Samuha n.d.) is located, quite a few women still carry on the tradition of annual foraging. Bagyamma, Sarojamma(Fig. 15.5) and Sheelamma are friends. They are similarly aged, married with children and all graze cattle for a living. The cattle grazing are apart from the jobs that the other family members perform. It is their very own independent calling (Fig. 15.6). They feed the cattle, day in and day out, (not taking even a single holiday since this would mean the bovine missing its feed) and milk the animal and sell it too. The cattle take them on long walks and as they graze on sweet grasses and weeds, the women follow their lead and forage the weeds too (Fig. 15.7).

Fig. 15.5 Sarojamma showing Kuppeakku (*Acalypha indica*)

Fig. 15.6 Nagarathnamma foraging along with Sharadamma

Fig. 15.7 Suresh Kumar holding Kommeakku (*Boerhavia diffusa*)

15.5 Spreading the Knowledge Through Community Initiatives: Sarjapura Curries

Suresh Kumar has learnt foraging from Nagarathnamma and Rathnamma who happen to be his aunts. He also has a habit of regularly tagging along with the village women to go on foraging sprees in Volagerekallahalli (Fig. 15.8). He has been running a workshop for people to identify and eat weeds growing around their environs and has been propagating several of the species and handing them out to home gardeners and farmers alike(Tharayil n.d.).

An artist and organic farmer, since 2019 the foraging workshops are one among the many initiatives led by him to inculcate a sensibility toward the land and the food it offers. Foraging is how Suresh began his journey into farming and propagating native plants. Veritable strides were made in finding forgotten edible weeds during this year. Sarojamma's back is bolt-straight as she points out a weed. She shows *Kuppeakku* or *Kuppe soppu* (*Acalypha indica*), a weed that grows easily around sidewalks and roads

Fig. 15.8 Community of foragers from VolagereKallahalli

(hence the name Kuppe (garbage/throwaway)- Soppu (green)) is eaten regularly enough but is also known to cure coughs in children. She isn't going to reach for a cough syrup anytime soon. The same way, multiple cures present themselves in the weeds like a real-time magical story for woes small and big. Or that is how it appears to our senses habituated to procuring things from supermarkets.

Foraging for Suresh is one more way to get closer to a way of life that is quickly being forgotten. He despairs the way food has become stripped of all its meaning and antecedents to be packed in Styrofoam and plastic to travel wide distances in refrigeration, bought via an app on the phone and eaten listlessly in front of televisions. Surely there is a better way to meet our food?

Moreover, he worries that the current crop of women foraging in the village might be one of the last to do so. He doesn't want it to be that way. It is tricky to embark on a mission to teach the farming community about food and tradition but he is doing so and there is some change in the offing. Many of them are growing kitchen gardens and also come along on the foraging escapades. The surprising thing is the new variegated groups of foragers who have sprung up. They look nothing like Nagarathnamma or Bhagyamma. They rustle in jeans and cotton T-shirts and sport backpacks and sunglasses.

Suresh gets a large number of students, faculty, researchers, and people of all walks interested in biology come forth to learn foraging (Kumari 2022). During the isolating months of Covid, with more time on their hands, many took to gardening on their rooftops and there has been renewed interest in both native species and in wild edibles (Tripathi 2019). This group falters quite a bit and needs a lot of reassurance about what is edible and safe but it is a committed one. There is a lot of renewed interest in foraging and medicinal plants in recent years, and the foraging workshops are almost crowded and colorful with people and children, bending, plucking, smelling, and talking to each other.

The foraging meetups are very informal and look like gatherings rather than a lecture. The foraging meetups happen in both villages bordering suburban spaces and also in urban green belts like lakes and public parks. Once the group have assembled Suresh walks in a characteristic meandering way looking for weeds. It isn't possible to predict what one might find and there are always surprises. Once an edible weed has been spotted, Suresh calls the crowd and shows them the plant, names it, and describes its uses. Many take photographs of the plant and keep it for reference. Usually by the end of a 2-h walk, there are at least a half a dozen edible weeds that have been identified. In the village, they are collected and often made into a meal at the end of the walk and shared with everyone who has been present.

15.6 The Threat to Weeds and a Prehistoric Knowledge

Tractors and water tankers drive past Hosahalli every half hour. The economy is no longer only agrarian (Aiyappa 2023; Kodira 2019). Several large apartment complexes have cropped up in the last decade to house the growing population of the city (Times Property 2023). A prominent property website writes "*Alongside the towering office buildings, the locality* [Sarjapura] *is also home to lush green spaces, serene lakes, and a vibrant cultural scene.*" It also lists waterlogging as a setback to buying or renting property in the area. What it conveniently leaves out is the fact that these were all primarily farmland till very recent times.

The farmers themselves find it more and more difficult to find farmhands to work their lands. Several of them sell water from their wells to the apartments, to make a living. Many work in different capacities as gardeners, house help, etc. in the apartments. Some of the bigger landowners have managed to ride the boom by selling their land at a great profit to builders. These far and few farmers live cushy lives, preferring to not farm anymore. The village is teetering on the verge of change, and the change is not an ecologically solid one. It most probably will not take trees, weeds, or lakes into consideration. It will be one where elevators and car parks will get better consideration. The apartment complexes also choose to grow only selective ornamental plants in the small designated manicured spaces they choose to call gardens. There is no space for weeds, let alone edible ones. In the current climate, foraging here is also an act of building a bond with the earth and all that it has to offer. It is also a form of dissent at its very core. Toward urbanization, toward mono crop agriculture, toward the lack of awareness about where food originates and how it does so. Just with the plucking and eating of a weed (Plates 15.1 and 15.2).

Plate 15.1 Saralamma out foraging

Plate 15.2 Cattle grazers with who Suresh goes looking for forgotten plants

15.7 Calling Ethnobotany Detectives: Plants Needing Identification!

This section provides some insights to interesting wild edible plants that are preferred for foraging and food preparations as locals are known about their benefits and values. However, identification and insights from ethnobotanists can highlight more values that includes their medicinal as well as aromatic properties (Fig. 15.9).

Fig. 15.9 (**a**) Yanigemundlu. (**b**) Kundli kommu or rabbit horn. (**c**) Karkodlu. (**d**) Karkodlu bush also called Cape Dates, the yellow fruits are sweet

References

Aiyappa M (2023) Simpler land conversion will boost investments in Karnataka. Times of India. https://timesofindia.indiatimes.com/city/bengaluru/simpler-land-conversion-will-boost-investments-in-karnataka-experts/articleshow/96999072.cms. Assessed on 15 May 2023

Kodira PP (2019) How the upcoming Satellite Town Ring Road will affect lives and livelihoods around Bengaluru. Citizens Matters. https://bengaluru.citizenmatters.in/bengaluru-metropolitan-region-satellite-town-ring-road-strr-livelihood-agriculture-impact-39601. Assessed on 2 June 2023

Kumari B (2022) The joys of urban foraging. Deccan Herald. https://www.deccanherald.com/specials/the-joys-of-urban-foraging-1133556.html. Assessed on 15 May 2023

Samuha S @sarjapura_curries (n.d.) *Posts* [Instagram profile]. Instagram. Retrieved September 3, 2023, from https://www.instagram.com/sarjapura_curries/?hl=en

Sarjapura Curries [@sarjapura_curries] (2021). "Jedidiyah Alfred says" After two years of looking up at trees, it's now time to look down .. at weeds. So much food at our feet!" [Photograph]. Instagram. https://www.instagram.com/p/CS1yGJpl2z6/?hl=en. Assessed on 10 Mar 2023

Tharayil S (n.d.) Uncultivated greens: reawakening forgotten food traditions. Reawakened. Agrobiodiversity explained. https://www.thelexicon.org/reawakened/uncultivated-greens/. Assessed on 9 May 2023

Times Property (2023) Discover the charm and growth of Sarjapur, Bengaluru: a perfect blend of urban development and natural beauty. Times Property. https://timesproperty.com/news/post/sarjapur-bengaluru-locality-guide-blid4696#3. Assessed on 12 Mar 2023

Tripathi S (2019) As Bengaluru loses land to concrete, a small initiative wants to remind the city of its green past. Scroll.in. https://scroll.in/magazine/926156/as-bengaluru-loses-land-to-concrete-a-small-initiative-wants-to-remind-the-city-of-its-green-roots. Assessed on 17 Aug 2023

Dhanya Rajaram is a writer with a background in Art History. She writes about Art, Food, Travel, Literature, and Biology. She is also a citizen scientist with a keen interest in watching all the life around her.

Foraging in Urban India: Finding Food and Cure

16

Abhijit Sinha and Sharmila Sinha

Abstract

Communities the world over have been foraging since time immemorial. More so in the global South. As cities developed so did urban foraging. In India, foraging to meet nutritional needs has always been the norm.

The authors through this paper bring to fore the need to understand urban foraging through the lens of community storying, sharing, and thereby sensitising lay persons of its benefits. They, along with an urban ecologist, lead fellow walkers through Sundar Nursery (an arboretum) which lies at the heart of India's capital Delhi. During this walk, they are also accompanied by an unschooled young man, who carries with him knowledge of plants and their medicinal properties. Generational memories of food and foraging are exchanged with the community that joins in, while on this walk.

The paper also delves into the need for open green spaces while planning cities. Urban foraging is important to build this community where foraging is understood by city bred populace and determines how and what they eat. As the community of responsible foragers increases, it will meet the food needs of the increased population.

Some food stories and recipes have been shared in this paper too.

Keywords

Community-based knowledge · Sacred groves · Preserving weeds · Medicine in the wild · Food that heals · Urban planning

A. Sinha · S. Sinha (✉)
Delhi, India

S. Dhyani, M. Sardeshpande (eds.), *Urban Foraging in the Changing World*,
https://doi.org/10.1007/978-981-97-0345-6_16

16.1 It Always Starts with a Story

The little boy is walking, frolicking a bit ahead of his parents. He bends down and picks up the tiny berries from a bush with spiny thorns. He looks back with a smile and gobbles the orange berries. Walking on the sparsely populated roads of Ladakh in the western part of the Himalayas in India, at an altitude of about 3000 metre in the month of October, one can see these orange berries growing abundant on the roadside.

Sea buckthorn, (Fig. 16.1) a 'wonder berry', grows wild above the tree line in this cold desert and the boy knows so. His parents and neighbours pluck these berries, which are then processed and bottled after extracting the juice and trucked off far and wide. Leh Berry (Leh is the main town of this province), in local parlance, has been reaping benefits for the local communities, as well as for those who need it for medicinal purposes.

The orange and yellow berries that dot the dry, rough mountain terrain have various medicinal properties. The generic name *Hippophae* in Latin comes from horse because in ancient Greece it was fed to the horses for weight gain and a shiny coat (Ontario 2023).

For centuries, it has been used as food and for medicinal purposes. Talking to the Ladakh community, one gets to know that it lowers fever, reduces inflammation, counteracts toxicity and abscesses, cleans the lungs, and treats cold and cough, and is effective in arresting tumour growth, especially in the stomach and oesophagus (Elford 2013). The Ladakhis brew an herbal tea by adding a few berries with the young leaves, while the tender thorny stalks, leaves, and fruits are chewed by the animals. Travellers are advised to suck the fruits to keep altitude sickness at bay.

Housed in spiky thorned bushes, these berries are not so easy to forage—high in protein, vitamins C and E, and organic acids—that provide the much-needed nutrients to humans and their animals.

Fig. 16.1 Sea buckthorn—*Hippophae rhamnoides*

This is the reason why it is called wonderberry—growing at the high altitudes of the Himalayas, the Hindu Kush range, and other cold desert regions. The local populace swears by its medicinal properties—a sure cure for gout, skin rashes, and more (Elford 2013).

Today, the sea buckthorn's properties are being researched in the field of biotechnology, nutraceutical, pharmaceutical, cosmetic, and environmental sciences. The oil, juice, cream, shampoo, and several other products travel to Europe and the Americas—fetching good revenue for the community that forage, wash, dry, and then extract the oil, to sell them far and wide (Larmo et al. 2014).

The shrub can also be seen growing wild in Himachal Pradesh, Uttarakhand, Sikkim, and Arunachal Pradesh states of India—ranging from west to east of the Himalayas—where vegetation is scant, with an average annual rainfall of 50 millimetre. The thorny bushes arrest soil erosion, attract wildlife, and are good as farm boundaries warding off big predatory animals. During the late 1990s and early 2000s, sea buckthorn was used for greening the mountainous terrain under the Cold Desert Afforestation Programme (Parvaiz 2014).

16.2 Sea Buckthorn: Rising Potentials

Harvesting of these berries is done for 3 to 4 weeks a year, when approximately 5% of the fruits are plucked. Majority of the berry collectors are from the economically weaker section of the society, with women constituting 67.4% of the work force.

While mean annual berry harvest was 213.4 ± 83.9 metric tonne (MT) in the period 2004 to 2015, the average income generation of individuals in 2013 was Rs 18,375—plucking for 12.6 days (on an average) during the season. With the increasing trend in demand for sea buckthorn, the berry harvest in Ladakh is expected to increase to 1509 MT in 2030 (as against 361.3 MT in 2016).

The Forest Department of Ladakh estimates that 2500 ha of barren land can be brought under sea buckthorn plantation by 2030 without much investment in Leh district, with a projected net income estimated at Rs 491 crore (USD 72 million). Value-added consumer products, if manufactured in the region, will increase the earnings of the extremely hardy people of the region (Stobdan et al. 2017).

16.3 Why Foraging

To forage, one needs to walk off the beaten path—literally and figuratively. First, to find, between the bushes and under the shade of large trees—and then, identify what is edible and what isn't. As human history developed, so did foraging. To meet food and nutritional demand, from the days of the hunter-gatherers to settlers and then the advent of civilisations, humans have been picking, scanning, tasting, and eating weeds and grasses (from which came the cereals), fruits and vegetables—as well as hunting animals, fish, and insects.

Fig. 16.2 Amrul (*Oxalis corniculata* L.)

This is what made them bipedal (Wayman 2012) up on their 2 feet. From being arboreal and foraging for food among the treetops, to coming down on the grasslands and other open areas. Foraging also made them discover and develop food as per their need. The Fertile Crescent, on the banks of the Rivers Euphrates and Tigris, is where the first remnants of agriculture have been reportedly found. Foraged barley is supposedly the first grain that was domesticated.

Humans then, and till date, forage from forests, open spaces in cities, and various other places for survival—not only to fill their stomach, but also their mind. As the world steps out of the COVID-19 impact, humans have realised that they have to get back to nature. To embrace food that is local, seasonal, and healthy (as much without pesticides as possible). Foraging meets these needs.

Micronutrients from Amrul (*Oxalis corniculata L.*) (Fig. 16.2) and Punarnava (*Boerhavia diffusa*) can never be gained as a tablet or capsule (though they are sold in bottles). This remains true for so many other weeds, grasses, and fruits, which are only available to a person who can recognise them as affordable nutrients that build immunity.

Foraging is therapeutic too—walking, picking up what one needs, and interacting with those who are doing the same. This walk, talk, and pluck combine to contribute to community building, much needed for one's health and well-being, especially during the COVID crisis, and supporting human and social capital.

16.4 Walking to Learn About the Herbs and Medicinal Plant: Sundar Nursery, Delhi

A wealth lies unheeded on the sides of the car path inside Sundar Nursery, an arboretum, in Delhi. In the heart of capital of India, adjacent to a locality with a high density of population, we walk to experience the plethora of trees, shrubs, herbs, birds, bees, butterflies, and insects. The 69-acre green space is vibrant and thriving with life, perhaps unknown to the visitors who drive past and park under the shade of the majestic trees that have been standing tall for long.

Formerly known as Azim Bagh or *Bagh e Azeem*, the sixteenth-century garden which had become derelict, lies adjacent to the Humayun's Tomb (mausoleum of the second Mughal ruler of India). Of the 15 restored monuments within the park, six are listed on the UNESCO World Heritage Sites after being revitalised and landscaped.

Originally, this garden was laid out beside the Mughal-era Grand Trunk Road that once ran from Peshawar in the northwest to Bengal in the east of India. Spread over 90 acres (36 hectare), the magnificent garden also had a lake, for the travellers to rest and slake their thirst.

During the last phase of the British rule in India, an arboretum was created with plants brought from various parts of India and other British colonies. Two years after Independence, in 1949, the nursery gates were opened to the public. Besides trees, herbs, shrubs, this place has been home to birds and butterflies too (CPWD 2023).

In this nursery, there is a Cocoa tree and a Brazilian Ironwood tree, brought here nearly a century ago by the British colonisers. Among them stand tall native trees of this land like Chamrod (*Ehretia laevis*), Ber (*Ziziphus mauritiana*), Arjun (*Terminalia Arjuna*), and Peelu (*Salvadora persica*).

Strolling under these trees, one can see Doob or Durva grass (*Cynodon dactylon*). Popularly known as Bermuda grass, it is a perennial herb that spreads through stolon and rhizomes, and readily takes roots at the nodes in the plains and hills all over India. A native of Asia, North Africa, southern Europe and Australia, this grass has found its place of importance in India, more so for its medicinal value, but cleverly inducted (to preserve and protect) as an essential religious requisite by our ancestors having deep knowledge of nature. No ritual, from birth to death, in a Hindu household, can do without Doob grass.

The herb contains crude proteins, carbohydrates, calcium, sodium, and potassium and has β-sitosterol and β-carotene, triterpenoids, vitamin C, and more (Reshu et al. 2018). Fresh juice is used as a diuretic and astringent to stop bleeding. The primordial Santhal tribe of the prehistoric Chhota Nagpur plateau use this to cure parasitic attack too (Krishna and Amrithalingam 2014).

Ber (*Ziziphus mauritiana*) (Fig. 16.3), the Indian Jujube, has been cultivated as a fruit for so long that no one knows now where it was found originally wild. But it is a fair guess that the dry Delhi region formed part of its native range (Krishen 2008). This evergreen deciduous tree with a pair of spines has several myths surrounding it. In the Hindu mythology, Lord Rama was given *Ber* by a woman from the marginalised community when he was wandering in the jungles, after being banished from Ayodhya.

Ber also has a place of importance in the cultural lore of Bengal (in today's parlance, West Bengal of India and Bangladesh), where it is called Kool. On the day of Basant Panchami in the month of *Magh* (from 15th January to 14th February), the people of Bengal offer this fruit to the goddess of education, Saraswati. One variety

Fig. 16.3 Ber (*Ziziphus mauritiana*), the Indian Jujube (Source: Google)

Recipe

In a pan, add a dash of mustard oil. When hot, add a dried red chilli and a few mustard seeds. As they start to crackle, add the Jujube fruits and stir for a minute. Add a pinch of salt. Stir. Then add a glass of water. Cover and boil till the fruits are tender. Add jaggery as sweetener according to your taste. Bring to boil. Your tangy sweet and sour after meal is ready. Just before removing from pan, throw in a few green chilies. Cover. The green chillies will give that extra zing to the chutney. You could also add a quarter teaspoon of powdered fennel seeds.

of the small, round and creamy brown *Kool*, native of the land, is made into a tangy chutney too. Each part of the tree is used by humans and animals alike. The leaves and the tender stalks are used as fodder, while the bark that gives a red hue is used as a dye agent.

Another tree that is prominent in Sunder Nursery is the Arjun (*Terminalia Arjuna*) that peeps out from here and there. The tall tree can usually be seen alongside rivers, streams, and dry water bodies throughout the Indo-sub-Himalayan and Aravalli foothills. It is also found in the forests of Sri Lanka and Mauritius and grows almost in all types of soil—but prefers humid, fertile loam, and red lateritic soils.

As for its medicinal properties, the bark of Arjun tree helps to cure rheumatic heart disease. This was found after a study in which 30 patients of rheumatic vascular heart disease with congestive heart failure (CHF) were given 200 mg Arjuna bark juice, thrice daily. Result was a significant improvement in the left ventricular ejection fraction (LVEF) and a significant reduction in heart size (Dwivedi and Chopra 2014).

Further down the road, Peelu (*Salvadora persica*), also known as *Salvadora indica* or toothbrush tree, stands tall. Called Miswak in Arabic, its twigs have been used for centuries as a natural toothbrush, and the World Health Organisation has officially recognised the fibrous branches for oral hygiene use.

Research suggests that it contains a number of medically beneficial properties, like being antiseptic, besides being an astringent and detergent. Miswak, according to *The Journal of Health*, can be as effective or more in reducing plaque and gum disease than a regular toothbrush (Sagar 2015). It is unique because of a combination of silica and natural antiseptics found inside the sap that cleans the same way as baking soda or manufactured toothpaste does.

In fact, the word Miswak means tooth-cleaning stick. It is a wild tree that grows in the regions across north-western India to Africa. It is used to treat gingival problems and has proved more effective than tooth brushing for removing plaque. One can chew the tip at either end to create bristles, which also releases oils within the wood that naturally fight plaque and protect teeth.

16.5 Don't Miss the Undergrowth

The robust trees catch the eyes of the passerby. But the overgrown foliage of the ground below is often missed. People trample over them, admiring the big trees and walk on. But beneath their feet lie medicines—mostly in food form for birds and animals as well as for the well-being of humans.

Discerning people come to forage and take them to plant in pots lining their shaded balcony gardens. Hardy, deep rooted, and hugging the ground, the foliage under the Nursery trees sit pretty, as if waiting to be noticed. These sturdy plants, called weeds, grow in various kinds of soil—often creating a microclimate of their own, clinging on to roots or barks of the larger trees.

There are some that green the floor below, peeping from nooks and crannies, from gaps in the rocks, or spreading and covering the wire mesh atop the boundary walls. In the nursery, we spot the *Giloy* that has crept over the fat main stem of the neem (Azadirachta indica) tree, and then spread all over the wire mesh.

16.5.1 Giloy (*Tinospora cordifolia*)

Giloy (*Tinospora cordifolia*) (Fig. 16.4) is called Amrita in Tamil. The creeper is revered all over the country. It is believed to be an immunity booster. During COVID-19, it was being foraged extensively (Kumari 2020) to recover from COVID-related symptoms. The leaves and stem are used as medicine against jaundice and other water-borne ailments, while the starch extracted out of the roots is used to treat dysentery (Krishna and Amrithalingam 2014).

Sharmistha Mukherjee, a housewife who juggles between gardening and singing also loves her plants. She is also walking with us through Sundar Nursery today. 'During COVID, as my fever rose, I started to drink giloy juice. Three times a day about half a cup each time. Within 2 days fever decreased. By the third day, I had stopped giloy juice. You see, excess of everything is not good. Only a measured amount is needed'.

Fig. 16.4 Giloy (*Tinospora cordifolia*)

Several case studies (Varshney 2021) have proved that Giloy is a miracle as long as it is taken in measured amount (Pandey et al. 2020).

16.5.2 Kalmegh (*Andrographis paniculate*)

Kalmegh (*Andrographis paniculate*) (Fig. 16.5) is a seasonal herb native to India and Sri Lanka. It is one of the most popular medicinal plants used traditionally for the treatment of diseases such as cold, fever, diabetes, hypertension, ulcer, bronchitis, skin diseases, diarrhoea, and malaria. Kalmegh powder is haemostatic, hence useful for bleeding piles and has blood purifying properties (Okhuarobo et al. 2014).

16.5.3 Bhang (*Cannabis sativa*)

Bhang (*Cannabis sativa*) (Fig. 16.6) is known in this subcontinent as an ancient medicinal herb, as famous as the Indian hemp Ganja *aka* Marijuana. This plant belongs to the Cannabaceae family. A wonder weed that has travelled from the Himalayas to other parts of the world, it grows wild, and every part of this plant is

Fig. 16.5 Kalmegh (*Andrographis paniculate*)

Fig. 16.6 Bhang (*Cannabis sativa*)

used for various purposes. The paste of fresh Bhang leaves is used to dress wounds and cure sores, while its juice is applied to cure lice infestation and dandruff (Sinha 2012).

16.5.4 Vasaka Leaves (*Justicia adhatoda*)

Vasaka leaves (*Justicia adhatoda*) (Fig. 16.7) is a one-stop medicine for cough and other lung-related diseases. The plant is not tall but bushy and has beautiful white flowers. The sturdy stalks and the leaves, pointed forward, make for a pretty bouquet. Wash off the dust, crush the leaves, and drink the juice. It is surely going to bring relief from your bouts of cough and breathlessness—treating bronchitis, asthma, and dental ailments, cures pyorrhea and bleeding gums.

Fig. 16.7 Vasaka leaves (*Justicia adhatoda*)

16.5.5 Amrul (*Oxalis corniculata L.*) and Punarnava (*Boerhavia diffusa*)

Amrul (*Oxalis corniculata L.*) and **Punarnava** (*Boerhavia diffusa*) are spread all over. Worldwide, *Oxalis corniculata Linn.* is highly used for its antioxidant, antifungal, anti-inflammatory, analgesic, astringent, diuretic, anti-microbial, cardio-relaxant, and several other properties (Anika et al. 2020). Mix the two greens. Add in salads. Stir fry. Add in soups. The sour leaves of Amrul is used to clean brass utensils.

16.5.6 Punarnava (*Boerhavia diffusa*)

Punarnava (*Boerhavia diffusa*) has diuretic, antibacterial, antioxidant properties and is used for menstrual and kidney disorder, abdominal tumour, and much more. It is also good for the heart.

16.6 Sacred Groves

Madhav Gadgil in his essay Sacred Groves: An ancient Tradition of Nature Conservation (Gadgil 2018) writes—a legacy of prehistoric tradition of nature conservation, sacred groves are patches of forest that rural community in developing world protect and revere as sacrosanct. …… not a tree is felled, nor a creature is harmed within its boundaries. These ecological refugia is home to numerous species of trees, medicinal plants, animals, birds, lizards, and many more (Madhav Gadgil 2018).

Near India's capital Delhi lies Mangar Bani a forest grove that has shrunk over the recent decades as has the Aravalli hills, due to human encroachment (Fig. 16.8). Spread over 274 hectares, this grove is protected by the spirit of Gudariya Baba whose temples sit in the middle. Not a twig should be taken outside the grove, as locals believe that the curse of Baba will fall on that person. However, in times of dire need like famine or fire, villagers seek permission of the Baba at his shrine. While trees cannot be axed but fallen twigs and leaves, etc. can be foraged, especially those that are good fodder for the cows and other milch animals that are in every backyard.

In today's world with the shrinking sacred groves, communities are trying to protect as much as they can by cordoning off patches of land as areas of monitored foraging—places where once a week a few families are allowed to forage cattle feed and some medicinal and food plants that are available within the boundaries of the grove. This way all villagers get chance to forage at least once a month one by one, helping the land to replenish its greens.

16.6.1 In My Backyard

In the comic book stories of Asterix and Obelix, Getafix, the druid carries a silver sickle tucked onto his flowing white robe. The druid has a fix for all—from curing Vitalstatix's

Fig. 16.8 Conversation in a sacred grove

weight issues to Dogmatix's sneezes. He deftly climbs trees to pluck mistletoe, and mixing it with various herbs that he gets from the forest—and hey presto, the right concoction has been made for the patient, which will work, druid knows.

Conjuring concoctions from herbs and medicinal plants while walking in the forest has always been traditional knowledge amongst many communities. Even nowadays, we can find these women and men in the small vegetable shops or as our domestic cooks and helps within our localities. Ubiquitous and leading a simple life, these druids of yore have been brought up with such knowledge, learning from their elders. Over the generations, they have the eyes for picking herbs and shrubs that are edible and will find them in the most unlikely of places.

It's summer, **Jamun (Java plum or Indian blackberry)** fruits have ripened. Called *Syzygium cumini* in science books, these fruit trees line the avenues in and around the British built Lutyens's Delhi. So do Bael (*Aegle marmelos*), Ber or Indian Jujube (*Ziziphus mauritiana*), Cluster Fig (*Ficus racemose*), and Neem (*Azadirachta indica*) too can be found all about this part of Delhi.

Come the ripening season in summer, the collectors contracted by municipal authorities spread cloth sheets under the Jamun trees. The juicy blackish blue fruits are shaken off or fall on their own and collected in small wicker baskets to be sold to the car-borne passers-by who stop to buy the succulent seeded fruits.

Bael (*Aegle marmelos*) too grows on thorny trees in the wilder parts of the national capital, but the more edible ones are those planted by discerning druids from amongst the native, and not so native, dwellers of Delhi. The fleshy fruit is plucked or sometimes fall on its own, the shell broken, and the pulp squeezed and strained through a soft cloth.

The juice is refined with water, adding a bit of sugar and salt for taste. Street vendors roam around the city carting the juice in earth pots. Mothers and grandmother share with neighbours and also bottle and store the sweet (or salty as per preference) sherbet, or the family just gulp it down straight. In the dry summer winds of Delhi, the Bael sherbet acts as an antidote (Fig. 16.9).

Another foraged nature's product is **Neem** (*Azadirachta indica*) twigs, flowers, and fruits. The twigs are collected from broken branches and sold throughout the year by the rustic druids, to be used as natural toothbrush by many. The juice of the bitter Neem twigs, chewed as the first and last thing of the day, is healthy and improves digestion.

Neem flowers bloom in spring. Clusters of the tiny white dot bloom briefly and are collected, just like the falling Jamun fruits, on sheets of cloth placed beneath. Fruits (ripening in the months of May and June) of this tree are small and yellow. They are sweet with an after taste of bitterness. Our home-grown druids know the medicinal value of these short-lived flowers and fruits which are made into chutneys by adding slivers of small green mangoes that have fallen from the trees nearby (or bought from the roaming sellers) (Fig. 16.9).

Neem leaves too are in demand with the advent of summer. Chewed raw or sautéed to be eaten with a dab of rice—they act as antidote (Health 2010) to measles and chicken pox (Gopinath and Karthikeyan 2021) (Fig. 16.10).

All these foraged fruits, flowers, twigs, and leaves are a sustenance for a larger population of those whom we often call the marginalised communities. The

Fig. 16.9 Bael (*Aegle marmelos*)—the fruit, the pulp inside, and finally sherbet

Fig. 16.10 Neem flowers and fruits

roadside vendors go around selling their ware, interacting with people, striking a conversation and keeping a watchful eye on the neighbourhood as to who comes and goes. That's how a community stays safe and thrives, profiting from each other.

Wahid and Rakesh sell vegetables on their cycle carts in a colony where there is a lot of demand for such foraged greens. They also sell neem leaves and Bael fruit. A bunch of neem leaves weighing about 100 gms sells for INR 20 each and a Bael fruit for INR 80.

Bael leaves are sold by flower seller Prashant, along with the pooja flowers (flowers offered to Gods) each morning. His boys cycle across the locality to deliver the INR 100 costing packets each morning. This has become a ritual for many homes just like getting the morning newspaper.

Savitri the homemaker will pick up both the packet of flowers for pooja (the flowers will later go into one of the flowerpots adorning the entry gate of her home) and newspaper lying on the floor next to it for the man of the house.

The neighbourhood shop sells dried Kalmegh leaves and twigs for INR 50 each (about 50 gms) and a bundle of six or eight tender twigs of Neem (used as toothbrush by many) are sold for INR 30.

16.7 Building Long-Term Community Resilience in my City

Cities and development are synchronistic with landscapes in constant flux, and changes in socio-economic structures caused by demographic fluidity. Coupled with fluctuating temperatures in relation to climate change, this has had immense impact on the natural systems—water, ecosystems, fertility of land, and more. The loss of biodiversity disrupts the flow of greenhouse gases, soil nutrients, and water (among other things) in the ecosystems, making them less efficient at regulating weather, growing crops, holding fresh water, and so on.

This reduction in habitable land, potable water, and access to food is having grave consequences for the human society and on the quality of life, particularly for the poor (Sardeshpande et al. 2021). Planting of forageable species can improve urban living conditions by providing access to nutrition and recreation for humans, and support and regulation for the urban ecosystem.

Another impacting feature on our lives is the fact that food in cities travel kilometres through a circuitous route. It goes through a chain of industrial production processes, increasing food miles along with cost. This can be mitigated by producing, foraging, and eating local food.

Kitchen gardens in urban backyards were a norm not many years ago, bringing nutrition to the pot and benefiting the habitat too. A few leaves plucked from one or more plants, bushes, or creepers, along with a flower or two from the others, added that extra zest to the curry of fish or vegetable.

But even now, with crunched or no backyards, just a few vegetables and flowerpots can do the trick—lined up on the balcony strips to meet the needs of the everyday greens. Alongside a pumpkin (*Cucurbita*) creeper sprouting from an earth-filled garden pot, it is easy to grow a chilli plant (*Capsicum frutescens*) and butterfly pea (*Clitoria ternatea Linn*). While pumpkin and chillies are consumed as vegetables, tea brewed with an infusion of Aparajita (butterfly pea flower) and its leaves brewed with tea is good to pep one up. Again, the home druids use the warm juice of the leaves, mixed with a pinch of salt, and applied around the ears to relieve earache.

But all the knowledge and variety of such edible and medicinal plants are getting lost with each day. To weather these vagaries, we need to build small food spaces individually, as well as work as a community. The 'new normal'—the habit, and thinking, that food comes only from delivery boys from department stores—is passe. We need to move on. Urban foragers are stepping out to form communities. Walking...building groups of like-minded people who talk, eat, and share foraged food.

In Delhi, more than two decades ago, friends got together to form a walking group called *Knowing and Loving Delhi Better*. For the past 28 years, the group has been walking through the known—and not so known—parts of Delhi, bonding over conversations that satiate hunger of the soul, mind, and of course the stomach too. From this group, an offshoot has branched out that focuses on food, foraging, and farming.

Called *Vividhara*, this online (WhatsApp) group has connected with people through messages to hold small nature fairs, talks, and interactions on and off the field (Fig. 16.11).

Vallari Sheel, an urban ecologist pursuing her PhD, likes to take friends and soon-to-become one, for walks through the parks of Delhi— to show them the trees, shrubs, and the foliage that kisses the ground.

She likes to squat impromptu to pick up a leaf or two and speak about their beneficial aspects. Vallari, like many such passionate young people in Delhi, enjoys spending her time outside in the parks, gardens, and biodiversity zones—watching and reading about nature, and conversing and making friends with her walks.

Then there are accomplished cooks like Swapan Maity, unschooled but understanding the value of herbs and plants through folklore knowledge that probably is generationally imbibed and uses them in his regular cooking. This provides nutrition to the family he cooks for, and his own too.

Fig. 16.11 Vallari Sheel on far left speaks while Swapan Maity on far right (in blue) listens to her and gives his inputs

16.8 The World Is My Home

As the world shrinks in size because of social media, we learn about many social groups that promote foraging. One such *Wild Foodies of Philadelphia* (Sardeshpande et al. 2021; Wild Foodies.org n.d.), a group of 6600 people have a detailed online library and organise walks explaining to its members what to eat, and when. During the pandemic, as the world closed, this group organised several online meetings where the participants were acquainted with winter foraging. This example highlights the evolution of social capital linked to foraging, that adapts to new forms of connection and communication.

A growing number of such groups are making social media their means of communication.

16.9 Conclusion

The purpose of foraging is to create a positive budget that helps save money and gives ample nutrition. The allied purpose is to create and bond with a community. Cities need them—green open spaces in the midst of congested areas are a breather for the residents. A case in point is the Sundar Nursery. People from the densely crowded neighbourhood in Nizamuddin come to relax in the park. They, while returning, will pick up some herbs and leaves for their daily use. So will many others who can identify them.

This is where individual experts like Vallari Sheel help people get to know and understand the nutritional plants, leaves, flowers, barks, and twigs. Similarly, small, focused groups like *Vividhara* have the ability to lead conversations around food foraging and more using communication tools like WhatsApp. With the escalating cost of food, Swapan Maity can provide much-needed nutrition to his family.

Both Vallari, an urban ecologist, and Swapan, an unschooled young man, create green spaces around them to showcase what can be grown and shared and also bring people together to propagate that all weeds are not bad. The weeds that need to be identified, foraged, and eaten too.

16.10 Epilogue

Foraging has been a way of life in the Global South—be it in Asia, Africa or South America. The knowledge of herbs, their nutritional value and way to keep them replenished within communities, has been generationally handed down. These druids still live within communities, going about with their regular work besides engaging with local people and lived locale for their foraging needs.

Memories of food and culture are passed on, I would say genetically. They intertwine between food, feasting and fasting.

Every culture is driven by the landscape, its harvest, big and small. Sacred groves are the protectors of these landscapes, keeping intact their flora and fauna. Locals know how to tread softly for the upkeep of these. In India, each sacred grove will have a deity. The belief is that the divine is responsible for the upkeep of it. Seasonally, associated with some cultural connection, in short, some festive season, foraging of herbs and greens are done with utmost care from these groves. These are haven of greens that meet nutritional need of local population in times of crisis.

These memories are transitioned when humans migrate. They seek to find their small worlds and cultures in and around their new homes. They plant what they foraged. Those who do not have space nor the ability to plant they go looking for them in green spaces within the concrete jungle that they dwell in.

As an example, a house help may have an eye on the green discarded weed from the garden. This weed had been food for him and his family, meeting their nutritional need. So, he recognises it well. He will pick it up to incorporate it in his meal.

Urban foraging is thus hybridisation of knowledge systems—bringing in the science of it through research and including indigenous knowledge. It cannot be scientific alone. The twinning of two is the need of today. Intergenerational knowledge handed down through time tells us what and when to pluck and eat, bringing to fore the need for specific greens for specific season and thus festivity.

So, when future urban planners plan for a community, they need to keep green spaces in their plan process. These spaces give the community an important sociocultural landscape that provide priority to species and the terrain.

Poorly planned community space causes drastic change in geographical distribution, thereby causing degradation of key elements of ecosystems. Another policy bane is when green public spaces are privatised and are beyond reach of everyday person.

These negative aspects change the community's focus too. The rich buy from the shelves of e-markets and malls. The poor depends on cheap packaged food that satiate their hunger but not nutritional need.

The solution to this is to build community-based groups in urban spaces to bring about awareness about grass beneath one's feet which is discarded as weed (Fig. 16.12). These open green spaces where biodiversity is prioritised have multiple benefits besides being recreational, these are places of interactive commons also aid food security.

As mentioned above, food memories are an integral part of humanity. With these memories are intertwined local ethnobotanical and ethnoecological knowledge. Urban habitats when built unsustainably affect the ecology of organisms. The biodiversity of cities plays an important role in human health and well-being, besides providing several ecosystem services.

'All food cannot be bought off the shelf', notes Sharmistha, the home maker. 'That thought has to change. Nonetheless, indiscriminate foraging for personal benefit should not happen too. That will snatch away source of nourishment these plants provide, for many, especially the marginalised'.

Fig. 16.12 Community building while learning about plants, herbs, and more—community that converses lives happily

References

Anika AK, Shorna F, Upoma SSK, Singh T, Dash PR (2020) Therapeutic potentials of Oxalis corniculata Linn. As a medicinal plant: a review. Int J Pharm 7(4):87–95

Dwivedi S, Chopra D (2014) Revisiting Terminalia arjuna—an ancient cardiovascular drug. J Tradit Complement Med 4(4):224–231. https://doi.org/10.4103/2225-4110.139103

Elford E(2013) Introduction to Sea Buckthorn. ONspecialitycrops. Accessed online on 13 Mar 2023 at https://onspecialtycrops.ca/2013/05/21/introduction-to-sea-buckthorn/

Gadgil M (2018) Sacred groves: an ancient tradition of nature conservation. Sci Am. https://www.scientificamerican.com/article/sacred-groves-an-ancient-tradition-of-nature-conservation/

Gopinath H, Karthikeyan K (2021) Neem in dermatology: shedding light on the traditional panacea. Indian J Dermatol 66(6):706. https://doi.org/10.4103/ijd.ijd_562_21

Health (2010) Use neem to treat chicken pox the natural way. DNA. https://www.dnaindia.com/health/report-use-neem-to-treat-chicken-pox-the-natural-way-1462203

Krishen P (2008) Trees of Delhi: a field guide. Conserv Soc 6(1):102

Krishna N, Amrithalingam M (2014) Sacred groves of India—a compendium C.P.R. Environmental Education Centre, Chennai

Kumari J(2020) Benefits of Giloy: how Giloy fights COVID-19 and other deadly infections by building immunity. https://www.india.com/lifestyle/benefits-of-giloy-how-giloy-fights-covid-19-and-other-deadly-infections-by-building-immunity-4067283/

Larmo PS, Järvinen RL, Yang B, Kallio HP (2014) Sea buckthorn, dry eye, and vision. In: Handbook of nutrition, diet and the eye. Academic Press, pp 473–480. https://doi.org/10.1016/B978-0-12-401717-7.00048-4

Okhuarobo A, Falodun JE, Erharuyi O, Imieje V, Falodun A, Langer P (2014) Harnessing the medicinal properties of Andrographis paniculata for diseases and beyond: a review of its phytochemistry and pharmacology. Asian Pac J Trop Dis 4(3):213–222. https://doi.org/10.1016/S2222-1808(14)60509-0

Ontario KP (2023). Introduction to sea buckthorn. Ministry of Agriculture, Food and Rural Affairs, Canada. Accessed online on 13 Mar 2023 at https://www.ontario.ca/page/introduction-sea-buckthorn

Pandey V, Vaishya JK, Balakrishnan P, Nesari TM (2020) Nutritional aspects of Tinospora cordifolia (Giloe). Med Plants Int J Phytomed Related Industries 12(1):158–160. https://doi.org/10.5958/0975-6892.2020.00021.0
Parvaiz A(2014) Seabuckthorn could transform Ladakh economy. The Third Pole. Accessed online on 13 Mar 2023 at https://www.thethirdpole.net/en/nature/seabuckthorn-could-transform-ladakh-economy/
Reshu V, Tarun V, Satbir S, Geeta M, Girish M (2018) Hidden potential of doob grass-an Indian traditional drug. Res Pharm Health Sci 4:478–482. https://doi.org/10.32463/rphs.2018.v04i03.13
Sagar S (2015) Role of natural toothbrushes in containing oral microbial flora—a review. Asian J Pharm Clin Res 8(4):29–33
Sardeshpande M, Hurley PT, Mollee E, Garekae H, Dahlberg AC, Emery MR, Shackleton C (2021) How people foraging in urban greenspace can mobilize social–ecological resilience during Covid-19 and beyond. Front Sustain Cities 3:686254. https://doi.org/10.3389/frsc.2021.686254
Sinha S(2012).Beyond stoned. DowntoEarth https://www.downtoearth.org.in/news/beyond-stoned-36875
Stobdan, T, Dolkar, P, Chaurasia, O. P., Kumar, B. (2017). Sea buckthorn (Hippophae rhamnoides L.) in trans-Himalayan Ladakh, India. Defence Life Sci J, 2(1), 46–53.doi: https://doi.org/10.14429/dlsj.2.11074
Varshney, V. (2021). Slay those symptoms: how some giloy can help amid COVID-19. DowntoEarth. https://www.downtoearth.org.in/news/food/slay-those-symptoms-how-some-giloy-can-help-amid-covid-19-76935
Wayman, E. (2012) Becoming human: the evolution of walking upright. Smithsonian Magazine. Accessed online on 13 Mar 2023 at https://www.smithsonianmag.com/science-nature/becoming-human-the-evolution-of-walking-upright-13837658/#:~:text=The%20earliest%20hominid%20with%20the,%2Dyear%2Dold%20Ardipithecus%20ramidus
Wild Foodies.org (n.d.). http://www.wildfoodies.org/

Abhijit Sinha Abhijit Sinha is a journalist turned storyteller/writer, with a career span of over three decades in national and international media. His writings focus on Political and Socio-Economic aspects of SDG and Demographic trends. He has been associated with various education institutes and public bodies. He designs Behavioural Change Communication programmes for the government, corporate and professional institutions.

Sharmila Sinha is a writer and a food connoisseur who writes to connect the dots between climate, culture, and cuisine. She travels and trains people to make them aware of their inherited culture, tradition, and knowledge base and their interconnectedness. As a foodie, she organizes food pop-ups to acquaint people with why it is important to eat local and seasonal. A published author, she also conducts writing workshops to engage the young and old on environmental issues.

The Ubiquitous Dhekia and Its 'deep roots' to Nurture People's Lives: A Case of Urban Foraging from Assam

17

Oinam Sunanda Devi and Sonali Ghosh

Abstract

Northeast India is endowed with a bounty of wild natural resources, many of which are edible in nature and hence have been a part of the regular diets of the local communities for centuries together. Over 828 numbers of wild and tradeable resources have been recorded from the state of Assam, out of which 10.62% are wild edible plants, 7.37% are edible fruits and 82.01% as seeds and other derivatives. Many of these wild edible plants are found in rural areas and collected by forest fringe communities for a coordinated supply to the urban elite. The traditional ecological knowledge (TEK) that supports the collection of such wild plants is fast disappearing and at the same time, over-extraction to feed urban city markets has also made many of the practices unsustainable. We argue that the revival of TEK by formalizing the knowledge systems through training and linking it to sustainable harvesting practices would be a win-win situation for all. Further, scaling up TEK and linking it to Mission LiFE to reshape food systems to better support nutrient-rich diverse diets and environmental sustainability are the way forward.

Keywords

Wild edibles · Traditional ecological knowledge · Mission LiFE · Environmental sustainability

O. S. Devi (✉)
Assam State Biodiversity Board, Aranya Bhawan, Guwahati, Assam, India

S. Ghosh
Chief Conservator of Forests (REWP), Aranya Bhawan, Guwahati, Assam, India

S. Dhyani, M. Sardeshpande (eds.), *Urban Foraging in the Changing World*,
https://doi.org/10.1007/978-981-97-0345-6_17

17.1 Introduction

Assam being a part of the Eastern Himalaya Biodiversity Hotspot supports a wide variety of vegetation types, and the forests here are rich in natural resources which are a boon for the local people as a source of food and income for sustaining their livelihood. Since time immemorial, the local communities have an intrinsic relationship with forests which are depicted in their rich culture and traditions. The rich forest ecosystems of the state are a constant source of valuable Non-Timber Forest Produces (NTFP) in the form of wild edible fruits, flowers, leaves, roots, resins, rhizomes, spices, dyes, bamboo, cane, etc. for the local communities. The role of these NTFPs in providing food security and income for the forest-dwelling communities cannot be ignored. Not only forests, the state's rich riverine and wetland habitats and other non-forested areas also provide sufficient natural resources for the local communities for their sustenance and livelihood. Local communities such as Karbis, Bodos, Mishing, Nepalis, Tiwas, and Dimasas, among others, mainly women, travel to far-flung areas to collect these bioresources and sold them in the small- and medium-sized local markets or *haats* scattered all over the state (Devi 2019). The traders and private collectors buy these resources from the small local markets and supply these raw materials to urban communities and various industries and manufacturers operating in the state. A one-year study conducted by Assam State Biodiversity Board in 2019 documented 61 edible wild fruit species, 85 edible wildplants, 410 medicinal plant species, 114 Aromatic plants, 68 species of plant-based dyes and resins, 37 species of flavouring agents and spices, 30 species of Bamboo and grasses, 36 ornamental plant species and 20 staple cash crops which are traded in about 248 local markets of Assam (Devi 2019). A checklist of 85 wild and semi-wild edibles consumed by rural and urban populace as vegetables, commonly traded in local markets of Assam is given in Table 17.1. This data shows the richness of these wild/semi-wild bioresources in Assam which makes it a potential source of economic growth for the local communities if their trade linkages can be traced efficiently.

Table 17.1 List of wild and semi-wild edibles consumed by rural and urban populace as vegetables, commonly traded in local markets of Assam

Sl. No.	Family	Local name	Scientific name	Seasonal availability	Status
1	Agaricaceae	Mushroom	*Agaricus bisporus* Quell	Oct–Feb	R,O
2	Amaryllidaceae	Tlangpurun/ Maroi Napakpi	*Allium hookeri*Thwaites	Oct–Apr	R,O
3	Amaryllidaceae	Naharu	*Allium sativum* L.	All year round	C, A
4	Araceae	Mukhi / Panch Mukhi Kochu	*Alocasia cucullata* (Lour.) G.Don	Nov–Mar	C, A
5	Araceae	Mann Kochu	*Alocasia macrorrhizos* (L.) G.Don	Oct–Mar	C, A
6	Araceae	Dahi Kochu	*Alocasia odora* (Lindl.) K.Koch	Rainy season	C, A

(continued)

Table 17.1 (continued)

Sl. No.	Family	Local name	Scientific name	Seasonal availability	Status
7	Araceae	Koni/Horu Kochu	*Alocasia* sp.	Nov–Mar	C, A
8	Zingiberaceae	Tora (Pullei)	*Alpinia nigra* (Gaertn.) Burtt	Feb–Apr	C, A
9	Amaranthaceae	Matikaduri xak	*Alternanthera sessilis* (L.) R.Br. ex DC.	Oct–Mar	R,O
10	Amaranthaceae	Khutura xak	*Amaranthus viridis* L.	Oct–Feb	C, A
11	Amaranthaceae	Ronga Moricha	*Amaranthus tricolor* L.	Oct–Jan	C, A
12	Araceae	Ol Kochu	*Amorphophallus bulbifer* (Roxb.) Blume	Oct–Mar	C, A
13	Araceae	Mann Kochu	*Alocasia macrorrhizos* (L.) G. Don	Oct–Mar	C, A
14	Acanthaceae	SirataTita	*Andrographis paniculata* (Burm.f.) Nees	Dec–Feb	C, A
15	Auriculariaceae	Mushroom (Uchina)	*Auracularia delicata* P. Henn	Feb–Apr	R,O
16	Plantaginaceae	Brahmi xak	*Bacopa monnieri* (L.) Wettst.	Nov–Mar	C, A
17	Poaceae	Bhaluka bah gaaj	*Bambusa balcooa* Roxb.	Jun–Aug	C, A
18	Poaceae	Jaati bah	*Bambusa tulda* Roxb.	Jun–Aug	C, A
19	Basellaceae	Puroixak	*Basella alba* L.	Feb–Mar	C, A
20	Malvaceae	Tikoni Barual xak	*Byttneria aspera* Collebr. ex Wall.	Feb–April	C, A
21	Arecaceae	Raidang bet	*Calamus erectus*Roxb.	Apr–Jun	C, A
22	Leguminosae	Tebi	*Canavalia gladiata* (Jacq.) DC.	Oct–Mar	C, A
23	Apiaceae	Bor Manimuni	*Centella asiatica* (L.) Urb.	Mar–May	C, A
24	Amaranthaceae	Jilmil xak/Bathua	*Chenopodium album* L.	Nov–Mar	C, A
25	Lamiaceae	Nephaphu	*Clerodendrum glandulosum* Lindl.	Oct–Feb	R,O
26	Araceae	Kola Kochu	*Colocasia esculenta* (L.) Schott	Nov–Mar	C, A
27	Malvaceae	Mora paat	*Corchorus capsularis* L.	Feb–Mar	C, A
28	Cucurbitaceae	Rongalao Phul	*Cucurbita maxima* Duchesne	Dec–Mar	C, A
29	Poaceae	Kako bah gaj	*Dendrocalamus hamiltonii* Nees & Arn. ex Munro	Jun–Aug	C, A
30	Dilleniaceae	Outenga	*Dillenia indica* L.	Dec–Apr	C, A
31	Athyriaceae	DhekiaXak	*Diplazium esculentum* (Retz.) Sw.	Year Round	C, A
32	Dioscoreaceae	Kath Aalu	*Dioscorea alata* L.	Nov–Feb	C, A
33	Dioscoreaceae	Mowa Aalu	*Dioscorea esculenta* (Lour.) Burkill.	Nov–Jan	C, A
34	Brassicaceae	Helochi Xak	*Enhydra fluctuans* Lour.	Oct–Dec	R,O

(continued)

Table 17.1 (continued)

Sl. No.	Family	Local name	Scientific name	Seasonal availability	Status
35	Apiaceae	Mann Dhania	*Eryngium foetidum* L.	Oct–Dec	C, A
36	Nymphaeaceae	Makhana	*Euryale ferox* Salisb.	July–Aug	C, A
37	Moraceae	Dimoru	*Ficus hispida* L.f.	Year Round	C, A
38	Moraceae	TengaBor	*Ficus geniculata* Kurz	Year Round	C, A
39	Gnetaceae	Hanthu pat	*Gnetumgnemon*L.	Year Round	C, A
40	Zingiberaceae	Kor Phul (Spiked Ginger Lily)	*Hedychium spicatum* Sm.	Feb–Apr	R,O
41	Malvaceae	Ronga Tenga Mora	*Hibiscus sabdariffa* L.	Oct–Mar	C, A
42	Saururaceae	Masundari	*Houttuynia cordata* Thunb	July to August	C, A
43	Araliaceae	Horu Manimuni	*Hydrocotyle rotundifolia* Roxb.	Mar–May	R,O
44	Convolvulaceae	Kolmou xak	*Ipomoea aquatica* Forssk.	Jun–Aug	C, A
45	Convolvulaceae	Mitha Aalu	*Ipomoea batatas* (L.) Lam.	Oct–Feb	C, A
46	Leguminosae	Seem	*Lablab purpureus* (L.) Sweet	Oct–Mar	C, A
47	Araceae	Sengmora	*Lasia spinosa* (L.) Thwaites	Oct–Mar	R,O
48	Lamiaceae	Durun xak	*Leucas aspera* (Willd.) Link.	Oct–Dec	C, A
49	Solanaceae	Kon Bilahi	*Lycopersicon* sp.	Dec–Mar	C, A
50	Malvaceae	Lofa xak	*Malva verticillata* L.	Oct–Feb	C, A
51	Euphorbiaceae	Himolu Aalu	*Manihot esculenta* Crantz.	Jan–Apr	C, A
52	Lamiaceae	Pudina	*Mentha arvensis* L.	Oct–Feb	C, A
53	Leguminosae	Lajuki lota	*Mimosa pudica* L.	All year round	C, A
54	Moringaceae	Sajina	*Moringa oleifera* Lam.	Nov–Mar	C, A
55	Cucurbitaceae	Bon Tita Karela	*Mukia maderaspatana* (L.) M.Roem.	Oct–Feb	C, A
56	Rutaceae	Narasingha	*Murraya koenigii* (L.) Spreng.	Nov–Feb	C, A
57	Musaceae	KachKol	*Musa paradisiaca* L.	Oct–Feb	C, A
58	Musaceae	Malbhog Kol Dil	*Musa sanguinea* Hook.f.	Oct–Feb	C, A
59	Nymphaeaceae	Podumguti	*Nelumbo nucifera* Gaertn.	Jun–Jul	C, A
60	Nymphaeaceae	Ronga bhet	*Nymphaea rubra*Roxb. ex Andrews	Aug–Oct	C, A
61	Lamiaceae	Bon Tulshi (Mayang ton)	*Ocimumamericanum* L.	Oct–Feb	C, A
62	Bignoniaceae	Bhatghila (Dingdinga)	*Oroxylum indicum* (L.) Kurz	Feb–June	C, A

(continued)

Table 17.1 (continued)

Sl. No.	Family	Local name	Scientific name	Seasonal availability	Status
63	Oxalidaceae	Sengeri Tenga/ Tengeshi	*Oxalis corniculata*L.	Oct–Feb	C, A
64	Rubiaceae	Bhedailata	*Paederia foetida* L.	Oct–Mar	C, A
65	Rubiaceae	Ronga bhedai lota	*Paederia scandens* L.	Oct–Jan	C, A
66	Leguminosae	Manipuri Urahi (Yonchak)	*Parkia timoriana* (DC.) Merr.	Oct–Mar	C, A
67	Polygonaceae	Bon Ghehu/ Yelang	*Persicaria barbata* (L.) H.Hara	Oct–Feb	C, A
68	Acanthaceae	Nongmagkha/ RongaTitaPhul	*Phlogacanthus thyrsiflorus* Nees	Oct–Apr	C, A
69	Pleurotaceae	Mushroom	*Pleurotus squarrosulus* R. Singer	Feb–Apr	C, A
70	Pleurotaceae	Mushroom (Uyen)	*Pleurotus ostreatus*	Feb–Apr	C, A
71	Polygonaceae	Modhuxuleng	*Polygonum microcephalum* D. Don	Oct–Feb	C, A
72	Polygonaceae	Singju/Phak Pai/ Knotweed	*Polygonum posumbu*Buch.-Ham. ex D. Don	Oct–Feb	C, A
73	Papillionaceae	Tengnoumanbi/ Chari SiriaUrohi	*Psophocarpus tetragonolobus* (L.) DC.	Oct–Mar	R,O
74	Polygonaceae	TengaPaleng	*Rumex vesicarius* L.	Mar–June	R,O
75	Schizophyllaceae	Mushroom (Kanglayen)	*Schizophyllum commune* E. Fries	Feb–Apr	R,O
76	Pedaliaceae	Teal paat	*Sesamum indicum* L.	Aug–Oct	R,O
77	Leguminosae	Bokful	*Sesbania grandiflora* (L.) Pers.	Feb–Apr	R,O
78	Solanaceae	Deori Tita/ BhekuriGuti	*Solanum anguivi* Lam.	Oct–Apr	R,O
79	Solanaceae	BhotBengena/ Titabhekuri	*Solanum ferox* L.	Oct–Apr	C, A
80	Araliaceae	Bhotola	*Trevesia palmata (*Roxb. ex Lindl.) Vis.	Oct–Feb	R,O
81	Lyophyllaceae	Termite mushroom	*Termitomyces eurrhizus*	Apr–Jun	C, A
82	Cucurbitaceae	Snake gourd	*Trichosanthes cucumerina* L.	Feb–June	C, A
83	Leguminosae	Methi guti xak	*Trigonella foenum-graecum* L.	Oct–Mar	C, A
84	Compositae	Ogora	*Xanthium strumarium* L.	Oct–Dec	R,O
85	Rutaceae	Mejenga	*Zanthoxylum oxyphyllum*Edgew.	Aug–Sep	C, A

Abbreviations: C:- Common; R:- Rare; A:- Abundant; O:- Occasional; Status as per frequency of occurrence in the markets

17.2 Importance of Edible Wild Resources in Improving the Diet and Nutrition of the Urban Communities

Urban foraging is a global phenomenon that is gaining attention especially in terms of the benefits it can provide to foragers like additional income and nutritional supplements. However, urban foraging is largely understudied and the types of species foraged, patterns of foraging, socio-economic profile of foragers, and impact of urbanization on foraging are still being investigated (Somesh et al. 2021). Different communities have their own culture, traditions, and food habits. These food habits are again intrinsically linked with the changing seasons (Komor and Devi 2016). Each season is marked by different festivals which call for some specific traditional food preparation, viz. during the celebration of Rongali Bihu festival in Assam, the Assamese community, prepares elaborate dish of different leafy wild or semi-wild edible plant species such as *Dhekiaxak* (*Diplazium esculentum*), *Brahmi xak* (*Bacopa monnieri*), *Kolmou xak* (*Iopmoea aquatica*), *Bhedai lata* (*Paederia foetida*), *Durun Xak* (*Leucas aspera*), *Bor Manimuni* (*Centella asiatica*), *Horu Tengesi* (*Oxalis corniculata*), *Ronga Tenga Mora* (*Hibiscus sabdariffa*), *Morapat* (*Corchorus capsularis*) Jilmil xak (*Chenopodium album*), etc. The Karbi people also consumed two special wild plants, namely *Hanthu pat* (*Gnetum gnemon*) and *Mehek* (*Rhynchotechum ellipticum*) which are part of their sacred dish called '*Hanmi-Hanmoi*' and '*Thekar*' (Terangpi et al. 2013). Likewise, similar traditional uses of wild edible plants are also prevalent among other communities of the region. Some notable works on wild edible plants used by different commmunities in Assam are dietary uses of wild plants of Karbi Anglong (Kar and Borthakur 2008); Mishing tribe (Dutta et al. 2016); Tai Phake tribe of Upper Brahmaputra Valley in Assam (Borgohain et al. 2016); Dimasa tribe (Bose et al. 2018); Bodo Tribe (Baro et al. 2015; Dutta 2012), among others. These non-conventional edible bioresources are not only a source of traditional delicacies but also an important aspect of cultural traits that have been passed through generations (Komor and Devi 2016). It not only provides nutrition to the economically poor but is also an important facet of providing food security to the rural and urban populations (Plates 17.1, 17.2, and 17.3).

a) Dhekia (*Diplazium esculentum*)

b) Tora (*Alpinia nigra*)

c) Makhana (*Euryale ferox*)

d) Ronga Morisha (*Amaranthus tricolor*)

e) Ronga Tengamora(*Hibiscus sabdariffa*)

f) Roga Puroi Xak(*Basella alba*)

Plate 17.1 (**a**–**f**) Wild and semi-wild edible leafy vegetables. (**a**) Dhekia (*Diplazium esculentum*). (**b**) Tora (*Alpinia nigra*). (**c**) Makhana (*Euryale ferox*). (**d**) Ronga Morisha (*Amaranthus tricolor*). (**e**) Ronga Tengamora (*Hibiscus sabdariffa*). (**f**) Roga Puroi Xak (*Basella alba*)

a) Bor Manimuni (*Centella asiatica*)

b) Duron Hak (*Leucas aspera*)

c) Bhot Bengena (*Solanum ferox*)

d) Khutura xak (*Amaranthus viridis*)

e) Bamboo Shhots(*Dendrocallamus hamiltoni*)

f) Hanthu Pat(*Gnetum gnemone*)

Plate 17.2 (**a**–**f**) Wild and semi-wild edible leafy vegetables. (**a**) Bor Manimuni (*Centella asiatica*). (**b**) Duron Hak (*Leucas aspera*). (**c**) Bhot Bengena (*Solanum ferox*). (**d**) Khutura xaak (*Amaranthus viridis*). (**e**) Bamboo Shoots (*Dendrocalamus hamiltonii*). (**f**) Hanthu Pat (*Gnetum gnemon*)

a) LanghinBozar, Karbi Anglong District

b) Kothiatoli Bozar, Nagaon District

c) Gandhi Maidan Market,Hojai District

d) Sitajakhala Market, Morigaon District

e) Rowta Market, BTC

f) Sonai Market, Cachar District

Plate 17.3 Women vendors selling wild and semi-wild edible leafy vegetables at different local market of Assam. (**a**) Langhin Bazar, Karbi Anglong District. (**b**) Kathiatoli Bozar, Nagaon District. (**c**) Gandhi Maidan Market, Hojai District. (**d**) Sitajakhala Market, Morigaon District. (**e**) Rowta Market, BTC. (**f**) Sonai Market, Cachar District

17.3 Traditional Ecological Knowledge (TEK) Towards the Conservation of These Resources

Tribal dominated tracts of Northeast India are the storehouse of knowledge about the multiple uses of plants (Arora 1981). Traditional communities living here for thousands of years have built a precious knowledge base about the use of the rich bioresources of the region. Studies undertaken by various researchers have documented 1350 species of plants which are used in ethnomedicinal preparations from the entire NE India (Dutta and Dutta 2005). Some notable researchers who have worked on traditional and ethnomedicinal uses of wild plants by different communities in Assam include Borthakur 1996; Das et al. 2006, 2013; Chakravarty and Kalita 2012; Teron and Borthakur 2013; Basumatary et al. 2014; Bailung and Puzari 2016; Bora et al. 2016, among others. These wild edible plant resources be it in the forms of roots, leaves, stems, flowers, fruits, or bark are also used for traditional medicinal purposes besides being consumed as food items. The local traditional healers (variously known as *bez, ozah, hakims*) used these resources in making traditional medicines to heal various types of ailments, broken bones, etc. and such knowledge has been passed down to generations. Even though modern medical facilities are available in all the districts of Assam, in remote far-flung areas, these traditional healers are the sole saviour of many lives in emergency situations. Of the 85 wild and semi-wild edibles consumed by rural and urban populace as vegetables documented in this study, 32 of them are being used in traditional medicines and AYUSH industries (Devi 2019). The list of 32 wild/semi-wild resources consumed as vegetables used in traditional medicinal system are given in Table 17.2. However, more efforts are needed for documenting all the wild plant species which are used to make traditional medicines, consumed orally or applied externally and the knowledge associated with them so that such knowledge can be protected from theft or biopiracy.

As a step for the conservation and protection of such resources and associated knowledge, the Government of India enacted the Biological Diversity Act in 2002 and a three-tier structure was created at National, State, and Grassroot level for implementation of the Biological Diversity Rules 2004 in India. Accordingly, every State constituted its own State Biodiversity Board and Biodiversity Management Committees at local all levels for fulfilling the mandate of the Biological Diversity Act 2002. These Biodiversity Management Committees (BMC) have been entrusted to document all biological resources and associated traditional knowledge of their jurisdiction in the 'Peoples Biodiversity Register' (PBR), which is a legal document for safeguarding and regulating access of the biological resources for research and commercial utilization. These steps check overexploitation of the resources by commercial entities and also provide baseline information for planning sustainable management of these resources.The PBR also contains information of local traditional healers and their knowledge of different uses of plants and animals in traditional medicine, so that it can be protected from theft or biopiracy. This step will also enable them to register their knowledge for patents, if required. At present, there are 2549 Biodiversity Management Committees (BMC) at all local bodies and also at the three Autonomous Councils of Assam. These BMCs have documented their PBRs with assistance from Assam State Biodiversity Board, and they have initiated preparation of management plans for safeguarding the local biological resources by

Table 17.2 List of wild/semi-wild resources consumed as vegetables used in traditional medicinal system

Sl. No.	Family	Scientific name	Trade name	Associated traditional knowledge/medicinal uses	Medicinal system tag
1	Zingiberaceae	*Alpinia nigra* (Gaertn.) Burtt	Galangal	It is reported to be a stimulant and is used for treatment of rheumatism. The roots are also used for treatment of fungal infections such as ringworm and melasma	A, F
2	Araceace	*Alocasia macrorrhizos* (L.) G. Don		It is reported to be useful in curing abdominal diseases, abscess, and burns	F
3	Amaranthaceae	*Amaranthus tricolor* L	Red Spinach	It is considered to be a high fibre diet useful for treatment of bowel problems. It is also reported to be useful for treatment of external inflammation and bladder distress	F
4	Amaranthaceae	*Amaranthus viridis* L	Cholai	It is reported to be useful for treating urinary problems. A decoction of the plant is useful in stopping dysentery and inflammation	A, F, S,
5	Acanthaceae	*Andrographis paniculata* (Burm.f) Wall. Ex Nees	Kalmegh	The plant is consumed by local people as a bitter tonic useful for treatment of chronic fever, ulcers, skin diseases, intestinal worms, and dysentery	A,F,H,S,U
6	Plantaginaceae	*Bacopa monnieri*(L.) Wettst	Brahmi, Jal Brahmi, Nirbrahmi	It is considered as tonic for brain. It is reported to be an effective and powerful herb helpful for memory and combating stress. It is also reported to be useful for treatment of epilepsy and asthma	A,F,H, S

(continued)

Table 17.2 (continued)

Sl. No.	Family	Scientific name	Trade name	Associated traditional knowledge/medicinal uses	Medicinal system tag
7	Basellaceae	*Basella alba* L	Malabar Spinach	It is considered to have high iron content and reported to be useful in preventing osteoporosis and anaemia. The leaves are used in ayurveda for treatment of boils and sores	A, F
8	Apiaceae	*Centella asiatica (L.) Urb*	Brahmi booti	It is reported to be a liver tonic useful for treatment of stomach problems. It is also useful for treatment of wounds and cuts	A,F,H,S
9	Amaranthaceae	*Chenopodium album* L		The plant is generally consumed by the local people for its rich contents of vitamin A, calcium, potassium, and phosphorous. It is reported to be useful for treating kidney stone	F
10	Lamiaceae	*Clerodendrum glandulosum* Coleb. Ex Wall		It is reported to be useful for curing high blood pressure. Mishing tribes of Assam used this plant for making local wine. It is also useful for treatment of dysentery and abdominal pain	F
11	Dilleniaceae	*Dillenia indica* L		It is reported to have anti-dandruff and anti-hair fall properties. It reported to be useful in treatment of fever, cough, and dysentery. It is also considered to have aphrodisiac properties	A, F
12	Dioscoreaceae	*Dioscorea alata* L	Sewali Kodi	It is reported to be useful in relieving pain related to gout and rheumatism	A,F,S

(continued)

Table 17.2 (continued)

Sl. No.	Family	Scientific name	Trade name	Associated traditional knowledge/medicinal uses	Medicinal system tag
13	Dioscoreaceae	*Dioscorea esculenta* Burk		It is considered effective in curing abscess, anorexia, beriberi, menopause, rheumatism, ulcers, and swelling	F
14	Araceae	*Diplazium esculentum* (Retz.) Sw		It is reported to be useful for treatment of urinary complaints. It is also considered to be a rich source of iron by local communities	F
15	Brassicaceae	*Enhydra fluctuans* Lour		It is reported to be useful for treatment of cough, skin diseases, and gonorrhoea	A, H, F
16	Apiaceae	*Eryngium foetidum* L	Long Coriander/ Culantro	It is reported to be useful in traditional medicine for burns, earache, fevers, hypertension, constipation fits, asthma, stomachache, worms, infertility complications, snake bites, and in treatment of malaria	F
17	Nymphaeacae	*Euryale ferox* Salisb. Ex K.D. Koenig	Phoolmakhana	It is reported to be useful for treatment various ailments such as chronic diarrhoea, spleen disorders, diabetes and kidney problems. It is also reported to be an aphrodisiac	A,S,U, F
18	Gnetaceae	*Gnetum gnemon* L		It is reportedly used in traditional medicine for purification of blood and is consumed as a traditional delicacy by local people specially the Karbi tribes of Assam	A, F

(continued)

Table 17.2 (continued)

Sl. No.	Family	Scientific name	Trade name	Associated traditional knowledge/medicinal uses	Medicinal system tag
19	Zingiberaceae	*Hedychium spicatum* Sm	Kapoor kachri	It is reported to be useful for treatment of liver complaints, fevers, vomiting, diarrhoea, inflammation, and pains	A,F,U, F
20	Malvaceae	*Hibiscus sabdariffa* L	Roselle	It is reported to be useful in treatment of dysentery and fever. It is also considered to have antihypertensive, diuretic, and laxative properties	A, F
21	Araliaceae	*Hydrocotyle rotundifolia* DC	Water Pennywort	It is reported to be used in traditional medicine for treating insomnia, cardiac debility, and epilepsy. The juice of the plant is used locally over the affected portion in case of elephantiasis	F
22	Lamiaceae	*Leucas aspera* (Willd.) link	Dron pushpi	It is reported to be useful for treatment of liver ailments, snake bite, scorpion sting and sinusitis, etc. it is also considered to have anti-helminthic, laxative, and diaphoretic properties. Traditionally, the leaves are applied externally to cure boils	A,F,H,S
23	Euphorbiaceae	*Manihot esculenta* Crantz	Cassava	In traditional medicine, the fresh rhizomes are made into a poultice and applied to sores. It is also reported that eating the boiled tubers cure diarrhoea	A, F

(continued)

Table 17.2 (continued)

Sl. No.	Family	Scientific name	Trade name	Associated traditional knowledge/medicinal uses	Medicinal system tag
24	Lamiaceae	*Mentha piperita* L	Menthol, peppermint	It is used traditionally for relieving spasms. It is reported to be a stimulating herb, both cooling and warming. It is also reported to be effective in relieving digestive problems and soothing fevers	H,U, F
25	Moringaceae	*Moringa oleifera* Lam	Sahenjana, Sajina	The stem juice of the plant is reported to be useful in treatment of high fever, jaundice, rheumatic fever, and gout. In folk medicine, the plant is also used for curing bone fracture	A,F,S,U
26	Rutaceae	*Murraya koenigii*(L.) Spreng	Kariveppila, curry pata	Traditionally, it is considered to have anti-diabetic, antioxidant, and hepato-protective properties. It is reported to be useful in treatment of nausea, vomiting, digestive problems, diarrhoea, and bloated stomach	A,F,S
27	Oxalidaceae	*Oxalis corniculata* L	Puliyaarila	It is reported to be useful in treatment of dysentery and blood pressure. The leaves are acrid, bitter, and mildly astringent. Infusion of the leaves is used in treatment of scurvy. Juice of the leaves mixed with milk is used in treatment of jaundice	A,F,S,U

(continued)

Table 17.2 (continued)

Sl. No.	Family	Scientific name	Trade name	Associated traditional knowledge/medicinal uses	Medicinal system tag
28	Rubiaceae	*Paederia scandens* (Lour.) Merr	Prasaarani	It is reported to be useful for treatment of infertility and paralysis. A decoction of the plant is used for treatment of abdominal pain, abscess, arthritis, and overheating	F
29	Pedaliaceae	*Sesamum indicum* L	Til	It is reported to be useful in treatment of piles, dysentery, constipation, and cough. The leaves are used for treatment of skin diseases and consumed by local people for increasing secretion of milk of nursing mother	A,S,F
30	Solanaceae	*Solanum ferox* L	Night shade	It is considered useful for treatment of worm infection and skin diseases. Also useful for treatment of sore throat and rashes in tongue, etc	F
31	Compositae	*Xanthium strumarium* L		It is reported to be a cooling, laxative, fattening, anti-helminthic, tonic, digestive, antipyretic in ayurveda. It also improves appetite, complexion, and memory	F
32	Rutaceae	*Zanthoxylum oxyphyllum* Edgew		It is reported to be useful in treatment of dental problems. Traditionally, it is considered to be an astringent, digestive, and stimulant	F

Abbreviations: A:- Ayurveda; F:- Folk; S:- Siddha, U:- Unani; H:- Homoeopathy

a) A traditional healer with his customer in the Goroimari local market, Sonitpur District

b) A traditional healer with his wares in the Sitajakhala local market, Morigaon District

Plate 17.4 (**a, b**) Traditional healers encountered in local markets of Assam. (**a**) A traditional healer with his customer in the Goroimari local market, Sonitpur District. (**b**) A traditional healer with his wares in the Sitajakhala local market, Morigaon District

way of promoting sustainable utilization of the resources. Beside these, the Government of India also initiated the Traditional Knowledge Digital Library (TKDL) where the age-old traditional knowledge of different communities was translated, codified, and kept for future references (Plate 17.4).

17.4 Key Threats to Wild/Semi-Wild Bioresources

With rapid urbanization, our dependencies on the natural resources have also escalated. The increasing demands of these natural products from consumers worldwide have led to the overexploitation of these resources especially from the wild. To meet the demand, the local people have moved away from their earlier lifestyles and developed new ways of accessing the resources which differs from our ancient understandings of sustainability that ascribed limits to use of biological resources. Some of the factors threatening the existence of biodiversity of the NE region are habitat fragmentation and destruction due to deforestation, developmental activities, shifting cultivation, poaching, trade in wild flora and fauna, introduction of exotics, and rapid widespread of invasive species (Behera 2001). The absence of modernized socio-economic and public healthcare systems along with lack of employment and good infracstructure facility has compelled the poor people of this region to depend on bioresources for their livelihood generation (Devi 2019). A total of 234 local vendors were interviewed from 78 local weekly markets of Assam during this study conducted in 2019 with structured questionaire on their dependency to wild bioresources and the possible threats to these resources. Of the total, about 33.33% responded that the key threats to bioresources are from habitat degradation, 52.99% of the vendors believed that the main reason is overexploitation and about 13.67% of them responded that both habitat degradation and overexploitation equally contribute to the decline of rare and endemic wild/semi-wild bioresources which are abundantly available few years back (Fig. 17.1). Of the total 234 local vendors interviewed, 58.54% revealed that selling bio-resoures in the local market or *Haat Bozar* is their primary occupation and they rely solely on it for their survival, whereas 24.78% of them responded that agricultural labour is their main occupation, 11.96% revealed that agriculture is their primary occupation and 4.70% respondents reveal that they do petty jobs like carrying stones and sand during non-market days (Fig. 17.2).

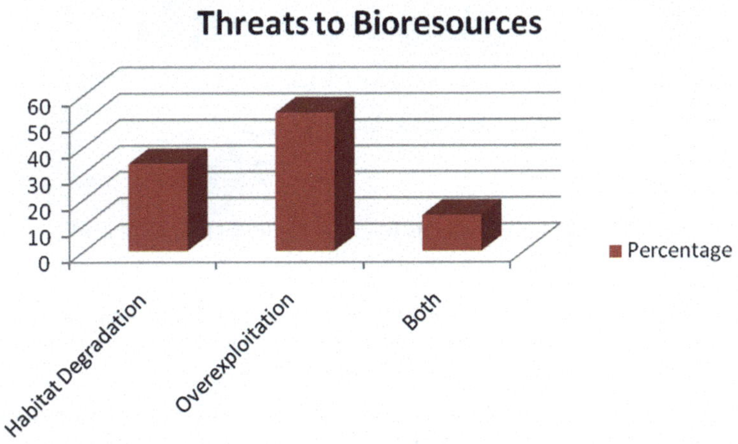

Fig. 17.1 Threats to bioresources as per the perception of local vendors

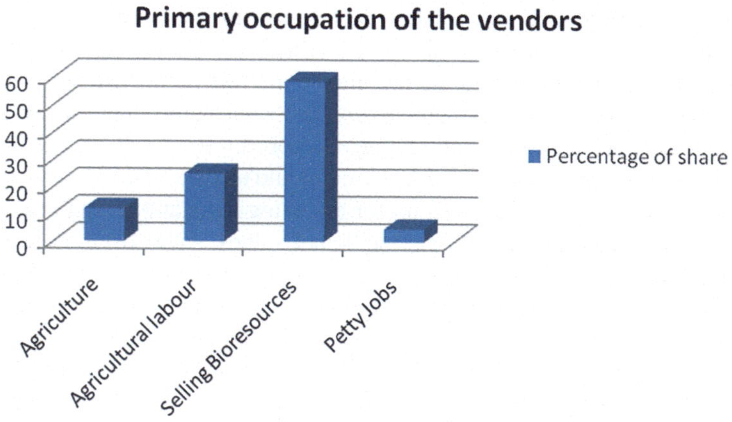

Fig. 17.2 Source of income of the vendors in the study area

Even though no primary data could be collected directly from the field for presenting the actual rate of decline of the bioresources commonly traded in local markets of Assam, it is somewhat evident from the perception of the local vendors that unsustainable extraction of the bioresources continues unchecked in all forested and non-forested areas. The rapid increase in population leading to unemployment also adds to the increasing dependency of marginal populations on this trade for livelihood as depicted in Fig. 17.2. These factors directly or indirectly aid in the rapid loss of rare and endemic bioresources including those species on which we depend for our nutritional needs. Unsustainable harvesting is reported to be main reason for overexploitation of bioresources in Assam till date, and the primary collectors who are basically local indigenous people inhabiting in and around the forests collect these resources for their own consumption as well as for selling them in local markets. Due to lack of awareness on sustainable harvesting, the local people tend to extract all the available resources they encountered, including rare and endemic species, which they thought will fetch good price in the market. Lack of proper storage and processing units in the state also aids to overexploitation of bioresources as many of the collected resources gets damaged before it could be sold in the market at the anticipated price or send them to outside state traders who supply these resources to manufacturing units. This further leads to unsustainable exploitation of the resources in their natural habitat in order to fulfil the demands from consumers and bioresources based industries.

17.5 The Way Forward: Mission LiFE and Bringing in Environmental Sustainability

The present scenario is becoming a major cause of concern for us, and the time has come to curb our greediness and think about the sustainability of these resources for posterity. For this, we need to take up steps for conservation of the bioresources without affecting the livelihood of many forest and fringe village dwelling communities. Therefore, initiative should be taken up to aware the local people specially the primary collectors of the value chain of these bioresources so that they are better rewarded for their hard work. Suitable innovative mechanisms should be developed for sustainable harvest, processing, and value addition of these resources for better yield and additional employment opprotunities. Value chain analysis study of certian potential tradable bioresources could also be taken up subsequently. Research initiatives should be taken up for assessing the status of high value tradable bioresources and their traditional medicinal values. Apart from these, conservation parks or arboretum should be set up at suitable locations for propagation of rare and endemic resources. Emphasis should also be given on spreading awareness about the potential role of the bioresources in livelihood generation by the concerned government agencies and public sector organizations of the state. The government may also adopt means in the state action plans of respective departments for involvement of Biodiversity Management Committees (BMC) and Joint Forest Management Committees (JFMCs) for sustainable management of these forest resources.

References

Arora RK (1981) In: Jain SK (ed) Native food plants of Noth eastern tribals. Glimpses of Indian ethnobotany. Oxford & IBH, New Delhi, pp 91–106

Bailung B, Puzari M (2016) Traditional use of plants by the Ahoms in human health management in upper Assam, India. J Med Plants Stud 4(2):48–51

Baro D, Baruah S, Borthukar SK (2015) Documentation on wild vegetables of Baksa district, BTAD (Assam). Arch Appl Sci Res 7(9):19–27

Basumatary N, Teron R, Saikia M (2014) Ethnomedicinal practices of the Bodo-Kachari tribe of KarbiAnglong District of Assam. Int J Life Sci Biotechnol Pharma Res 3(1):161–167

Behera B (2001) Forest degradation in India: extent & determinant. Indian J Agric Econ 56(4)

Bora D, Mehmud S, Das KK, Medhi H (2016) Report on medicinal plant practices for dysentery, diarrhoea and cholera in different parts of Assam, India. J Med Plants Stud 4(6):208–212

Borgohain MP, Teron R, Tamuli AK (2016) Ethnomedicinal plants of the Tai Phake tribe of upper Brahmaputra Valley in Assam, India. East Himalayan Soc Spermatophyte Taxonomy 10(1):123–133

Borthakur SK (1996) Wild edible plants in markets of Assam, India-an ethnobotanical investigation. Ethnobiol Hum Welf:31–34

Bose SD, Teron R, Tamuli AK (2018) Wild plant resources used as food by the Dimasas of Karbi—Anglong District, Assam, India. Int J Pure Appl Biosci 6(4):158–163

Chakravarty S, Kalita JC (2012) An investigation on anti-diabetic medicinal plants used by villagers in Nalbari district, Assam, India. Int J Pharm Sci Res 3(6):1693

Das NJ, Saikia SP, Sarkar S, Devi K (2006) Medicinal plants of north-Kamrup district of Assam used in primary healthcare system. Indian J Tradit Knowl 5(4):489–493

Das AK, Choudhury MR, Sharma GC (2013) Medicinal plants used by Koch Rajbangshi of north Salmara subdivision, Bongaigaon, Assam, India. Our Nat 11(1):45–53

Devi OS (2019) Tradable bioresourcse of Assam. Assam State Biodiversity Board, 284 p

Dutta U (2012) Wild vegetables collected by the local communities from the Chirang reserve forest of BTAD, Assam. Int J Sci Adv Technol 2(4):116–125

Dutta BK, Dutta PK (2005) Potential of ethnobotanical studies in north East India: an overview. Indian J Tradit Knowl 4(1):7–14

Dutta G, Baruah G, Devi A (2016) Wild food plants of Mishing tribe—an ethnobotanical survey. Trop Plant Res 3(2016):221–223

Kar A, Borthakur SK (2008) Wild vegetables of Karbi—Anglong district, Assam. Indian J Nat Prod Resour 7(5):448–460

Komor P, Devi OS (2016) Edible bioresources & livelihoods. Assam State Biodiversity Board, 233 p

Somesh D, Rao R, Murali R, Nagendra H (2021) Patterns of urban foraging in Bengaluru city. Urban For Urban Green 57:126940

Terangpi R, Engtipi U, Teron R (2013) Utilization of less known plants, *Gentum gnemon* L. & *Rhynchotechum ellipticum* (Dietr.) A. DC. among the Karbis. Northeast India. J Sci Innov Res 2(5):943–949

Teron R, Borthakur SK (2013) Folklore claims of some medicinal plants as antidote against poisons among the Karbis of Assam, India. East Himalayan Soc Spermatophyte Taxonomy 7(2):346–356

Oinam Sunanda Devi (PhD) is working as a Scientific Officer at the Assam State Biodiversity Board. She has a PhD on "Diversity and ecology of frugivorous birds in Eastern Assam," and she has extensively worked in the field of wildlife conservation.

Sonali Ghosh (PhD) is an Indian Forest Service Officer with years of work experience in the field of wildlife and natural heritage management in India. She is currently serving as first woman Field Director of the renowned Kaziranga National Park (KNP), Assam. Her research interest includes Agroforestry, Ex-situ conservation, PA governance and exploring nature-culture linkages.

Exploring Urban Foraging in Cities: Cultivating Sustainable Urban Environments for Food Security and Social Cohesion

18

Jayshree Shukla

Abstract

Urban green spaces provide multiple benefits to society, and one of the major benefits is the provision of foraging wild edibles. Inclusion of wild edibles as a dietary source can provide food security to cities in the face of impending crises, including pandemics and natural calamities. This chapter emphasizes the need for a comprehensive understanding of urban foraging, which includes its diverse dietary value, complex spatial dynamics, policy regulations, and accessibility concerns. The chapter builds on recently published perspectives regarding foraging habits and patterns in cities, highlighting the importance of conscious foraging behavior and offers an overview of the potential of instrumentalizing urban foraging for the future of sustainable cities. The integration of wild edible species and the preservation of traditional knowledge enhance the multifunctionality and benefits of urban foraging. The need for awareness and knowledge transfer among future foragers to ensure responsible and successful integration of foraging into urban planning and development is emphasised.

Keywords

Urban foraging · Edible landscapes · Conscious foraging · Food security

J. Shukla (✉)
National Environmental Engineering Research Institute (CSIR-NEERI), Nagpur, Maharashtra, India

University of Adelaide, Adelaide, Australia

S. Dhyani, M. Sardeshpande (eds.), *Urban Foraging in the Changing World*, https://doi.org/10.1007/978-981-97-0345-6_18

18.1 Introduction

Urbanization is the primary driver of the transitions in food supply patterns and the continuously increasing global food demand. The current projections regarding the prevalence of undernourishment indicate that approximately 600 million people could face chronic undernourishment by 2030, leading to the possibility of potential failure to meet the UN Sustainable Development Goal 2, which aims to eradicate hunger by 2030 (FAO et al. 2023). The struggles faced by impoverished populations in producing, accessing, consuming, and disposing of sufficient food highlight the challenge of meeting growing demand for safe and sustainable food, while changes in global trade relationships suggest a growing vulnerability to food insecurity even in prosperous economies (Calicioglu et al. 2019). Approximately one billion people worldwide harvest wild food as a way to improve their food security and dietary diversity (Garekae and Shackleton 2020). Over the past few years, there has been a growing recognition that cities and urban areas serve a dual purpose, not only as places where food is consumed, but also as spaces with the potential to be productive food environments (Ritchie and Roser 2020). This shift in perspective has been driven by practical initiatives aimed at reshaping food supply chains and promoting local food production within urban settings, along with creative, political, and theoretical reconsiderations of food systems and urban dynamics (Gebre and Gebremedhin 2019). Certain perspectives acknowledge that cities possess the potential to contribute significantly to addressing the challenges of supplying environment friendly, functional, and healthy food to a growing population amidst the continuously altering landscapes (Nyman 2019). Urban green spaces (UGS) are comprehensively defined as the land covered by all types of vegetation, regardless of their size and function in the urban areas. These spaces supply a wide range of ecosystem services that contribute to human well-being and support sustainable urban growth (Pinto et al. 2022). The UGS, including parks, playgrounds, and gardens, increase human–nature interactions and enhance nature-relatedness among city dwellers. The potential of UGS to provide provisioning services, particularly related to wild food foraging, is being increasingly recognized (Kaika 2017; Colléony et al. 2020; Hurley et al. 2022).

Urban foraging refers to the practice of individuals gathering natural resources from the urban environment that are not commercially cultivated or farmed. These resources encompass various items such as wild foods, fuelwood, materials for crafts, and medicinal herbs (Mino et al. 2021). In urban ecosystems, the presence of foraging associated species and spaces plays a vital role in providing essential services to both human and animal species, such as provisioning, regulation, and support of ecosystems. This, in turn, can aid in enhancing the ability of cities to adapt to climate change and build resilience (Schunko and Brandner 2022). Urban foraging can serve as a means of subsistence and generate income for those engaged in it. Foraged products have the potential to enhance food security in urban households, alleviate poverty, and contribute substantially to the economic development of the society (Chakona and Shackleton 2019). Urban foraging not only has cultural and recreational significance for individuals and communities, but it also enables

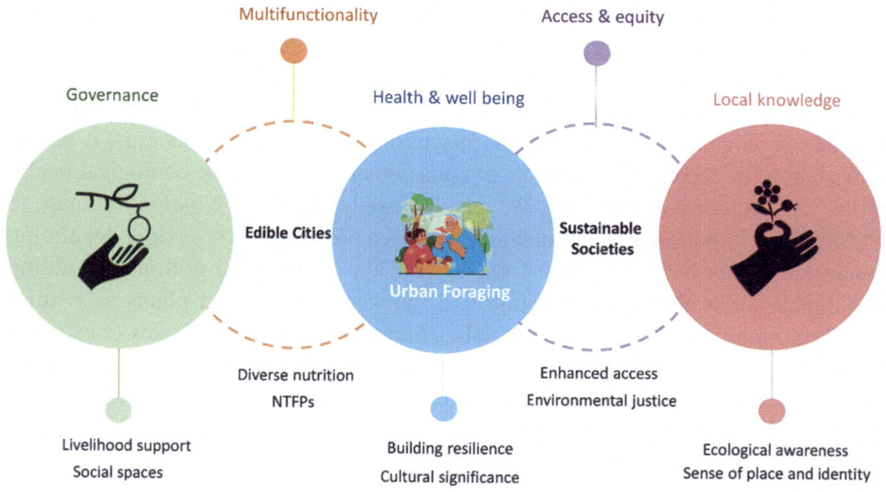

Fig. 18.1 Overview of the multiple aspects of urban foraging (Dotted lines: comprehensive outputs from including urban foraging in policy and infrastructure planning; Leader lines: Co-benefits of urban foraging)

them to partake in traditional customs and contribute to the creation of a sense of place (Somesh et al. 2021). Foraging becomes a valuable strategy for promoting participatory co-management and stewardship of urban ecosystems by embracing a commons governance framework, allowing citizens to actively engage in the sustainable management of shared resources (Fig. 18.1).

Urban foraging for wild plants and fungi challenges conventional notions of the origin, development, distribution, and consumption of food. It creates different experiences that gives a new significance to the edibles. By exposing the diverse and uncertain nature of food, informal methods of food acquisition in specific urban spaces challenge our collective perceptions of food (Sardeshpande and Shackleton 2020). Urban foraging entails spontaneous harvesting of resources as and when found and includes harvesting of both native and non-native species. This process facilitates potential management of greenspaces and plants that are otherwise not acknowledged as cultivated species, thereby creating habitats for wild plants (Guenat et al. 2023). It also enhances the local ecological knowledge and promotes knowledge transfers, fostering a sense of co-management and ownership amongst foragers. Despite the increasing studies on the importance of urban foraging of wild food, the potential contribution of wild food as an alternative food source for improving dietary diversity and food security has seldom been acknowledged (Garekae and Shackleton 2020). The inadequate acknowledgment of the importance of wild edibles leads to gaps in the food security policies, interventions, and regional and national food composition data. Therefore, it is imperative to understand the role of wild plants in urban food systems for informed urban planning. Studies have reported multiple constraints faced by urban foragers during selection of suitable UGS for foraging, including accessibility, potential pollution of UGS,

UGS management, and regulations (Fischer and Kowarik 2020; Brandner and Schunko 2022). Addressing these constraints that lead to caution exercised by the foragers can help in improving the foraging habitats and enhance nature connectedness among the city dwellers, in turn strengthening the local biodiversity conservation and ecological restoration. While discussing the foragers' concerns, it is also important to underscore the impacts of conscious foraging on the ecological health of the UGS (Guenat et al. 2023). We use the insights from the existing literature to highlight the importance of urban wild plants as a dietary source worldwide, aiming connect the dots between urban foraging and the widespread notion of building sustainable cities through cultivation and management of wild plants as nature-based solutions. We then discuss the policy barriers and pollution as key constraints against foraging practiced by the city dwellers. Finally, we discuss the need for inclusion of wild species in UGS planning for resilient cities and infrastructure, while emphasizing on good foraging practices for potential foragers to eliminate any negative ecological impact due to foraging.

18.2 Foraging in Urban Areas

Urban foraging has been extensively investigated in metropolitan regions throughout the world, including India, Germany, Uganda, Brazil, and United states (Synk et al. 2017; Landor-Yamagata et al. 2018; Ojelel et al. 2019; Gaither et al. 2020; Guenat et al. 2023). The collection of numerous species from urban environments has been documented in both the Northern and Southern hemispheres, encompassing temperate and tropical regions (Mollee et al. 2017; Somesh et al. 2021). Foraging takes place in different locations within cities, including public and private lands, as well as communal lands. The types of land used for foraging are diverse, ranging from formally managed parks to abandoned lots and spaces in between (Palliwoda et al. 2017). In any given city, numerous species, both native and non-native, appear to be harvested across urban, suburban, and peri-urban landscapes (Synk et al. 2017). This includes different types of land ownership, such as public, state, and private lands. Kujawska and Łuczaj (2015) discuss the fact that foraging is a prevalent exercise, with foragers often traveling to several locations in search of their desired species and experiences. It is widely recognized that fewer wild edibles are consumed by the people in urban areas as compared to those in the rural settings as the wild edibles are not as accessible to the urban residents as they are in the rural areas (Kujawska and Łuczaj 2015).

Urban foragers are people who search for food or provisions growing in the wild for consumption (Linnekin 2017). Foragers come from diverse social, cultural, and economic backgrounds, and they experience a multitude of advantages that arise from foraging, which often occur simultaneously (Brandner and Schunko 2022). Some of these advantages include the opportunity to connect with nature, gain a greater understanding of the environment, and develop a deeper appreciation for the food they consume. The significance of particular benefits of urban foraging may shift over time due to individual and collective situations, such as the accessibility of natural areas and species, the presence of job opportunities or occurrences like natural calamities, and instances of armed conflict (Short Gianotti and Hurley 2016). For example, in areas prone to natural disasters such as hurricanes or earthquakes, foragers may have an advantage over those who rely on supermarkets for their food supply. It has been suggested that in regions experiencing economic downturns, foraging can provide a low-cost source of nutrition and a potential income source for those willing to sell their harvest (Kumar et al. 2017). A study by Sachdeva et al. (2018) reports the relevance of foraging during the great recession of 2007–2009 wherein the self-provisioning strategies were highlighted in the print and digital mass media articles, suggesting foraging as an economically viable practice during heightened food insecurity and financial instability. In line with the previous finding, a study by Clouse (2022) highlighted the resurgence of urban foraging during the COVID-19 pandemic in America. The research explored the incidence of citizens' interest in acquiring wild edibles from their vicinity, emphasizing the need to incorporate this trend in urban planning and organizational frameworks to improve food security in similar events of disasters. Thus, wild edibles act as nature-based solutions (NbS) that can improve resilience and provide food to the inhabitants, while improving connectedness and ecological awareness among city residents.

18.2.1 Nature-Based Solutions for Sustainable Urban Transformation

The rapidly vanishing agricultural spaces designed for food production due to increased population pressures necessitate the creation of alternative landscapes to realize food demands in urban settings (Oliver et al. 2018). Wild edibles are widely discussed in numerous scholarly articles and have been referenced in case studies across the globe (Säumel et al. 2019; Scharf et al. 2019; Mino et al. 2021; Sartison and Artmann 2020; Russo and Escobedo 2022). Nevertheless, there remains a lack of a definitive, unified explanation for what constitutes cultivation of wild edibles in cities, and numerous authors endeavor to provide their own perspectives, incorporating additional factors that contribute to the idea of a resilient city (Säumel et al. 2019). Cultivation of food in common spaces within urban areas acts as NbS, yielding numerous concurrent ecological, communal, and economic advantages. Furthermore, these practices are regarded as localized, action-oriented strategies aimed at achieving sustainable transformation to resilient cities through desirable objectives (Russo and Cirella 2019) of fostering social unity and improving quality of life, establishing a connection between humans and their food, as well as encouraging pro-environmental behavior (Sartison and Artmann 2020). The NbS encompass various elements, including community gardens, which are defined as shared spaces cultivated collectively by local members with vegetables or flowers such as *Abelmoschus esculentus* (Okra), *Bauhinia purpurea* (Bahuni), *Hibiscus rosa-sinensis* (Gudhal), *Matricaria recutita* (Camomile), and *Ocimum basilicum* (Basil). Additionally, green roofs, vertical gardens, urban forests, residential gardens, kitchen gardens, historical gardens, informal green spaces, etc. all contribute to the sustainable network of edible food components and structures within a city. Initiatives aimed at establishing sustainable cities give rise to the creation of nourishable landscapes, which are spaces that incorporate edible plants and other types of landscaping flora to form multifunctional environments (Świąder et al. 2023). These landscapes are closely associated with urban settings that are perpetually productive, with the production of food being an integral component of the urban fabric, empowering native stakeholders to address social challenges, stimulate the emergence of new businesses and employment opportunities, drive economic growth, and enhance social cohesion (Plassnig et al. 2022). Encouraging citizens to actively participate in the co-creation of sustainable development paths through promoting their interactions with the local biodiversity, imparting nature consciousness, and facilitating responsible foraging activities can proactively transform the urban surroundings and establish a more interconnected urban way of life that brings benefits to all (Scharf et al. 2019).

18.2.2 Exploring Foraging in Rapidly Altering Urban Spatial Dynamics

Urban studies increasingly argue that urbanization and urban areas have expanded to a global scale (Kidokoro et al. 2022). This phenomenon calls for a better understanding of the differentiation between the urban environment and the distinctive physical structure of the city (Fang and Yu 2017). When examining the potential of cities that produce food, it becomes evident that population density plays a significant role. It can emerge as a consequence of the strain on infrastructure and as a response to limited land availability and resource inefficiency (Nyman 2019). In exploring global, urbanized food systems, the physical structure of cities can uncover the hidden uncertainties related to food that are not always acknowledged directly. Cities combine various cultural and material elements that sometimes conflict with the realm of food. Neglected and vacant areas can give rise to spontaneous groupings of urban vegetation, which can be edible based on their physical and cultural status, as well as how we interact with and navigate through such landscapes (Poe et al. 2013; Gandy 2016). The intricate nature and consequences of population density, the uneven distribution of urban forest cover, and growing trend of higher built density and vertical stratification in urban food production can contribute to the development of environmental friendly cities that appear to be more productive. Nonetheless, these advancements can present challenges such as restricted access, difficulties in navigating spaces, and limitations on informal food collection or provision, as suggested by Sachdeva et al. (2018).

Over time, the disconnect between city dwellers and the sources of their food, within a capitalist urbanization context, deeply ingrains a significant cultural and psychological perception regarding how food should be organized spatially (Heynen et al. 2006). Land use regulations and zoning restrictions can dictate where and how urban foraging activities can take place. Certain areas may be off-limits or inaccessible for foraging due to private property ownership or legal restrictions (Hanna and

Wallace 2022). These boundaries can limit the opportunities for urban residents to engage in foraging practices. Factors such as the presence of parks, green spaces, and edible plantings in public areas can encourage and facilitate urban foraging (Nagendra et al. 2018). Arrington et al. (2017) suggest the use of geo-tagged and cross-referenced satellite data to enhance our understanding of the complex spatial dynamics of urban foraging. The study explores community-based environmental monitoring and citizen science data sets to create unique data repositories. These data repositories can be used for subsequent qualitative and quantitative interpretations of the community data to understand the landscape level and demographic characteristics of foragers and foraging grounds. This information can be instrumental in planning and management of UGS to impart resilience to vulnerable cities resulting from rapid urbanization and altering land use patterns.

18.3 Balancing Foraging Practices, Regulations, and Ecological Awareness

Urban foraging generally involves acquiring wild food without cost and often sharing it rather than selling it. However, this does not mean that there is no involvement of capital or labor. Tsing (2012) describes gathered sustenance as a generous, unearned abundance, emphasizing its value as something unexpected and unrequested. Wild, underutilized species are abundant in micronutrients, resilient to climatic extremities, and necessitate minimal resources for cultivation (Hunter et al. 2019). Embracing and incorporating these species into our food systems not only preserve vital genetic resources for the future but also provide habitats for wildlife, including fruit-eating animals and insects that depend on these species for survival (Fehlmann et al. 2021). Foraging and scavenging of wild edibles are often restricted by the local governments as a method of organization (Shackleton et al. 2017). Although foraging is recognized as a right in many countries, it is often limited by the land owners and public authorities that hold the jurisdiction of the land. A study by Lee and Garikipati (2011) highlighted the restraints held by the British law on wild food gathering to limit excessive exploitation and habitat conservation. An article by Cunningham (2018) also mentions the restrictions set by the authorities in Seattle, US against foraging in municipal parks as a result of destruction and exploitation of the biodiversity in the parks by the foragers. However, the article presents hope for the foragers as it reports that some cities in the US have started acknowledging the foraging communities as conscious gatherers, allowing the harvest to certain fruits and plant species. Similar research has been published by Brito and Borelli (2020) and Guenat et al. (2023) revealing that foraging practices were discouraged and forbidden in some cities of Brazil. The reports indicate that these regulations are often unclear, subject to contention, and are ambiguous in their interpretation. These laws are conveniently negotiated, circumvented, and interplayed through personal agreements with the regulating bodies. However, the unevenness of these provisions call for better understanding of the foraging dynamics by the land managers to facilitate transparent policy formation (Short Gianotti and Hurley 2016). Under this perception, Shortly and Kepe (2021) suggest a major ideological shift on the stakeholders' side to embrace the foraging practices while monitoring the foragers' demographics, their motivation, their understanding of the local biodiversity, and develop foraging grounds to promote local ecological knowledges and peoples' interconnectedness with the nature.

18.4 Increasing Pollution: A Major Barrier to Foraging in Cities

Understanding and navigating the constraints of urban foraging are essential for individuals engaged in the activity, as it allows them to make informed decisions, minimize risks, and maximize benefits. These risks arise not only from the possibility of misidentifying species (Shortly and Kepe 2021), but also from encountering chemical contaminants present in the soil, on the surfaces of collected items, or within consumed tissues when gathering or consuming materials from urban settings (von Hoffen and Saumel 2014). Synk et al. (2017) suggest that minimizing exposure to pollutants for individuals involved in collecting foraged materials and other plants grown in urban areas can be achieved through careful evaluation of environmental pollution, including factors such as nearby traffic density, as well as thorough cleaning practices.

Pollution poses a significant barrier to urban foraging and is a growing concern for creating healthy cities overall. A recent inquiry by Amato-Lourenco et al. (2020) has expounded that air pollution stemming from traffic in the urban sector of Sao Paulo, Brazil, exhibits a strong correlation with significant contamination found in wild plants. Consequently, it deduces that engaging in foraging activities within cities should preferably occur in low-traffic areas or parks. In the case of horticulture, the specific locations, produce, and fruits become important in identifying unsafe levels of contaminants (Brandner and Schunko 2022). Conversely, crops grown on urban rooftops have minimal levels of heavy metals, suggesting that these areas are less susceptible to contamination from vehicular traffic (Grard et al. 2018). However, studies also suggest that while crops grown in urban gardens may be free of contaminants, there are risks associated with the inadvertent consumption of soil in the same locations (Brandner and Schunko 2022). Studies have reported that due to intensive land use and waste production, trace metal contamination has been prevalent in urban soils (Gupta et al. 2019). A study conducted in California found elevated levels of lead and cadmium in soil from sites with wild edible plants. However, based on tissue testing, the study concluded that thoroughly washed greens were tested safe for consumption (Khalid et al. 2018). Nevertheless, there is limited verified evidence regarding safety concerns associated with wild edible plants in general (Stark et al. 2019). The uncertainties surrounding the risks imposed by pollution and contamination underline the necessity for supplementary studies, particularly within the realm of environmental education, to enhance awareness regarding potential risks associated with specific species or locations, as well as the handling and consumption of forageable resources (Palliwoda et al. 2017).

To minimize potential health risks associated with foraging, it is advisable to engage in the collection of wild mushrooms, herbs, and vegetables within expansive green spaces or parks, while maintaining a distance from roadways (Amato-Lourenco et al. 2020). Furthermore, it is crucial to consider the historical land use patterns of the designated foraging area, as highlighted by McLain et al. (2014). In addition to individual precautions, community-based approaches can play a pivotal role in strengthening urban foraging practices. Collaborative efforts among foragers, local authorities, and environmental organizations can help identify safe foraging zones, conduct soil testing, and educate the public about the potential risks and benefits of foraging in certain areas. Building a network of informed foragers and establishing guidelines for responsible foraging can contribute to safer and more sustainable urban foraging experiences. This collaborative approach also fosters a sense of community and shared responsibility for the well-being of both foragers and the urban environment.

18.5 Practicing Conscious Foraging

The responsible institutions for management of urban green spaces may face aesthetic, environmental, and public health issues when deciding whether to allow foraging in cities (Clouse 2022). It is crucial to recognize that the influence of foraging is overshadowed by the alterations in land use and land cover that transpire during the process of urbanization (Säumel et al. 2016; Pinto et al. 2022). Foraging is often discouraged in cities due to low levels of biodiversity and higher proportions of threatened species as compared to the surrounding landscapes, which can be endangered due to negative impacts of foraging (Bunge et al. 2019).

However, studies have reported negligible impacts of foraging on the threatened taxa (Fischer and Kowarik 2020). This indicated that the foragers prefer harvesting abundant species as compared to the rare ones. Interestingly, foragers in urban settings appear to possess a remarkable level of awareness and knowledge concerning the local natural environment. This awareness is often cultivated through active participation in community events during their developmental years, as observed by Guenat et al. (2023). Foragers who exhibit such conscientiousness and understanding of the ecosystem demonstrate a heightened sense of empathy toward biodiversity. This empathetic perspective translates into informed decision-making regarding the selection and quantity of species harvested (McLain et al. 2017). Beyond their role as harvesters, foragers also play a vital part in the co-management of urban foraging grounds. In addition to their harvesting activities, they engage in supplementary activities that contribute to the maintenance of these areas. They participate actively in tasks such as seed dispersal, weeding, pruning, and the creation of wildlife-friendly adventure trails (McLain et al. 2017). This multifaceted involvement showcases how foragers can positively impact the ecological balance of urban settings. Schunko et al. (2021) suggest that urban wild food harvesting is an intellectually demanding pursuit. Foragers must possess the ability to adapt their practices to the evolving availability of resources resulting from rapid urban transitions.

This highlights the cognitive flexibility and resilience required by foragers to navigate the dynamic urban ecosystem successfully.

In summary, the narrative surrounding urban foraging is evolving, driven by the discovery of unexpected positive attributes of foragers. These findings highlight the potential for foraging to serve as a bridge between urban residents and their surrounding natural environments, fostering a deeper understanding of, and connection to, the complex ecosystems that exist within cities.

18.6 Conclusion

The chapter highlights the need for inclusion of wild species as dietary sources in the regional and national food composition data. This process would entail recognition of the wild edibles by the land managers as an important ecosystem benefit, rendering dedicated management of these edibles to prevent overexploitation and conduct sustainable foraging practices. Foragers should strive to reduce their negative ecological impact on the foraging habitats by supporting biodiversity restoration and maintenance initiatives. As urban foraging continues to gain popularity, community-based initiatives can be instrumental in preserving the practice while mitigating its associated challenges. This would aid in bridging the gap in perceptions held by public authorities against urban dwellers, leading to the creation of sustainable foraging spaces and realization of the optimum potential of urban foraging.

References

Amato-Lourenco LF, Ranieri GR, de Oliveira Souza VC, Junior FB, Saldiva PHN, Mauad T (2020) Edible weeds: are urban environments fit for foraging? Sci Total Environ 698:133967. https://doi.org/10.1016/j.scitotenv.2019.133967

Arrington AB, Diemont SA, Phillips CT, Welty EZ (2017) Demographic and landscape-level urban foraging trends in the USA derived from web and mobile app usage. J Urban Ecol 3(1):jux006

Brandner A, Schunko C (2022) Urban wild food foraging locations: understanding selection criteria to inform green space planning and management. Urban For Urban Green 73:127596. https://doi.org/10.1016/j.ufug.2022.127596

Brito VV, Borelli S (2020) Urban food forestry and its role to increase food security: a Brazilian overview and its potentialities. Urban For Urban Green 56:126835

Bunge A, Diemont SAW, Bunge JA, Harris S (2019) Urban foraging for food security and sovereignty: quantifying edible forest yield in Syracuse, New York using four common fruit- and nut-producing street tree species. J Urban Ecol 5(1). https://doi.org/10.1093/jue/juy028

Calicioglu O, Flammini A, Bracco S, Bellù L, Sims R (2019) The future challenges of food and agriculture: an integrated analysis of trends and solutions. Sustainability 11(1):222.8

Chakona G, Shackleton CM (2019) Food insecurity in South Africa: to what extent can social grants and consumption of wild foods eradicate hunger? World Dev Perspect 13:87–94

Clouse C (2022) The resurgence of urban foraging under COVID-19. Landsc Res 47(3):285–299

Colléony A, Levontin L, Shwartz A (2020) Promoting meaningful and positive nature interactions for visitors to green spaces. Conserv Biol 34(6):1373–1382

Cunningham J (2018) Hungry for change: urban foragers take the law into their own hands. The Guardian. Accessed 2024 Apr 5; Available from: https://www.theguardian.com/cities/2018/aug/07/hungry-for-change-urban-foragers-take-the-law-into-their-ownhands

Fang C, Yu D (2017) Urban agglomeration: an evolving concept of an emerging phenomenon. Landsc Urban Plan 162:126–136. https://doi.org/10.1016/j.landurbplan.2017.02.014

FAO, IFAD, UNICEF, WFP and WHO (2023) The State of Food Security and Nutrition in the World 2023. Urbanization, agrifood systems transformation and healthy diets across the rural–urban continuum. Rome, FAO https://doi.org/10.4060/cc3017en

Fehlmann G, O'riain MJ, FÜrtbauer I, King AJ (2021) Behavioral causes, ecological consequences, and management challenges associated with wildlife foraging in human-modified landscapes. Bioscience 71(1):40–54. https://doi.org/10.1093/biosci/biaa129

Fischer LK, Kowarik I (2020) Connecting people to biodiversity in cities of tomorrow: is urban foraging a powerful tool? Ecol Indic 112:106087. https://doi.org/10.1016/j.ecolind.2020.106087

Gaither CJ, Aragón A, Madden M, Alford S, Wynn A, Emery M (2020) "Black folks do forage": examining wild food gathering in Southeast Atlanta Communities. Urban For Urban Green 56:126860

Gandy M (2016) Unintentional landscapes. Landsc Res 41(4):433–440

Garekae H, Shackleton CM (2020) Foraging wild food in urban spaces: the contribution of wild foods to urban dietary diversity in South Africa. Sustainability 12(2):678

Gebre T, Gebremedhin B (2019) The mutual benefits of promoting rural-urban interdependence through linked ecosystem services. Glob Ecol Conserv 20:e00707

Grard BJP, Chenu C, Manouchehri N, Houot S, Frascaria-Lacoste N, Aubry C (2018) Rooftop farming on urban waste provides many ecosystem services. Agron Sustain Dev 38(1):2

Guenat S, Bailey-Athias JP, Fischer LK (2023) Urban foraging in Brazilian public greenspaces. Ambio 52:1248–1261. https://doi.org/10.1007/s13280-023-01847-y

Gupta N, Yadav KK, Kumar V, Kumar S, Chadd RP, Kumar A (2019) Trace elements in soil-vegetables interface: translocation, bioaccumulation, toxicity and amelioration—a review. Sci Total Environ 651:2927–2942. https://doi.org/10.1016/j.scitotenv.2018.10.047

Hanna C, Wallace P (2022) Planning the urban foodscape: policy and regulation of urban agriculture in Aotearoa New Zealand. Kōtuitui N Z J Soc Sci 17(3):313–335

Heynen N, Perkins HA, Roy P (2006) The political ecology of uneven urban green space: the impact of political economy on race and ethnicity in producing environmental inequality in Milwaukee. Urban Aff Rev 42(1):3–25

Hunter D, Borelli T, Beltrame DM, Oliveira CN, Coradin L, Wasike VW et al (2019) The potential of neglected and underutilized species for improving diets and nutrition. Planta 250:709–729

Hurley PT, Becker S, Emery MR, Detweiler J (2022) Estimating the alignment of tree species composition with foraging practice in Philadelphia's urban forest: toward a rapid assessment of provisioning services. Urban For Urban Green 68:127456

Kaika M (2017) 'Don't call me resilient again!': the new urban agenda as immunology … or … what happens when communities refuse to be vaccinated with 'smart cities' and indicators. Environ Urban 29(1):89–102. https://doi.org/10.1177/0956247816684763

Khalid N, Hussain M, Young HS, Boyce B, Aqeel M, Noman A (2018) Effects of road proximity on heavy metal concentrations in soils and common roadside plants in Southern California. Environ Sci Pollut Res 25:35257–35265

Kidokoro T, Matsuyuki M, Shima N (2022) Neoliberalization of urban planning and spatial inequalities in Asian megacities: focus on Tokyo, Bangkok, Jakarta, and Mumbai. Cities 130:103914

Kujawska M, Łuczaj Ł (2015) Wild edible plants used by the polish Community in Misiones, Argentina. Hum Ecol 43(6):855–869

Kumar SA, Lakhran HA, Meena RS, Jangir CK (2017) Current needs of sustainable food and forage production to eliminate food and forage insecurity under climate change era. Forage Res 43(3):165–173

Landor-Yamagata JL, Kowarik I, Fischer LK (2018) Urban foraging in Berlin: people, plants and practices within the metropolitan green infrastructure. Sustainability 10(6):1873

Lee J, Garikipati S (2011) Negotiating the non-negotiable: British foraging law in theory and practice. J Environ Law 23(3):415–439
Linnekin BJ (2017) Food law gone wild: the law of foraging. Fordham Urb LJ 45:995
McLain RJ, Hurley PT, Emery MR, Poe MR (2014) Gathering "wild" food in the city: rethinking the role of foraging in urban ecosystem planning and management. Local Environ 19(2):220–240
McLain RJ, Poe MR, Urgenson LS, Blahna DJ, Buttolph LP (2017) Urban non-timber forest products stewardship practices among foragers in Seattle, Washington (USA). Urban For Urban Green 28:36–42
Mino E, Pueyo-Ros J, Škerjanec M, Castellar JA, Viljoen A, Istenič D et al (2021) Tools for edible cities: a review of tools for planning and assessing edible nature-based solutions. Water 13(17):2366
Mollee E, Pouliot M, McDonald MA (2017) Into the urban wild: collection of wild urban plants for food and medicine in Kampala, Uganda. Land Use Policy 63:67–77. https://doi.org/10.1016/j.landusepol.2017.01.020
Nagendra H, Bai X, Brondizio ES, Lwasa S (2018) The urban south and the predicament of global sustainability. Nat Sustain 1(7):341–349. https://doi.org/10.1038/s41893-018-0101-5
Nyman M (2019) Food, meaning-making and ontological uncertainty: exploring 'urban foraging' and productive landscapes in London. Geoforum 99:170–180. https://doi.org/10.1016/j.geoforum.2018.10.009
Ojelel S, Mucunguzi P, Katuura E, Kakudidi EK, Namaganda M, Kalema J (2019) Wild edible plants used by communities in and around selected forest reserves of Teso-Karamoja region, Uganda. J Ethnobiol Ethnomed 15:1–14
Oliver TH, Boyd E, Balcombe K, Benton TG, Bullock JM, Donovan D, Feola G, Heard M, Mace GM, Mortimer SR, Nunes RJ (2018) Overcoming undesirable resilience in the global food system. Global Sustainability 1:e9
Palliwoda J, Kowarik I, von der Lippe M (2017) Human-biodiversity interactions in urban parks: the species level matters. Landsc Urban Plan 157:394–406. https://doi.org/10.1016/j.landurbplan.2016.09.003
Pinto LV, Inácio M, Ferreira CSS, Ferreira AD, Pereira P (2022) Ecosystem services and well-being dimensions related to urban green spaces—a systematic review. Sustain Cities Soc 85:104072. https://doi.org/10.1016/j.scs.2022.104072
Plassnig SN, Pettit M, Reichborn-Kjennerud K, Säumel I (2022) Successful scaling of Edible City solutions to promote food citizenship and sustainability in food system transitions. Front Sustain Cities 4:190
Poe MR, McLain RJ, Emery M, Hurley PT (2013) Urban forest justice and the rights to wild foods, medicines, and materials in the city. Hum Ecol 41:409–422
Ritchie H, Roser M (2020) Environmental impacts of food production. Our world in data
Russo A, Cirella GT (2019) Edible urbanism 5.0. Palgrave Commun 5(1):1–9. https://doi.org/10.1057/s41599-019-0377-8
Russo A, Escobedo FJ (2022) From smart urban forests to edible cities: new approaches in urban planning and design. Urban Plan 7(2):131–134
Sachdeva S, Emery MR, Hurley PT (2018) Depiction of wild food foraging practices in the media: impact of the great recession. Soc Nat Resour 31(8):977–993
Sardeshpande M, Shackleton C (2020) Urban foraging: land management policy, perspectives, and potential. PloS One 15(4):e0230693. https://doi.org/10.1371/journal.pone.0230693
Sartison K, Artmann M (2020) Edible cities—an innovative nature-based solution for urban sustainability transformation? An explorative study of urban food production in German cities. Urban For Urban Green 49:126604. https://doi.org/10.1016/j.ufug.2020.126604
Säumel I, Weber F, Kowarik I (2016) Toward livable and healthy urban streets: roadside vegetation provides ecosystem services where people live and move. Environ Sci Policy 62:24–33. https://doi.org/10.1016/j.envsci.2015.11.012
Säumel I, Reddy SE, Wachtel T (2019) Edible city solutions—one step further to foster social resilience through enhanced socio-cultural ecosystem services in cities. Sustainability 11(4):972

Scharf N, Wachtel T, Reddy SE, Säumel I (2019) Urban commons for the edible city—first insights for future sustainable urban food systems from Berlin, Germany. Sustainability 11(4):966

Schunko C, Brandner A (2022) Urban nature at the fingertips: investigating wild food foraging to enable nature interactions of urban dwellers. Ambio 51:1168–1178. https://doi.org/10.1007/s13280-021-01648-1

Schunko C, Wild A-S, Brandner A (2021) Exploring and limiting the ecological impacts of urban wild food foraging in Vienna, Austria. Urban For Urban Green 62:127164. https://doi.org/10.1016/j.ufug.2021.127164

Shackleton CM, Hurley PT, Dahlberg AC, Emery MR, Nagendra H (2017) Urban foraging: a ubiquitous human practice overlooked by urban planners, policy, and research. Sustainability 9(10):1884. https://doi.org/10.3390/su9101884

Short Gianotti AG, Hurley PT (2016) Gathering plants and fungi along the urban-rural gradient: uncovering differences in the attitudes and practices among urban, suburban, and rural landowners. Land Use Policy 57:555–563. https://doi.org/10.1016/j.landusepol.2016.06.019

Shortly A, Kepe T (2021) Consuming the city: challenges and possibilities for foraging in Toronto's parks. For Trees Livelihoods 30(2):75–89

Somesh D, Rao R, Murali R, Nagendra H (2021) Patterns of urban foraging in Bengaluru city. Urban For Urban Green 57:126940

Stark PB, Miller D, Carlson TJ, de Vasquez KR (2019) Open-source food: nutrition, toxicology, and availability of wild edible greens in the East Bay. PloS One 14(1):e0202450. https://doi.org/10.1371/journal.pone.0202450

Świąder K, Čermak D, Gajewska D, Najman K, Piotrowska A, Kostyra E (2023) Opportunities and constraints for creating edible cities and accessing wholesome functional foods in a sustainable way. Sustainability 15(10):8406. https://doi.org/10.3390/su15108406

Synk CM, Kim BF, Davis CA, Harding J, Rogers V, Hurley PT, Emery MR, Nachman KE (2017) Gathering Baltimore's bounty: characterizing behaviors, motivations, and barriers of foragers in an urban ecosystem. Urban For Urban Green 28:97–102. https://doi.org/10.1016/j.ufug.2017.10.007

Tsing A (2012) Unruly edges: mushrooms as companion species: for Donna Haraway. Environ Humanit 1(1):141–154. https://doi.org/10.1215/22011919-3610012

von Hoffen LP, Säumel I (2014) Orchards for edible cities: cadmium and lead content in nuts, berries, pome and stone fruits harvested within the inner city neighbourhoods in Berlin. Ger Ecotoxicol Environ Safety 101:233–239

Jayshree Shukla is a Ph.D. candidate at the University of Adelaide, Australia. She has worked as a Project Associate at the CSIR-National Environmental Engineering Research Institute, Nagpur, India. Her primary areas of research include urban ecology management, forest conservation and restoration, and ecosystem modeling. She is an active member of the IUCN Commission on Ecosystem Management.

Part VI

Looking to the Future

Foraging and Consumption of Cerrado Biome Species in the Brasilia Alternative Urban Food System

19

Janaína Deane de Abreu Sá Diniz, Stéphane Guéneau, Jessica Pereira Garcia, and Barbara Loureiro Borges

Abstract

Brasilia, the capital of Brazil, is considered a park-city, precisely because it is heavily wooded and has a large population of fruit trees—both native to the Cerrado and exotic. In this sense, our chapter seeks to understand how initiatives involving foraging of native, exotic, neglected and underutilized species in natural areas or agroforestry systems have contributed to the consolidation of alternative urban food systems, as well as to encourage greater access to green areas to carry out the practice of foraging in the city of Brasília. The text presents a brief theoretical approach on urban foraging and alternative urban food systems, followed by the context of valuing local biodiversity in restaurants and school meals to promote conservation in the Cerrado, as well as a description of urban and peri-urban initiatives of foraging native and exotic species, such as an application to map fruit trees and a plan for planting trees launched by the Landless Rural Workers' Movement. This set of innovative initiatives demonstrates the potential for urban food systems transformation toward more sustainable and inclusive practices, where foraging for food is an important and possible practice that must be stimulated and supported by public policies and by the society.

J. Deane de Abreu Sá Diniz (✉)
University of Brasília, Brasilia, Brazil
e-mail: janadiniz@unb.br

S. Guéneau
Cirad – Moisa (French Agricultural Research Centre for International Development – Montpellier Interdisciplinary Center on Sustainable Agri-Food Systems), Montpellier, France
e-mail: stephane.gueneau@cirad.fr

J. Pereira Garcia
Center for Sustainable Development, University of Brasília, Brasilia, Brazil

B. Loureiro Borges
Department of Forest Sciences, University of Brasília, Brasilia, Brazil

S. Dhyani, M. Sardeshpande (eds.), *Urban Foraging in the Changing World*,
https://doi.org/10.1007/978-981-97-0345-6_19

Keywords

Park-city · Fruit collection · Healthy food · Planting trees

19.1 Introduction

Foraging is a practice carried out by a large number of people around the world, mainly with the purpose of supplying fruits, mushrooms, roots, and leaves, whether for self-consumption, exchange, or commercial purposes (Svizzero 2016). In countries with conserved forest areas, such as Brazil, despite high rates of deforestation (Escobar 2020), there are still many regions where foraging of food products is practiced, especially in the Amazon, contributing to the generation of income and food and nutritional security of communities involved in the activity (Porro and Sousa 2023), as well as in urban contexts (Borelli et al. 2020).

Also known as extractivism[1] or collection among traditional rural communities, foraging represents an important subsistence activity and/or serving local markets or supplying national and/or global chains (Porro and Sousa 2023; Jerin et al. 2022). Unlike agricultural activity, which requires planning between planting and harvesting, extractivism is defined by Drummond (1996) as any goods-producing activity in which useful natural resources are taken directly from their area of natural occurrence. Foraging, not only practiced in forest areas, includes products that are available in nature and that do not follow a high-yelding logic from planting to harvest.

Considering that in Brazil there are many cases of plant collection, both for subsistence (Porro and Sousa 2023) and for serving consolidated supply chains (Superti et al. 2018), in the present text we address both cases, in the context of Brasília and surrounding areas. We also discuss situations of urban and peri-urban collection, in places that do not necessarily constitute native forests (because these have been suppressed), but also with other planted species (including exotic ones), or because they are urban areas where some plant species occur, mainly for landscaping purposes (Lima 2009).

Brasilia, the capital of Brazil, was inaugurated in 1960 in the middle of the Cerrado biome. Initially, there was a great deal of deforestation to build its infrastructure, including buildings (residential, administrative, and commercial) and roads, but today the city is considered a park-city, precisely because it is heavily

[1] The interpretation of extractivism in the literature generally refers to the unsustainable extraction of natural resources. Recently, however, several authors have stated that extractivism cannot simply be used to describe the destructive forms of logging, mining, fishing, among others, but that this term also has a strong political-economic connotation that structures the processes of production and reproduction of capital (Ye et al. 2019). A part of Brazilian literature, on the other hand, is part of the current of thought of socio-environmentalism (Santilli 2005), which associates extractivism with sustainable practices, as a counterpoint to predatory practices of agricultural colonization and conversion of native vegetation. This concept is widely used to describe systems based on the collection of non-timber forest products (NTFP) from the Amazon rainforest for commercialization (Pinton and Emperaire 2001).

wooded and has a large population of fruit trees—native to the Cerrado (cagaita—*Eugenia dysenterica*, jatobá—*Hymenaea courbaril*, pequi—*Caryocar brasiliense*), from other Brazilian biomes (cashew—*Anacardium occidentale, jenipapo—Genipa Americana, jabuticaba—Plinia cauliflora, guava—Psidium guajava and Brazilian cherry—Eugenia uniflora*) and exotic (mango—*Mangifera indica*, jackfruit—*Artocarpus heterophyllus*, rose apple—*Syzygium jambus*, lemon—*Citrus limon*, tamarind—*Tamarindus indica* and date palm—*Phoenix dactylifera*). Most of these trees are accessible in parks and other green areas, or even on public roads in urban areas (Agostini-Costa et al. 2006). Until a decade ago, native fruits and their derivatives were little commercialized, being consumed *in natura* (fresh) mainly by residents immediately after being foraged.

The context of foraging in the region of Brasília is also marked by the sustainability issues of the Cerrado biome. Indeed, on the one hand, this biome is the tropical savanna with the highest biodiversity in the world, with endemic species which represent approximately 32% of all terrestrial and aquatic plants and vertebrates in the biome (Mittermeier et al. 2004). On the other hand, the Cerrado is one of the biomes where native vegetation is most threatened by the expansion of agribusiness, with high rates of deforestation and various socio-environmental damages (Eloy et al. 2018; Sauer and Oliveira 2021). In areas where native vegetation has been preserved, many local communities derive their livelihood from agro-extractivist systems which combine small-scale agriculture and the collection of biodiversity resources (Guéneau et al. 2019).

Brasília also stands out as a young city that is currently developing its gastronomic identity, excelling as the third largest gastronomic center in Brazil (Congresso em Foco 2017). The city is a place that concentrates public power, as it is the country's capital, and also has a population endowed with high financial and intellectual capital, favoring the concern of a large part of the population with issues related to sustainability. For several years now, we have been witnessing the emergence of movements seeking to enhance the value of native Cerrado species and neglected and underutilized species in Brasilia's gastronomy, by associating environmental values relating to the conservation of the biome to which the city's consumers are sensitive (Guéneau et al. 2022; Duarte et al. 2020, 2021).

This chapter seeks to understand how initiatives involving foraging of native, exotic, neglected and underutilized species in natural areas or agroforestry systems have contributed to the consolidation of alternative urban food systems, as well as to encourage greater access to green areas to carry out the practice of foraging in the city of Brasília.

Alternative food comprises a set of practices that indicate possibilities to resolve the limits and contradictions of the hegemonic agro-food model (Deverre and Lamine 2010), observed through the homogenization of the dietary pattern and the invisibility of production and producers, with impacts on human health, such as the increase in cases of obesity and the disconnection of human beings with nature (Schneider et al. 2016). In the case of alternative urban food systems, practices such as fair trade, short marketing channels, appreciation of local production, differentiated foods, and quality seals can be mentioned (Garcia et al. 2022). The hypothesis in this text is that foraging is a central piece of the emergence and consolidation of alternative urban food systems in Brasilia that contributes to the conservation of the endangered Cerrado biome.

The text is divided into two sections, in addition to this introduction, where we bring a brief theoretical approach on urban foraging and alternative urban food systems, and a short methodology topic. In the first section, we discuss the current context of valuing local biodiversity in restaurants and school meals to promote conservation in the Cerrado, while in the second section, we detail three other urban and peri-urban initiatives that involve foraging native and exotic species. In the last topic, we present a discussion on the challenges for maintaining the initiatives presented, as well as perspectives for continuing research and actions.

19.2 Methodology

The research methodology was qualitative, including a literature review, interviews, and the organization of a participatory workshop.

The first stage included interviews with key actors and the second consisted of holding a participatory workshop to encourage dialogue between actors, organized in Brasilia in June 2019. The objective of the workshop was the dialogue between chefs, agroextractivists and other local producers, representatives of cooperatives, professionals from different areas, consumers, and other people interested in the subject. It aimed to learn about innovations in the area of sustainable food systems (Valette et al. 2019), with the specific case of using fruits from the socio-biodiversity of the Cerrado in restaurants in Brasilia.

The third stage was carried out between October/2022 and March/2023, through a literature review and consultation of related documents and websites to collect information on other initiatives concerning sustainable food systems, leaded by initiatives conducted by enterprises, civil society, and public authorities. The first initiative reports from the introduction of native species of the Cerrado in school meals through the establishment of public policies at the national and regional levels. The second initiative reports on the creation and operation of an application to identify fruit species in urban areas and their geographical location, through the fruit-map application. And the third initiative is about the Plan "Plant trees, produce healthy food" elaborated by the Landless Rural Workers Movement—MST. In this sense, materials published by these initiatives were consulted and contact was made with their proposers or those responsible for confirmation and detailing of some information.

19.2.1 Valuing Cerrado Biodiversity Through Conscious and Educational Consumption

The Cerrado is the second biggest biome of Brazil, after Amazon, but, differently from the Brazilian rainforest, it is a neglected biome by governments and society in general, since it is not a forest, but a savannah landscape and, therefore, it is usually orientated by public and private projects in the direction of large agricultural activities, such as grain production and livestock than to conservation projects that allow sustainable collection (Diniz et al. 2013).

The construction of Brasilia, the capital of Brazil contributed to the transformation of the Center-West region, whose landscape was originally composed of the Cerrado biome. The Cerrado biome is made up of a savanna vegetation that occupies the Central Plateau area in Brazil, with the presence of more than 12,000 cataloged native plant species (CEPF 2017), and approximately with a rate of 32% of endemic species. With this, it is considered the most biodiverse tropical savanna in the world (Mittermeier et al. 2004).

Despite having been deforested for the construction of the capital's infrastructure, Brasilia is known for being a very wooded city—especially the Pilot Plan. There are also parks and rural properties that conserve remaining or recovered areas of the Cerrado. Some of these areas have recently started providing foraging products to supply restaurants in the capital. Also underway is the implementation of a local public policy aimed at supplying school canteens with plant species native from the Cerrado biome—main of them originated from collection—following a national trend toward regionalization of school meals.

19.2.1.1 Acquisition of Cerrado Species and Non-Conventional Food Plants by Restaurants in Brasilia

There is a growing number of chefs, all over the world, who choose ingredients that refer to sustainability characteristics, because they are produced in an organic, artisanal way and because they are typical or native to the biodiversity of their regions (Zaneti 2017; Pereira et al. 2019; Araújo and Finocchio 2020). In Brazil and in Latin American countries, such as Peru and Mexico, the kitchen appears as a space for the insertion of native species—most obtained from foraging—which are little used and/or do not have a structured value chain, but with great potential to be (re)integrated to the local food culture, if available for more accessible to the population. These actions are led by chefs who, concerned with local socio-environmental issues, promote the appreciation of regional food with the use and dissemination of food prepared from native plants, integration and strengthening of family farming and traditional communities, and the relationship to the countryside and city (Zaneti 2017; Pereira et al. 2019; Niederle and Schubert 2020).

In the case of products from the Cerrado biome, some authors seek to understand the appreciation of regional products in gastronomy through the process of building markets for these products (Zaneti and Balestro 2015; Guéneau et al. 2017; Duarte et al. 2020, 2021), that is, creating demand based on innovation in products, processes and differentiated relationships between producers or collectors and consumers or users. Encouraging the use of socio-biodiversity products can help in the conservation of the Cerrado, as agro-extractivism in the Cerrado is associated with the conservation and maintenance of the biome (Guéneau et al. 2020; Nogueira and Fleischer 2005). If the socio-biodiversity products are demanded and valued, the actors involved will be interested in maintaining collecting areas.

One practice which the actors interviewed saw as innovative and sustainable was the commitment to systematically using Cerrado fruits in the composition of the dishes of the establishments surveyed. Apart from a few fruit varieties that are part of the regional culinary heritage, such as the pequi fruit (*Caryocar brasiliense*), most fruits native to the Cerrado biome are consumed very little in Brazilian cities, and are even unknown to most Brazilian consumers (Garcia 2017). For instance, baru (*Dipteryx alata*), now known as the Cerrado nut, was hardly consumed a few years ago, even by traditional peoples. The inclusion of Cerrado fruits in the dishes of the establishments surveyed appears to evidence that for chefs, sustainable gastronomy involves not only an appreciation of products from the region where the restaurant is established, but also chefs playing a key role in preserving the local foraged biodiversity through the use of specific local foods in fine dishes.

However, the use of Cerrado fruits in the gastronomy does not guarantee that the fruit is grown sustainably (Guéneau et al. 2020). For this reason, some chefs seek to combine their local supply of Cerrado fruit with a direct relationship with the collectors, which enables them to visit the gatherers and collecting areas to observe the sustainability of the production systems.

Another important innovative practice witnessed among chefs in Brasília is the use of Neglected and Underutilized Species (NUS).[2] These edible wild plants and ruderal species (Neto et al. 2016) usually grow spontaneously and are classified as "weeds." They are consumed on a small scale or are no longer used by most of the population, who are not aware of the plants' nutritional and economic value and potential uses. More than half of the interviewees confirmed that their menus featured NUS, such as ora-pro-nóbis (*Pereskia aculeate*)[3] and taioba (*Xanthosoma sagittifolium*).[4] The frequent use of these plants demonstrates strong adherence to a movement driven by the publication of the NUS guide (Kinupp and Lorenzi 2014), which seeks to revive the consumption of neglected wild plants in Brazil, in clear opposition to the dominant species grown in industrial agriculture.

In few years, the offer by NUS has thus shifted from products collected in wild green or urban areas to more productive systems, such as agroforestry (spontaneously or intentionally) (Bredariol 2015; Herrmann et al. 2020), in cases of species with more consolidated demand. Demand is still restricted to a very limited public, mainly the population with greater purchasing power in Brasilia and/or consumers who prefer an alternative diet, associated with healthier products both in terms of human and environmental health.

In addition to restaurants and alternative businesses, in Brazil there are public policies that encourage the acquisition of native species by government bodies. In the next example, we address one of these policies, which has just had legislation approved in the Federal District, but which in other Brazilian municipalities and states is already more consolidated.

[2] In Portuguese, the Neglected and Underutilized Species is being used as non-conventional food plants (*Plantas Alimentícias Não Convencionais*, with PANC as acronym).

[3] Common English names for which include Barbados gooseberry leaves and leaf cactus.

[4] Known as arrow leaf or elephant ear in English.

19.2.1.2 Acquisition of Native Cerrado Species in Chools in Brasília

The Brazilian National School Feeding Program (PNAE) is a public policy that has existed since the 1950s. Among the changes already made in the policy, an important leap was the creation of Law 11,497 in 2009, when, in search of a healthy and quality food supply to public-school students, the program started to determine that at least 30% of the resources allocated to the purchase of food products were used in the acquisition of foodstuffs from family farming (Sonnino et al. 2014; Sidaner et al. 2013).

The percentage of acquisition of products from family farming varies greatly between municipalities and still shows a slow evolution in terms of fulfilling at least 30% of purchases from family farming (IPEA 2019). Among the products acquired in this category, there are still few that come from collection, however, in regions where extractivism is more practiced or organized, a greater insertion in school feeding is observed (Bispo 2014).

In the case of the Federal District, it was only in 2023 that the intention to acquire native species was officially signaled through Law 7228/2023, which provides priority for the purchase of fruits and products native to the Cerrado biome in school meals in the public network of the Federal District.[5] In addition to valuing local production and the biome, the law strengthens policies aimed at food and nutritional security and guarantees a more nutritious, healthy and diversified diet for students (Distrito Federal 2023).

In a newspaper article widely replicated in local and national media, the intention of the local government to acquire native fruits of the Cerrado, such as buriti (M*auritia flexuosa)*, gabiroba *(Campomanesia xanthocarpa)*, jatobá *(Hymenaea courbaril)*, umbu *(Spondias tuberosa)*, cagaita *(Eugenia dysenterica), and* pequi *(Caryocar Brasiliense)*, was signaled, in order to use them in recipes for school meals in the public-school system in the Federal District in 2023, which would offer juices, sweets, and cakes from different plants (Metrópoles 2023). However, in conversation with the team responsible for technical assistance to family producers, which include agroextractivists, it was informed that this action still requires better planning, given the complexity represented by the acquisition of products mostly from foraging, often scattered in different areas of the Federal District and surrounding areas, originating from places of difficult access and mostly perishable or very sensitive to transport. If the collected products were processed and/or frozen close to the collection sites, it would be necessary to maintain the cold chain, which entails additional transport costs. However, in some of these collection areas, mainly in the peri-urban areas, there is no adequate electrical energy network for processing, which makes it difficult to maintain the quality of processed products.

Before the law was signaled, in the Federal District, the only case of acquisition of products from the Cerrado occurred during the Covid-19 pandemic, in a Technological Institute. In this school, since most of the students come from poor families, they started to send baskets with products from a Central cooperative that gathers other 30 cooperatives from the Cerrado biome to the house of the students, to provide healthy

[5] In addition to native Cerrado fruits, the law also prioritizes the purchase of organic food, directly from family farming, from family entrepreneurs or their organizations, from agrarian reform settlements, from traditional communities, and from organic agricultural producers.

food and to keep the organizations supplying a demand, since other market channels—such as free fairs and points on the roads—were inaccessible. Although the scholars were receiving products originated from collection, few of them were from the city of Brasília, since there are no cooperatives that participate in this Central cooperative and the National Law of School Feeding allows the purchase of products from small producers in nearby regions if there are none in the same municipality.

In the following section, we address other experiences that we consider relevant to be presented in this chapter on foraging in Brasília, as examples of broader movements around the expansion of urban green spaces, or at least bringing the population closer to nature and expanded access to healthy and low or no cost food.

19.3 Urban and Peri-Urban Initiatives to Expand Green Areas in Brasília

As we have already described, the construction of Brasília provided the suppression of native vegetation and the inclusion of exotic species. Currently, there is a process of revaluation of native species. We will only describe the changes that made it possible for the urban structure to favor the collection of exotic fruit species, such as mango, avocado, blackberry, raspberry, and also native ones, such as cagaita (*Eugenia dysenterica*), cashew (*Anacardium occidentale*), jabuticaba (*Plinia cauliflora*), guava (*Psidium guajava*), pequi (*Caryocar brasiliense*), and pitanga (*Eugenia uniflora*).

Most conservation units in Brasilia do not allow collection for commercial purposes, because they are in the category of Integral Protection Units, such as Ecological Stations, Biological Reserves and National Park (BRASIL 2000). But in areas of legal reserve for urban and peri-urban agriculture, in agrarian reform settlements and private properties with agroforestry systems, it is possible to collect native and exotic species, which has been increasing in recent years due to the interest of a still restricted group of consumers.

While the older initiatives, with the advent of the pandemic, ended up weakening, others were created and/or strengthened precisely during the pandemic, showing the potential of urban commons for resilience in neighborhoods (Sardeshpande et al. 2021; Clouse 2022). The first situation is the case of the fruit-map application and some restaurants, while as examples of actions that have been strengthened we have the Landless Rural Workers Movement Plan (MST), the inclusion of products collected in Community Supported Agriculture baskets and the expansion of native species in the Agroforestry Systems that are generally associated with Community Supported Agriculture (CSA), that with this can expand the diversity of products to be offered in their baskets.

19.3.1 Fruit-Map Application Inspired by the City-Park

Brasilia is considered an open-air garden and one of the most wooded Brazilian cities. The tree planting plan originally envisaged in the report by urban planner Lucio Costa had the landscape purpose of configuring a garden city, with separate blocks and blocks framed by a green strip of trees (Lima 2009). At the time of its foundation, between 1955 and 1960, plant species common in other Brazilian regions were planted to create a visual identity that referred to other capitals in Brazil; however, there were many problems in adapting these species to the conditions of the region. It was only from the 1970s onwards that preference was given to native species. Currently, the Federal District has about 160 species of trees and palm trees, as well as shrubs, herbs, and flowers.

About 15% of the trees in the Brasília are fruit trees, of 35 different species, and there are 950,000 of them in public areas. Of the 35 species, nine are native to the Cerrado biome: araçá (*Psidium cattleyanum* Sabine), baru (*Dipteryx alata*), cagaita (*Eugenia dysenterica*), cajá (*Spondias mombin L*), ingá-mirim (*Inga fagifolia*), jatobá-da-mata (*Hymenaea courbaril*), jatobá-do-cerrado (*Hymenaea stilbocarpa*), and pequizeiro (*Caryocar brasiliense*). Among the most common are mango and jackfruit trees, which make up green spaces in residential areas, avenues, and public parks.[6] Figure 19.1 presents a simplified map of the Pilot Plan of Brasilia and an aerial photograph of a set of residential blocks in the North Wing (Asa Norte), where it is possible to find fruit species, such as jackfruit, blackberry, avocado, mango, and cagaita (*Eugenia dysenterica*), located next to an urban park.[7]

To facilitate the access of the population of Brasilia to this natural urban richness of fruit trees, in 2015 two young developers created the Fruit-map application[8], which allows to find and register public fruit trees and community gardens in the city and in its surrounding areas.

Although the application base is available for all the countries, initially it signaled the location of fruit trees only in Brasília (Fig. 19.2). And over time, as more people downloaded the app, which is collaborative, they added locations outside the Federal District. Currently, there are population of registered fruit trees and users in other cities, such as São Paulo. In 2019, the app had 50,000 users and more than 60,000 fruit trees and community gardens registered.[9]

During the pandemic, when a large number of people started to prefer outdoor activities, the Fruit Map was in high demand. The application is now having its versions updated, to improve functions. Figure 19.3 shows other screens of fruit-map application, where users can select a fruit to search or to mark anywhere in the city, know about the conditions of access[10] and see its distribution in the neighborhood and surroundings.

[6] https://agenciabrasilia.df.gov.br/2022/10/07/com-950-mil-arvores-frutiferas-de-35-especies-df-e-pomar-a-ceu-aberto/. Accessed in 01/28/2023.

[7] This park even corresponds to 2 Superblocks that did not have their construction project consolidated thanks to the mobilization of the local population, which defended the preservation of a water source, hence the name of the park (*Olhos d'água* in Portuguese).

[8] https://www.fruitmap.com.br/.

[9] https://casapark.com.br/novidade/brasilia-e-mais-verde/. Accessed in 01/26/2023.

[10] Where "Easy Level" would be collection just by stretching the arms, "Middle Level" would need a support or some instrument, and "Difficult Level" would need a ladder.

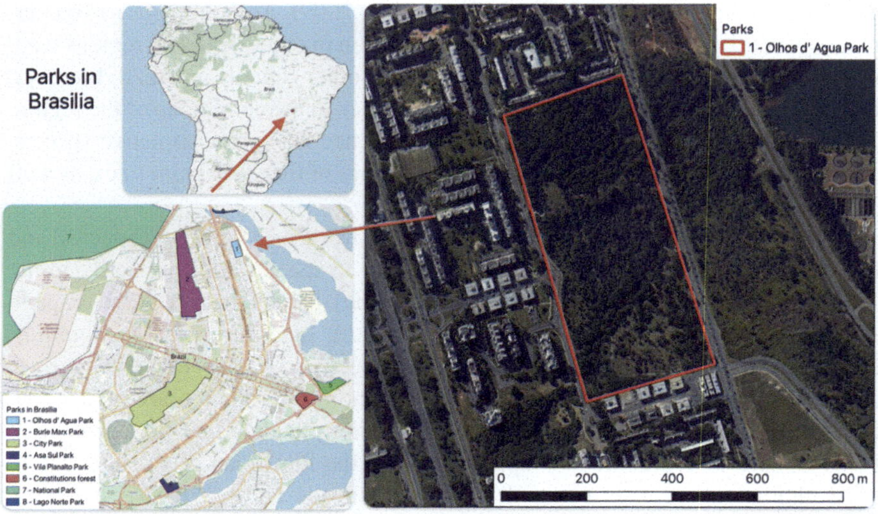

Fig. 19.1 Location of Brasilia and example of Superblocks from Plano Piloto, showing trees near buildings and a green area (park). Fonte: Adapted from QGIS Openstreetmap

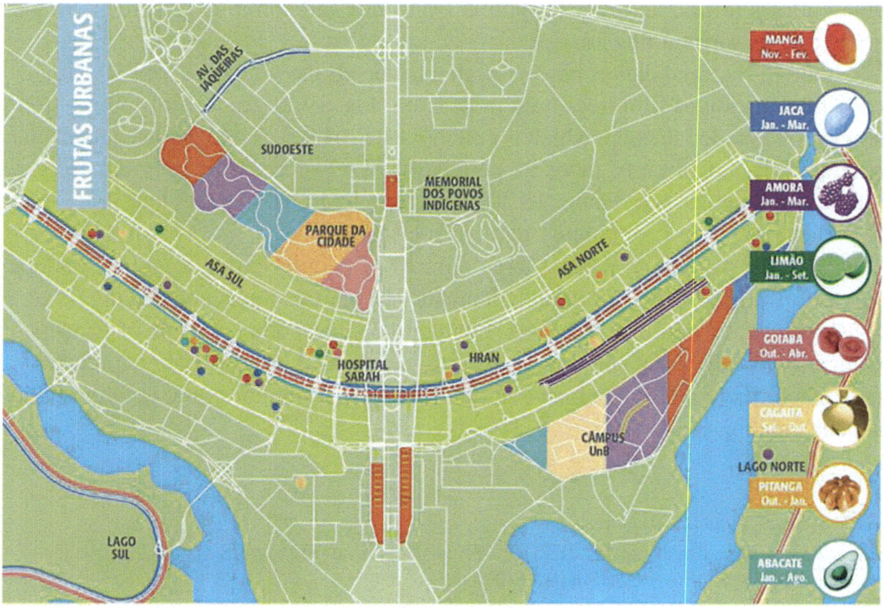

Fig. 19.2 Example of distribution of urban fruits from the Pilot Plan of Brasilia in the fruit-map app. Source: https://bolachinhacomovo.wordpress.com/2014/04/10/brasilia-cidade-pomar/

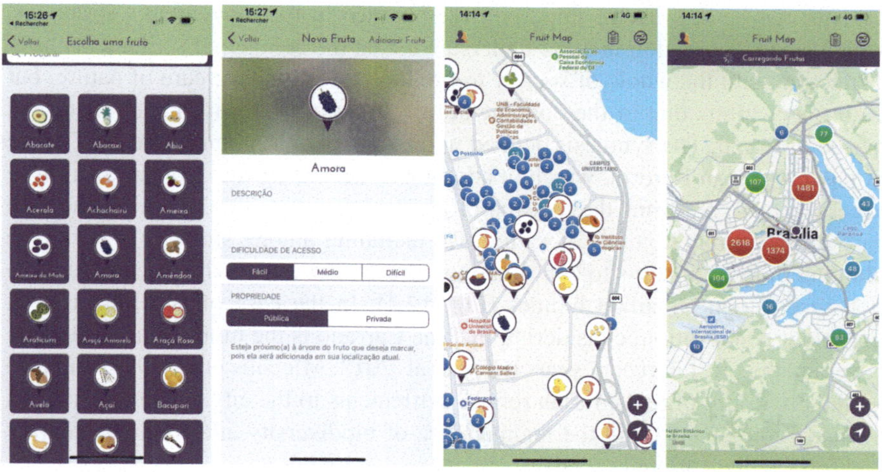

Fig. 19.3 Screens from fruit-map with information to choose a fruit to search location in the city (1), levels of access to the fruit (2), trees distribution on a university campus (3), and aggregation of marked pins in each region of the city (4)

The following initiative is nationwide, but as there is a large number of agrarian reform settlements (generally occupying degraded areas) in the Federal District, it is appropriate to consider this example as we are addressing possibilities for expanding urban and peri-urban green spaces for restoration, leisure, and also for food foraging (Siqueira et al. 2022). This is the case of several Agrarian Reform settlements that, despite often being threatened by real estate speculation, have recovered green areas, either to create legal reserves or to implement agroforestry systems to offer more food diversity, including native fruit trees (Cechi et al. 2020; Dourado and Eloy 2018).

19.3.2 Plan "Planting trees, producing healthy food" from the Landless Rural Workers' Movement

The Movement of Landless Rural Workers (MST) launched in 2020 the Tree Planting Campaign to plant 100 million trees in 10 years across the country and produce healthy food. It consists of the massive planting of trees linked to the organization of agroecological production systems, from the organization of seedling nurseries, seed collections, agroforestry, and productive backyards. Since the Plan is a space of articulation, training, political organization, and debate to reaffirm land reform, food sovereignty, agroecology, and care of the commons, the MST also planned to launch materials on agroecology, promote movements on social networks and national events, as well as training courses carried out online and face-to-face, addressing theoretical and practical issues.[11]

[11] https://mst.org.br/2020/09/17/mst-holds-national-tree-planting-campaign-throughout-brazil/. Accessed in 01/22/23.

This can be seen as an important reassurance in the movement's strategies, which since its creation in 1984 understands that the conquered settlements have a function with the whole of society to produce food and take care of nature, but with many limits due to the conjuncture of Agrarian Reform in each historical period in Brazil. The synthesis defined in the Agrarian Program of the MST is the Popular Agrarian Reform, which in addition to defending the democratization of access to land, reaffirms the mission of the settlements to produce food and take care of the common goods of nature. By including among strategic actions, the planting of trees, it starts to contribute to the recovery and environmental restoration in different Brazilian biomes (Fig. 19.4). In the case of Brasilia, it is an important contribution, considering that the Cerrado is the biome that has lost its coverage the most in recent years (Eloy et al. 2018; Miccolis et al. 2017) and also by the proximity of the agrarian reform settlements to the city, which determines a socio-ecological role in the maintenance of biodiversity and food sovereignty and food and nutritional security.

Fig. 19.4 "Planting trees, producing healthy food" Campaign poster and photos of planting actions by MST activists in different Brazilian states (https://mst.org.br/especiais/plantar-arvores-produzir-alimentos-saudaveis/#:~:text=Produzir%20Alimentos%20Saud%C3%A1veis-,Apresenta%C3%A7%C3%A3o,seus%20impactos%20ao%20meio%20ambiente)

The MST aims for the 100 million trees to be planted in rural schools, cooperatives, training centers, squares, avenues, and cities, in order to strengthen the production of healthy food in the areas of settlements and camps of the movement. The MST is present in 24 Brazilian states and has 450,000 settled families, but throughout Brazil 100,000 families still need land (MST 2023). Considering this number of families that still need land, a more effective agrarian reform, in addition to providing land, income, and food to landless families, it could contribute to the recovery of green areas, including urban and peri-urban ones, as long as there was support from the Government for settlers to reforest the land and not just practice agriculture in generally degraded areas.

19.4 Conclusions

This set of innovative initiatives, arising from restaurant chef groups, local government, app developers, or social movements in the city of Brasília, demonstrates the potential for urban food systems transformation toward more sustainable and inclusive practices, where foraging for food is an important and possible practice that must be stimulated and supported by public policies and by the society.

In Brasília, although the great efforts of the involved actors, the innovative actions in the restaurants, and in the application for mapping fruits available for foraging and urban gardens are still restricted to higher income consumers or people living in nobler areas in the city. On the other hand, initiatives to include Cerrado fruits in school meals and from MST can promote greater access to native and health food and to areas where they can be foraged, as pointed by Ferigollo et al. (2017) concerning the purchasing of family farmers by schools in the cities of the Brazilian Southeast state of Rio Grande do Sul.

Chefs' limited adoption of sustainable practices, such as serving seasonal local menus, reflects an important gap between their discourse and their practices. Moreover, it reveals the shortcomings of public policies dedicated to the transformation of urban food systems, with a lack of technical assistance and rural extension services, for instance, to support farmers and gatherers of Cerrado fruits to access these emerging markets in order to meet innovative chefs in their efforts to build sustainable food markets (Silva et al. 2017).

The same situation is observed in the support to gatherers to supply the School Feeding Program with foraging products. The participation of those products (mainly fruits, herbs, and flowers) is increasing in restaurants and street markets, but mostly by efforts from some entrepreneurs, groups of producers, and other collectivities rather than with an articulated policy. A diagnosis on these bottlenecks and possible changes in the flow, processing, and supply settings for schools and the local population in general, inspired in sustainable impact assessment tools, as proposed by Schmutz et al. (2017), could promote greater articulation between the different initiatives presented here.

We recognize that initiatives for greater access to wild species and expansion of green areas come from different sectors. A better integration between them could promote more effective and lasting results that would contribute to the goals of sustainable development. In this sense, it is expected that the understanding of foraging can be increasingly applicable to production systems that cover urban and peri-urban areas, which are also key elements for food and nutritional security and environmental sustainability.

Acknowledgments The authors would like to thank the developers of the Fruit Map application and the representatives of the Landless Rural Workers Movement for the information and images provided.

References

Agostini-Costa TDS, Silva DD, Vieira RF, Sano SM, Ferreira FR (2006) Espécies de maior relevância para a região Centro-Oeste. In: Vieira RF et al (eds) Frutas nativas da região Centro-Oeste do Brasil. Embrapa – Recursos Genéticos e Biotecnologia, Brasília, pp 12–25

Araújo AM, Finocchio CPS (2020) Farm-to-chef: produção local, proximidades e gastronomia na literatura internacional. Anais. Sociedade Brasileira de Economia, Administração e Sociologia Rural – SOBER. Foz do Iguaçu – PR

Bispo TW (2014) Agroextrativismo no vale do Rio Urucuia – Minas Gerais: formas de organização da produção e da distribuição. [Masters Thesis, Universidade de Brasília]. UnB Repository. https://repositorio.unb.br/handle/10482/16565

Borelli T et al (2020) Born to eat wild: an integrated conservation approach to secure wild food plants for food security and nutrition. Plan Theory 9(1299). https://doi.org/10.3390/plants9101299

BRASIL (2000) lei n° 9.985, de 18 de Julho de 2000. Regulamenta o artigo 225, § 1°, incisos I, II, III e VII da Constituição Federal, institui o Sistema Nacional de Unidades de Conservação da Natureza e dá outras providências. Ministério do Meio Ambiente, Brasília. http://www.planalto.gov.br/ccivil_03/leis/l9985.htm

Bredariol LR (2015) Levantamento e caracterização de Plantas Alimentícias Não Convencionais (PANCs) espontâneas presentes em um sistema agroflorestal no município de Rio Claro, São Paulo. Trabalho de Conclusão de Curso. Universidade Estadual Paulista, Instituto de Biociências de Rio Claro. Available at: https://repositorio.unesp.br/handle/11449/139038

Cechi A, Araújo VS, Amand L (2020) Exploring the synergy between community supported agriculture and afroforestry: institutional innovation from smallholders in a Brazilian rural settlement. J Rural Stud 81:3. https://doi.org/10.1016/j.jrurstud.2020.10.031

CEPF (2017) Perfil do ecossistema hotspot de biodiversidade do Cerrado. Critical Ecosystem Partnership Fund – ISPN – Conservação Internacional. Brasília

Clouse C (2022) The resurgence of urban foraging under COVID-19. Landsc Res 47(3). https://doi.org/10.1080/01426397.2022.2047911

Congresso em Foco (2017) Terceiro polo gastronômico, Brasília atrai restaurantes consagrados do eixo SP-Rio. https://congressoemfoco.uol.com.br/opiniao/colunas/terceiro-polo-gastronomico-Brasília-atrai-restaurantes-consagrados-do-eixo-sp-rio/

Instituto de Pesquisa Econômica Aplicada (IPEA) (2019) O desempenho recente das políticas de compras públicas da produção da agricultura familiar. Políticas sociais: acompanhamento e análise, Brasília n. 26

Deverre C, Lamine C (2010) Les systèmes agroalimentaires alternatifs. Une revue de travaux anglophones en sciences sociales. Économie rurale. Agricultures, alimentations, territoires, n 317:57–73. https://doi.org/10.4000/economierurale.2676

Diniz JDAS, Barbosa-Silva D, Souza C, Figueiredo AS, Wehrmann MES, Costa FMP (2013) Agregação de valores a espÉcies do Cerrado como oportunidade de inserção da agricultura familiar em mercados diferenciados. In: Conterato MA, Niederle PA, Triches RM, Marques

FC, Shultz G (Org.) Mercados e agricultura familiar: interfaces, conexões e conflitos. Porto Alegre: Via Sapiens (Difusão IEPE/UGRGS), pp 268–289

Distrito Federal [Diário Oficial do Distrito Federal] (2023) Lei N. 7,228 (23th January, 2023). Available at: https://dodf.df.gov.br/index/visualizar-arquivo/?pasta=2023|01_Janeiro|DODF%20017%2024-01-2023|&arquivo=DODF%20017%2024-01-2023%20INTEGRA.pdf. Accessed 28 Jan 2023

Dourado BF, Eloy L (2018) O cultivo de árvores entre agricultores que participaram de projetos de incentivo a sistemas agroflorestais no Cerrado. Cadernos de Agroecologia 13(1). Available at: http://cadernos.aba-agroecologia.org.br/cadernos/article/download/1026/543. Accessed 1 July 2023

Drummond JA (1996) A extração sustentável de pro- dutos florestais na Amazônia Brasileira. Estudos-Sociedade e Agricultura 6:116–137

Duarte LMG, Guéneau S, Diniz JDAS, Passos CJS (2020) Sistemas agroalimentares alternativos, construção social de mercados e gastronomização de produtos agroextrativistas do Cerrado brasileiro. In: Guéneau S, Diniz JDAS, Passos CJS (eds) Alternativas para o bioma Cerrado: agroextrativismo e uso sustentável da sociobiodiversidade. Mil Folhas, pp 405–446

Duarte LMG, Guéneau S, Diniz JDAS, Passos CJS (2021) Valorización de los patrimonios alimentarios y productos agroextractivistas del Cerrado brasileño en la gastronomía: un estudio sobre el Festival Gastronómico Cerrado Week. In: Rebaï N, Bilhaut AG, de Suremain CE, Esther E, Paredes M (eds) Patrimonios alimentarios en América Latina: recursos locales, actores y globalización. IRD-IFEA, pp 109–136. https://doi.org/10.4000/books.irdeditions.42804

Eloy L, Guéneau S, Nogueira MCR, Diniz JDAS, Leme A, Passos CJS (2018) Alternatives durables pour le biome Cerrado: occupations et usages des territoires par les producteurs agroextractivistes. Problèmes d'Amérique Latine 4(111):85–100. https://doi.org/10.3917/pal.111.0085

Escobar H (2020) Deforestation in the Brazilian Amazon is still rising sharply. Science 369:613–613. https://doi.org/10.1126/science.369.6504.613

Ferigollo D, Kirsten V, Heckler D, Figueredo OAT, Perez-Cassarino J, Triches RM (2017) Products purchased from family farming for school meals in the cities of Rio Grande do Sul. Rev Saude Publica 51:6. https://doi.org/10.1590/S1518-8787.2017051006648

Garcia JP (2017) Valorização de produtos agroextrativistas do Cerrado: comercialização e construção dos mercados da sociobiodiversidade. [Masters Thesis, Universidade de Brasília]. UnB Repository. https://repositorio.unb.br/handle/10482/24443

Garcia JP, Zaneti TB, Diniz JDAS, Guéneau S (2022) Alternative food dynamics and gastronomy: local products consumption in Brasilia restaurants. Grifos 31(47). https://doi.org/10.22295/grifos.v31i57.6729

Guéneau S, Diniz JDAS, Mendonça SD, Garcia JP (2017) Construção social dos mercados de frutos do Cerrado: entre sociobiodiversidade e alta gastronomia. SÉCULO XXI - REVISTA DE CIÊNCIAS SOCIAIS 7:130–156. https://doi.org/10.5902/2236672528133

Guéneau S, Capelari MGM, Diniz JDAS, Garcia JP, Zaneti, TB. (2022). The process of politicizing sustainability in the city of Brasilia: towards a transition of the local food system?. In: LeCoq JF, Grisa C, Guéneau S, Niederle P (eds) Public policies and food systems in Latin America. 1ed.Versailles: Éditions Quae, 2022, v. 1, p. 287-305.

Guéneau S, Diniz JDAS, Bispo TW, Mendonça SD (2019) Biodiversity-based supply chains of the Cerrado biome: opportunities and obstacles. Cirad, Iamm, INRA and Supagro. Montpellier: INRA, 27 p. Seminar of the European Association of Agricultural Economists: Governance of food chains and consumption dynamics: what are the impacts on food security and sustainability?. Montpellier, France, May 2019

Guéneau S, Diniz JDAS, Bispo TW, Mendonça SD (2020) Cadeias de produtos da sociobiodiversidade como opção de desenvolvimento sustentável no cerrado: o desafio da comercialização. In: Guéneau S, Diniz JDAS, Passos CJS (eds) Alternativas para o bioma Cerrado: agroextrativismo e uso sustentável da sociobiodiversidade. Mil Folhas, pp 329–367

Herrmann LK, Durigon J, Beskow GT (2020) O levantamento florístico de plantas alimentícias não convencionais (PANC) em uma agroflorestal no sul do Brasil. Cadernos de Agroecologia, v. 15,

n. 2. Anais do XI Congresso Brasileiro de Agroecologia, São Cristóvão, Sergipe. Available at: http://cadernos.aba-agroecologia.org.br/cadernos/article/view/2988

Jerin VA et al (2022) Role of non-timber forest products in income generation of the tribal population: a review. Asian J Agric Ext Econ Sociol 40(11):285–294., AJAEES 77936. https://doi.org/10.9734/AJAEES/2022/v40i111711

Kinupp VF, Lorenzi H (2014) Plantas alimentícias não convencionais (PANC) no Brasil: guia de identificação, aspectos nutricionais e receitas ilustradas. Instituto Plantarum de Estudos da Flora

Lima RMC (2009) Avaliação da arborização urbana no Plano Piloto. Masters Dissertation. Department of Forest Engineering, University of Brasília. Available at: https://repositorio.unb.br/handle/10482/5858

Metrópoles (2023) Sancionada lei que coloca frutos do Cerrado na merenda escolar do DF. https://www.metropoles.com/distrito-federal/sancionada-lei-que-coloca-frutos-do-cerrado-na-merenda-escolar-do-df

Miccolis A, Penereiros F, Vieira DLM, Marques HR, Hoffmann MRM (2017) Restoration through agroforestry: options for reconciling livelihoods with conservation in the Cerrado and Caatinga biomes in Brazil. Exp Agric 55(S1):208–225. https://doi.org/10.1017/S0014479717000138

Mittermeier RA et al (eds) (2004) Hotspots revisited: earth's biologically richest and most endangered terrestrial ecoregions. CEMEX – Agrupación Sierra Madre, Mexico City

MST (2023) "São 100 mil famílias em acampamentos, muitas há 10 anos debaixo da lona": MST cobra ações do governo Lula no Abril Vermelho. https://mst.org.br/2023/04/10/sao-100-mil-familias-em-acampamentos-muitas-ha-10-anos-debaixo-da-lona-mst-cobra-acoes-do-governo-lula-no-abril-vermelho/

Neto MJ, Maluf ACD, Boscaine TF (2016) Plantas ruderais com potencial para uso alimentício. Agroecol – Second Seminar in Agroecology in Latin America, Dourados, November

Niederle P, Schubert MN (2020) How does veganism contribute to shape sustainable food systems? Practices, meanings and identities of vegan restaurants in Porto Alegre, Brazil. J Rural Stud 78:304–313. https://doi.org/10.1016/j.jrurstud.2020.06.021

Nogueira MCR, Fleischer SR (2005) Entre a tradição e modernidade: potenciais e contradições da cadeia produtiva agroextrativista no Cerrado. Estudos Sociedade e Agricultura 13(1):125–157

Pereira, L. M., Calderón-Contreras, R., Norström, A. V., Espinosa, D., Willis, J., Guerrero Lara, L., . . . Pérez Amaya, O. (2019). Chefs as change-makers from the kitchen: indigenous knowledge and traditional food as sustainability innovations. Global Sustain, 2, e16, Article e16. https://doi.org/10.1017/S2059479819000139

Pinton F, Emperaire L (2001) Le manioc en Amazonie brÉsilienne: diversitÉ variÉtale et marchÉ. Genet Sel Evol 33(1):491–512

Porro R, Sousa RC (2023) Anatomy of babassu-nut value chain for policy guidance in support of traditional agroextractive communities in the Mearim Valley, Maranhão, Brazil. Rev Econ Sociol Rural 61(4):e263743. https://doi.org/10.1590/1806-9479.2022.263743

Santilli J (2005) *Socioambientalismo e novos direitos*: proteção jurídica à diversidade biológica e cultural. Peirópolis, São Paulo

Sardeshpande M, Rupprecht C, Russo A (2021) Edible urban commons for resilient neighbourhoods in light of the pandemic. Cities 109. https://doi.org/10.1016/j.cities.2020.103031

Sauer S, Oliveira KRA (2021) Agrarian extractivism in the Brazilian Cerrado. In: McKay B, Alonso-Fradejas A, Ezquerro-Cañete A (eds) Agrarian Extractivism in Latin America. Routledge. https://doi.org/10.4324/9780367822958

Schmutz U, Kneafsey M, Sarrouy Kay C, Doernberg A, Zasada I (2017) Sustainability impact assessment of different urban short supply chains: examples from London, UK. Renew Agric Food Syst 33(6):518–529. https://doi.org/10.1017/S1742170517000564

Schneider S, Cruz FTD, Matte A (2016) EstratÉgias alimentares e de abastecimento: desafios e oportunidades para as cidades e para o meio rural. In: Cruz TF da, Matte A, Schneider S. Produção, consumo e abastecimento: desafios e novas estratÉgias. Editora da UFRGS, Porto Alegre, pp 9–22

Sidaner E, Balaban D, Burlandy L (2013) The Brazilian school feeding programme: an example of an integrated programme in support of food and nutrition security. Public Health Nutr 16(6):989–994. https://doi.org/10.1017/S1368980012005101

Silva RRV, Gomes LG, Albuquerque UP (2017) What are the socioeconomic implications of the value chain of biodiversity products? A case study in northeastern Brazil. Environ Monit Assess 189:64. https://doi.org/10.1007/s10661-017-5772-2

Siqueira MM, Pachêco BS, Nascimento WJ, Santos PLL, Abreu VES (2022) Criando sinergias entre paisagismo e agroecologia: o uso de plantas nativas do Cerrado em jardins. In: Tullio L. Geração e difusão de conhecimentos de ciências agrárias. Atena, Ponta Grossa, pp 26–39

Sonnino R, Torres CL, Schneider S (2014) Reflexive governance for food security: the example of school feeding in Brazil. J Rural Stud 36. https://doi.org/10.1016/j.jrurstud.2014.06.003

Superti E, Pegler L, Araujo MMV (2018) The governance of emerging value chains and their impact on traditional communities. Int J Humanit Soc Sci 8(10). https://doi.org/10.30845/ijhss.v8n10p3

Svizzero S (2016) Foraging wild resources and sustainable economic development. J Econ Public Finance 2(1):132–153

Valette E, Schreiber K, Conaré D, Bonomelli V, Blay-Palmer A, Bricas N, Lepiller O (2019) An emerging user-led participatory methodology. Mapping impact pathways of urban food system sustainability innovations. In: Blay-Palmer A, Conaré D, Meter K, Di Battista A, Johnston C (eds) Sustainable food system assessment: lessons from global practice. Routledge, pp 19–41

Ye J et al (2019) The incursions of extractivism: moving from dispersed places to global capitalism. J Peasant Stud:1–29. Available at: https://www.tandfonline.com. https://doi.org/10.1080/03066150.2018.1559834

Zaneti TB (2017) Cozinha de raiz: as relações entre chefs, produtores e consumidores a partir do uso de produtos agroalimentares singulares na gastronomia contemporânea [PhD thesis, Universidade Federal do Rio Grande do Sul]. UFRGS LUME Digital Repository. https://www.lume.ufrgs.br/handle/10183/164708

Zaneti TB, Balestro MV (2015) Valoração de produtos tradicionais no circuito gastronômico: lições do Cerrado. Sustentabilidade em Debate 6(1):22–36. https://doi.org/10.18472/SustDeb.v6n1.2015.10709

Janaína Deane de Abreu Sá Diniz (PhD) is an Associate Professor in the University of Brasília/Brazil. Her research focuses on family farming organizations, non-timber forest products, and socio-biodiversity value chains, mainly in the Amazon and Cerrado biomes.

Stéphane Guéneau (PhD) Stéphane Guéneau is a socio-economist at CIRAD at the Montpellier Interdisciplinary Center on Sustainable Agri-food Systems (MOISA). His research topics encompass both quality in food supply chains and agri-food policy analysis.

Jessica Pereira Garcia is a PhD student at the Center for Sustainable Development, University of Brasília/Brazil.

Barbara Loureiro Borges is a PhD student in the Department of Forest Sciences, University of Brasília/Brazil.

More-Than-Human Urban Foraging: Designing Landscapes for Multispecies Sustainability in Shrinking Urban Japan

20

Christoph D. D. Rupprecht, Lihua Cui, Aoi Yoshida, Rika Shinkai, and Maximilian Spiegelberg

Abstract

Can a city be turned into a more-than-human feast? Edible landscapes are increasingly valued for their multiple socio-ecological benefits. Here we propose that making the urban fabric fertile grounds for foraging by humans, birds, bees, and other inhabitants alike expands the potential of edible cities as a paradigm for sustainability transformations to answer recent calls in sustainability research for meeting more-than-human needs. To this end, we reflect on ongoing landscape design research across shrinking Japanese cities: (1) a multispecies community garden concept, (2) bird and pollinator-oriented edible restoration of vacant lots, (3) wild-keeping of Japanese honeybees in urban contexts as entry to multispecies landscape stewardship.

C. D. D. Rupprecht (✉)
Department of Environmental Design, Faculty for Collaborative Regional Innovation, Ehime University, Ehime, Japan
e-mail: cr@multispecies.city; rupprecht.christoph.bq@ehime-u.ac.jp

L. Cui
Tokyo Metropolitan Research Institute for Environmental Protection, Tokyo, Japan
e-mail: sai-r@tokyokankyo.jp

A. Yoshida
AOI Landscape design LLC, Kanagawa, Japan
e-mail: yoshida@aoi-ld.com

R. Shinkai
Research Institute for Humanity and Nature & FEAST, Kyoto, Japan
e-mail: shinkai@chikyu.ac.jp

M. Spiegelberg
FEAST, Kyoto, Japan

S. Dhyani, M. Sardeshpande (eds.), *Urban Foraging in the Changing World*,
https://doi.org/10.1007/978-981-97-0345-6_20

Keywords

Human–nature relationship · Edible landscapes · Multispecies community garden · More-than-human edible landscapes

20.1 Introduction

Urban foraging as a practice is both widespread and diverse. Researchers are finding that benefits exceed nutrition alone (Sardeshpande and Shackleton 2019). Yet urban design providing opportunities for foraging remains limited. One reason may be that foraging implies a search for food without guarantee of success, rather than being associated with the certainty that a visit to a grocery store or home garden will lead to a meal. For some, the term foraging in an urban environment might evoke mental images of observing crows or stray cats and dogs on the lookout for treats. Positive acknowledgments of urban foraging as a ubiquitous activity, not necessarily the result of food insecurity or under provision, are conspicuously rare in the literature. Yet, recent historical research suggests that modern heavy reliance on what might be called formal agriculture, and especially on food being grown by specialists for profit to be sold across vast distances as a commodity, may be the outlier rather than the norm (Graeber and Wengrow 2021). For example, some indigenous people apparently chose to forgo cultivation involving labor-intensive human planting, harvesting and preparation in favor of strategic stewardship of the surrounding ecosystem to increase opportunities for foraging. Such practices are attracting increasing attention both in the context of considering the role of worldviews in biodiversity conservation (Fig. 20.1, Intergovernmental Science-Policy Platform on Biodiversity and Ecosystem Services 2022) and in relation to urgent calls in sustainability science for a radical re-thinking of agrifood systems to shift to a post-growth metabolism (Fig. 20.2, McGreevy et al. 2022).

Applying this way of thinking to a high-level approach of urban planning and design opens up conceptual space for reconsidering how cities might feed their inhabitants. Designating space for urban agriculture often leads to multi-faceted constraints, from regulatory hurdles to land prices and questions of what type of land use belongs in urban space. Concepts such as edible cities and edible landscapes instead aim to transform urban space to be edible by design (Creasy 2010). However, recent work on edible landscapes shows that these concepts have thus far mostly understood edibility to mean edible for humans (Rupprecht et al. 2023). In contrast, availability of food for non-human species has largely remained a question for urban ecology and biodiversity-supportive landscape design. In this chapter, we conceptually build on previous work on multispecies edible landscapes (Rupprecht et al. 2023) and edible urban commons (Sardeshpande et al. 2021) as a potential tool initiating a paradigm shift toward multispecies sustainability–meeting multiple species' needs as a requirement to create truly sustainable futures given the interdependence of human and non-human well-being (Rupprecht et al. 2020). We propose to maximize foraging opportunities for both humans and more-than-human urban inhabitants through an edible-by-design approach that fosters multispecies flourishing. Such an approach could be especially beneficial for shrinking cities, a phenomenon prevalent in Japan, but also being observed in countries such as Spain and New Zealand (Matanle and Sáez-Pérez 2019).

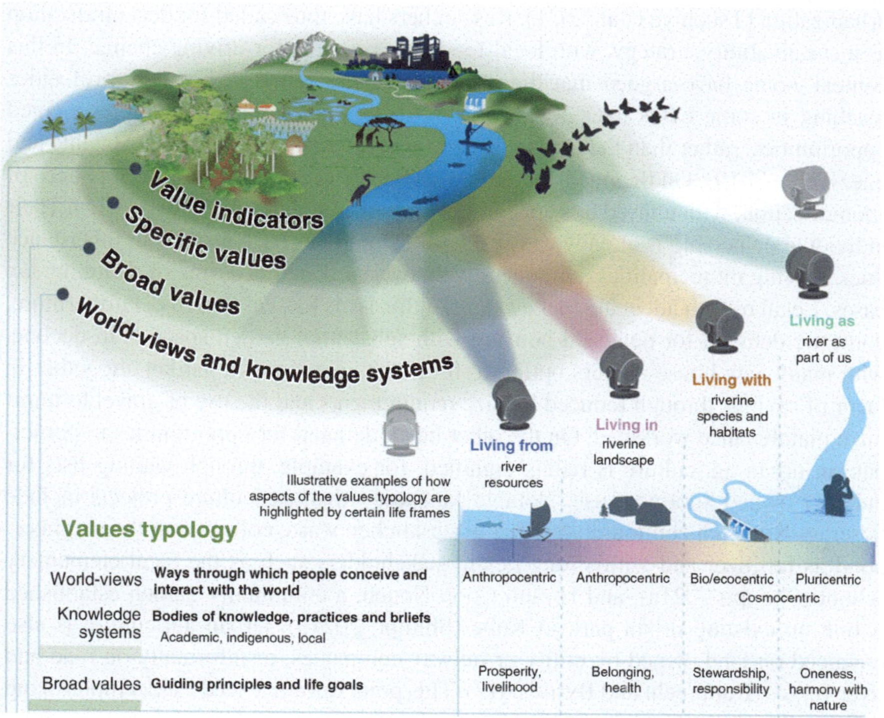

Fig. 20.1 Typology of human–nature relationship (modified from IPBES 2022, Figure SPM2)

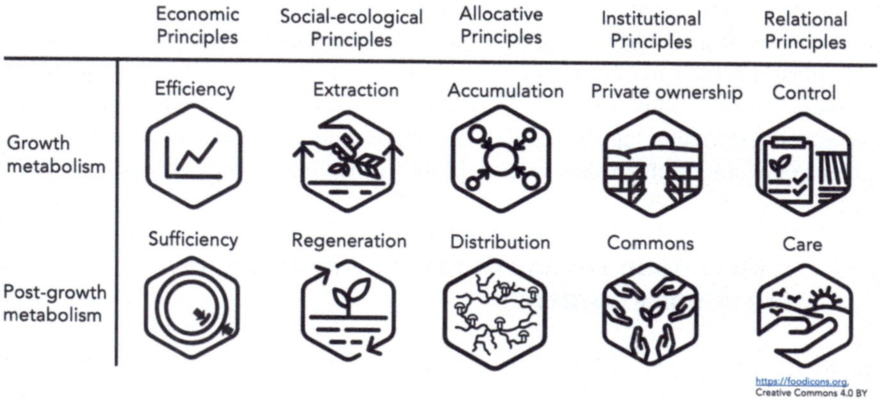

Fig. 20.2 Principles of growth and post-growth metabolisms (based on McGreevy et al. 2022)

Japanese cities tend to confuse foreign observers at first glance. Despite featuring some of the highest densities worldwide, tiny buildings persist next to towering apartment complexes, rice paddies in designated urban development zones, and informal gardens next to metropolitan train stations (Yokohari and Khew 2017). Beyond its idiosyncrasies, however, urban Japan faces the same urgent need for sustainable transformation as all other cities. Not only does the ecological footprint of Japanese cities far exceed the national biocapacity, per capita footprint increases with degree of urbanization (Tsuchiya et al. 2021). Researchers have thus called for decentralization as a sustainability strategy, with local food as a key area for driving change. In this context, some have argued that the projected steep decline in population of cities reaching in some cases over 30% in the next 20 years could offer unprecedented opportunities, rather than the crisis it is commonly understood to imply (Matanle and Sáez-Pérez 2019). On the one hand, urban shrinkage does usually follow a pattern of spongification, with unused lots appearing across the urban matrix (Aiba 2015). Their individual ownership is already causing significant issues such as unclear inheritance lines, leaving municipalities administratively overburdened in a time of shrinking tax bases. Legal reform adequate to the extent of this issue has yet to arrive. Furthermore, domestic demand for detached housing with substantial garden space is in decline, with many new house owners opting to forgo the creation of a garden altogether in favor of savings through reduced lot size requirements and the use of gravel to minimize maintenance workload. On the other hand, demand for opportunities to participate in urban agriculture is rising, signified, for example, through waiting lists for publicly owned allotment use. Notable Japanese urban agriculture projects include Seseragi Nouen, a community garden using kitchen waste collected in the neighborhood as fertilizer and with strong ties to stakeholders such as the local elementary school (Shimpo 2021a), and Hirano Coop Nouen, a community garden established within an existing urban park in Kobe (Shimpo 2021b). Urban agriculture is also practiced on land owned by utility or railway companies, or informally on road and river verges (Rupprecht and Byrne 2017). This point toward a need for deemphasizing individual and family-scale ownership and management of urban greenspace in general and food producing spaces in particular. Instead, a re-commoning and collective stewardship appear to be a strategy well-situated to simultaneously address large fluctuations in land ownership, changes in residents' preferences and lifestyles, and pressing needs for sustainability transformation (Sardeshpande et al. 2021).

Against this background, we offer three entry points to more-than-human urban foraging through edible-by-design cities based on ongoing landscape design research in shrinking Japanese cities. All three examples ultimately aim toward increasing overall multispecies sustainability by meeting multiple species' needs.

20.2 A More-Than-Human Gathering: Multispecies Community Gardens

Community gardens are well-known for providing a plethora of benefits to urban residents (Guitart et al. 2012). Could we extend the community at the heart of these gardens to include all species they consist of? This idea was the starting point for a landscape design proposal aiming to explore options for Kyoto's contexts of predicted depopulation and transition to a post-growth society. In a multi-stage project (KAKEN 2023), land use data analysis was combined with fieldwork to identify spatial unit types to consider where to best site a multispecies community garden. Focusing on Iwakura in Sakyo Ward, a suburban area surrounded by forest-covered mountains, suitable sites of abandoned former agricultural land were plentiful. Concept-wise, the design embraces changing urban land use dynamics while inviting multispecies agency to co-create space with support from trees and shrubs serving as wild-life support infrastructure (Fig. 20.3). As the city shrinks, sites are open for expansion with the potential to facilitate shifts from economic growth to post-growth lifestyles by providing meaningful engagement and more-than-human community interactions to improve well-being. Structurally, horizontal diversity aims to maximize habitat potential for a wide range of species, with special consideration in the form of fruit trees to local bird species to balance an increase in insects and open up opportunities for bird–human interactions. Spatially, the design integrates into a wider ecological network, with the garden functioning as a stepping stone while contributing through increased green and canopy cover to climate change adaptation.

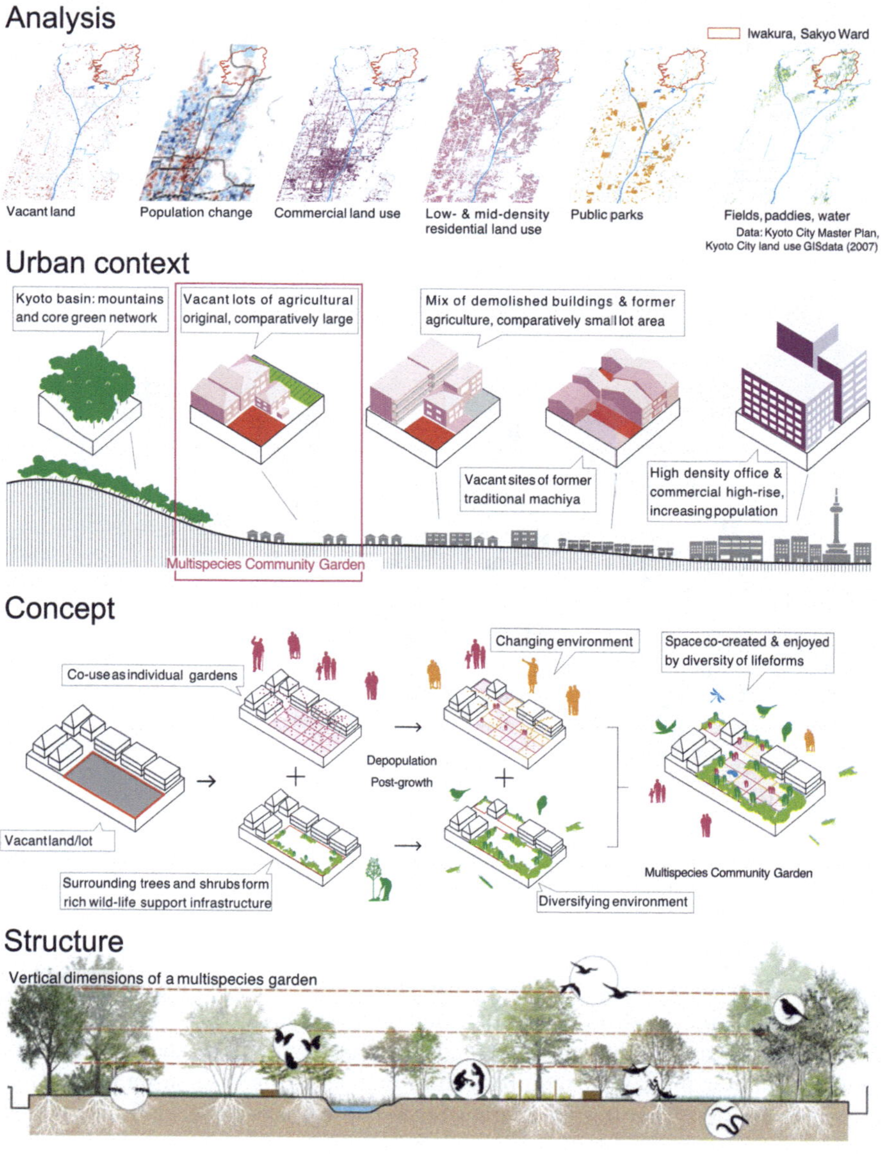

Fig. 20.3 Multispecies community garden design proposal process (design: AOI LANDSCAPE)

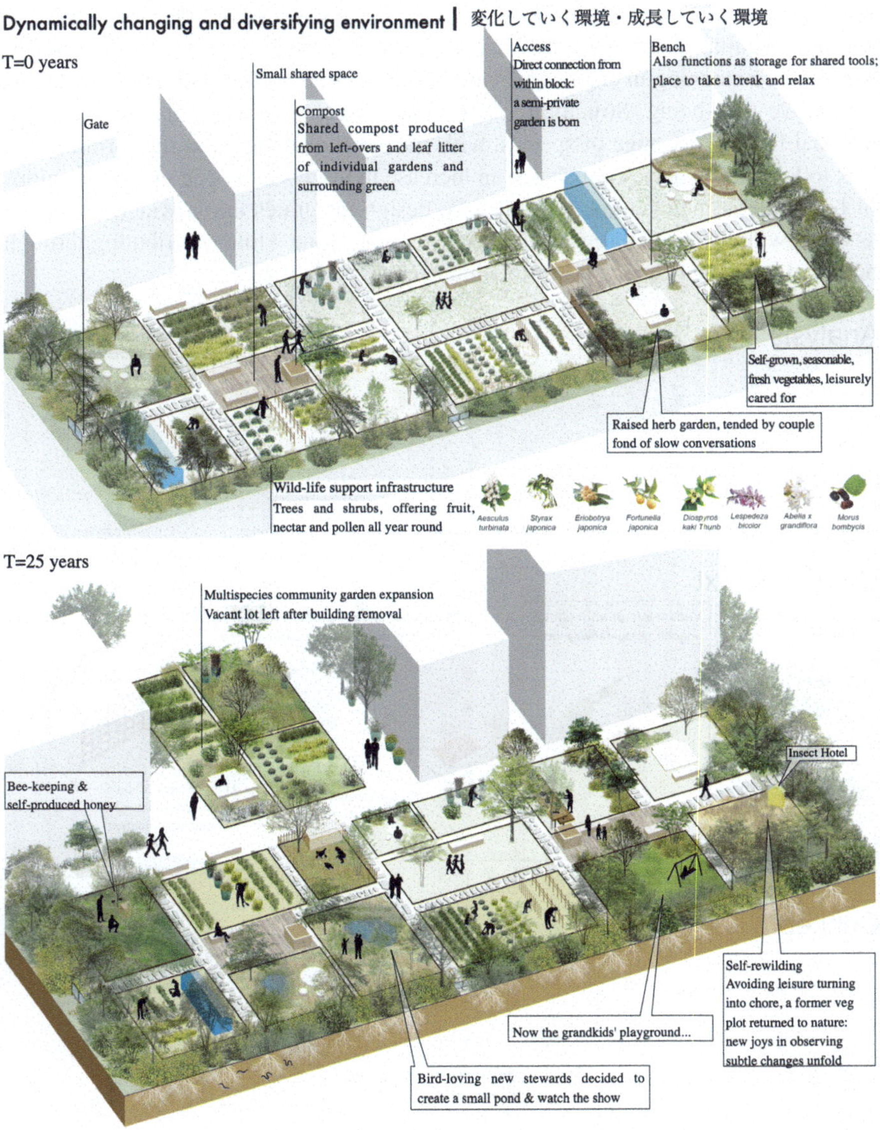

Fig. 20.4 Multispecies community garden design proposal envisioning dynamic changes in use over a 25-year time period (design: AOI LANDSCAPE)

The multispecies community garden design also responds to changes in its socio-ecology over time, allowing shifts from active gardening to hands-off stewardship (Fig. 20.4). For example, replacing vegetable beds with flowering perennials offering pollen and nectar to communally kept Japanese honeybees reduces maintenance work required while increasing more-than-human foraging opportunities. This in turn enables responsible beekeeping (and thus honey harvesting) that avoids simply depending on surrounding nectar and pollen sources, and also serves a wider community of pollinators. For periods of very low human engagement capacity, plots can be utilized through insect hotels and self-rewilding that creates space for multispecies agency to co-shape the community garden. Open questions remain around legal and procedural options for transitioning to a commons-oriented land ownership model for the garden, what approach to select for sustainable membership management and to what degree membership should be open to non-residents, and whether multispecies community gardens could be integrated into urban planning schemes for biodiversity conservation, disaster resilience, payments for ecosystem services, and carbon sequestration. Inclusion into such schemes, however, frequently requires initial stock take and application procedures as well as ongoing monitoring and quantification of services provided, possible deterrents that might be overcome through municipal or non-governmental organization-led support. Welcoming multispecies agency in shaping the space could also pose challenges in formulating or committing to specific management outcomes.

20.3 Foraging in More-Than-Human Edible Landscapes: Socio-Ecological Restoration of Vacant Lots

The example given above is only one option for addressing the increase in vacant urban land. Vacant lots come in all sizes and shapes, meaning diverse design options will prove vital in accessing their full potential (Rupprecht and Byrne 2014). Yet despite the clear need for ecological restoration, research on informal urban greenspaces shows that a classic biological conservation approach of excluding humans from spaces and leaving nature to its own devices is unlikely to succeed. For one, urban residents' perception of vacant lots changes drastically based on whether they are perceived to be neglected and abandoned or intentionally less manicured but still actively managed (Rupprecht 2017). Drawing upon insights from traditional Japanese satoyama landscape management and biocultural diversity theory, we were interested in whether fruit and nut trees and shrubs are viable options for socio-ecological restoration of vacant lots.

Transforming vacant lots into low-maintenance informal distributed community orchards for local consumption promises several benefits (Fig. 20.5). For human residents, they provide both physical health benefits in the form of added nutrients currently understood to be lacking in Japanese diets, and social benefits in strengthened community ties that have disintegrated as modern lifestyles leave people with less time and connection to place, but that are vital for community resilience in a disaster-prone nation. For more-than-human inhabitants, a careful selection of plants to maximize edibility of fruits as well as nectar/pollen production and habitat potential while taking into account site characteristics means a more densely knit habitat mosaic and expanded, more diverse food sources (Baldock et al. 2019; Wenzel et al. 2020). Socio-ecologically, a focus on indigenous plant species and varieties can not only help best meet the needs of local animal species, but also

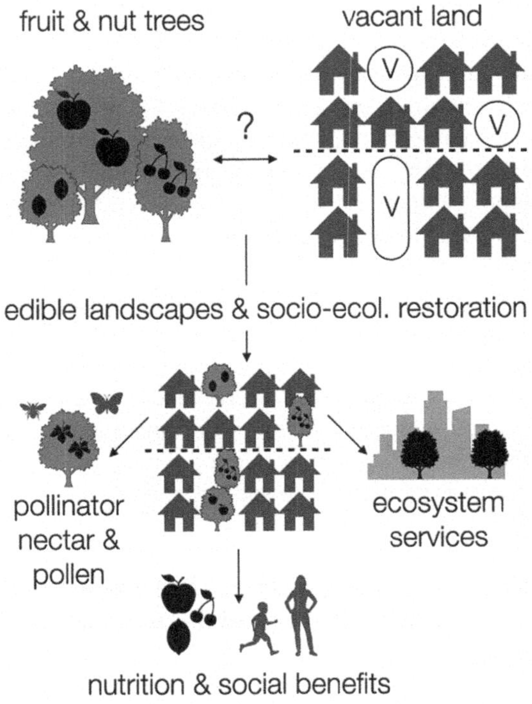

Fig. 20.5 Conceptual model for socio-ecological restoration of vacant lots into distributed communal orchards

contribute toward creating an increased sense of place through a regionally situated landscape characteristic. Efforts to preserve traditional plant varieties including fruit cultivars have experienced challenges in sourcing the necessary land and resources. A crowd-sourced approach through in-situ conservation (Raggi et al. 2022) could not only avoid an increasing homogenization of garden fruit cultivation driven by centralized sapling procurement, but opens up potential for (re-)establishing plants traditionally cultivated, including those grown by indigenous people such as the Ainu in Hokkaido. As a community endeavor, such community orchards might also be isolated from economic pressures that lead to productivity being prioritized over ecological function and conservation benefits.

However, in conducting fieldwork in Kyoto and Sapporo to explore the viability of this model, we encountered several constraints to be considered and addressed in the future. First, data availability for heterodox fruit types (e.g., *Lonicera caerulea*/haskap, *Myrica rubra*/red bayberry, *Elaeagnus multiflora*/gumi, *Actinidia arguta*/hardy kiwi) was highly limited both in regard to nutrition and yield, making it impossible to calculate nutritional provision gains for local populations without extensive basic research. Second, few producers of heterodox fruit and nut tree saplings (e.g. *Juglans ailantifolia*/Japanese walnut) exist in Japan, leading to a chicken-or-egg problem where residents interested in exploring alternative options have few choices to get started. An informant in Sapporo suggested the best, and possibly only, chance to access traditional indigenous plants would be to locate them in forests in the wild—an option, few residents are likely to consider given the logistic and legal questions involved. Sapporo's colonial origins and the devastating effects of cultural assimilation pressure on indigenous Ainu in Hokkaido likely contributed to this through loss in biocultural knowledge (Gayman 2020; Grunow et al. 2019). Providing municipal and prefectural support for efforts to revive an ecosystem of cultural knowledge and stewardship skills could be a powerful means to boost local engagement in biodiversity conservation efforts as well as sustainability transformations. We suggest the goal should be to democratize the whole knowledge and supply chain, integrating social, cultural, and ecological learning to build stewardship skills for managing a distributed, patchy network of lots forming a more-than-human edible landscape while providing residents with increased means to meet their own needs. In the spirit of food sovereignty (Patel 2009), experts could contribute through helping to select and curate a menu of locally viable plant species, while residents have the final choice and thus decision what kind of edible landscape to create.

20.4 From More-Than-Human Foraging to Multispecies Edible Landscape Stewardship: Urban Wild-Keeping of Japanese Honeybees

One issue prevalent in the literature on urban more-than-human co-existence is that of wild-life conflicts. In fieldwork on urban beekeeping practices conducted across sites in Japan, we learned about how beekeepers navigate their role as multispecies intermediaries. Having experienced a steep decline in beekeeping from 1979 to 2009, Japan has recently seen a resurgence in the practice with a shift from occupational to hobby beekeeping (Shinkai et al. 2020). We argue that this shift has several important implications for more-than-human urban foraging. What distinguishes hobby from commercial beekeeping in Japan are usually scale, intent, and the species of bee involved. Based on a survey of beekeeping practices, hobby beekeepers often keep only a few hives and do not harvest honey with the intention to sell it. Linked to this is their choice of the Japanese honeybee (*Apis cerana japonica*), which produces less honey per hive on average than Western honeybees (*Apis mellifera*), can better self-defend against natural predators in contrast to the Western honeybee, yet is also widely understood to be less aggressive. However, the main difference lies in how beekeepers are engaging in care practices.

To keep Japanese honeybees, the lack of control over their reproduction means that every swarm is necessarily sourced from the wild. Japanese honeybee keepers have developed highly diverse ways of attracting and enticing wild swarms to remain in the hive rather than abscond. These include using orchids, swarm catchers, numerous variations in hive design, harvesting practices and tools, but also a keen understanding for how to meet the bees' needs for shelter, water, nectar, pollen, protection from predators, and other environmental factors (Fig. 20.6). Queen gates, separators, and wing clipping common in industrial beekeeping to limit the queen's freedom of movement are not part of the practices. While this is evidence of an intricate relationship between keeper and bee, the Japanese honeybee

Fig. 20.6 NGO-led community beekeeping on the roof-top of the Kyoto Nakagyō Ward Office. In addition to the box hive, bees are provided with shade, shelter, water, and a highly biodiverse environment with 130 species of plants (e.g., biwa, cherry, plum, kiwi, lemon, loquat, kumquat, blueberry, strawberry, persimmon, etc.), many of them selected to provide nectar and pollen throughout the year (personal communication)

nevertheless remains undomesticated. Indeed, little effort is made to complete the domestication process, or domestication is intentionally rejected (Takekawa 2021). Instead, wild-keeping has been proposed as a form of beekeeping in which absolute control over reproduction and behavior is forgone in favor of an arrangement in which bees remain highly autonomous. The practice is also open for an alternative interpretation from a multispecies perspective. Whether a hive is used or not is no doubt highly influenced by knowledge, skill, and care the beekeeper invests, but the ultimate decision remains with more-than-human agency of the bee swarm. Beekeepers understand this, and our survey reflects that beyond the points raised above, the human–bee relationship and beekeepers' awareness of it also differ in comparison to commercial beekeeping. Rather than seeing the bees as livestock, beekeepers report a bond of partnership. Through their multispecies partners, beekeepers then learn new skills that do not end at the hive entrance, but rather extend to the landscape and its stewardship at large. The nature of knowledge and knowledge exchange are transformed as well, as best practices are no longer seen as a business secret but increasingly shared over open online platforms such as MitsubachiQ&A with over 10,000 users (MitsubachiQ&A 2023).

Against this background, we suggest that wild-keeping of Japanese honeybees may have the potential of serving as a model to revolutionize human–nonhuman relationships more broadly. Learning about and seeking to meet multispecies needs for a forage-able urban matrix could provide countless opportunities for new insights into urban ecosystems, including interspecies relationships between animals, insects, plants, and microbes we have only begun to grasp. More-than-human care relationships thus can also be seen as extending our sensory perception beyond the capabilities of our immediate body, opening access to multispecies perspectives on urban landscapes in which nectar and pollen sources are as relevant as traffic conditions.

20.5 Conclusion

In this chapter, we have argued that expanding urban foraging to make the city edible for a variety of species has the potential of increasing urban multispecies sustainability through better meeting more-than-human needs. Opportunities abound especially in shrinking cities with growing areas of vacant land, a phenomenon prevalent in but not unique to Japan. The shared need for food in this regard points to the importance of urban foraging as a concept in transforming cities to sustainably serve more-than-human needs. Whether designing a multispecies community garden, exploring options for socio-ecological vacant lot restoration through communal orchards, or considering the potential of transformed human–nonhuman relationships in fostering multispecies landscape stewardship, such approaches to landscape design provide not only immediate benefits on the ground, but also a chance to transform our thinking. Working with multispecies partners to develop landscape stewardship practices appears to be one promising pathway to transform not only the built environment, but shift from anthropocentric world views and value sets toward a focus on more-than-human relationships as outlined in a recent report on biodiversity conservation.

Acknowledgments We thank the editors for the invitation to contribute to this volume. Our deep appreciation goes to all human and non-human stakeholders and community members who generously provided time and advice throughout many stages of the research, including the Miyako Mitsubachi no Kai Non Governmental Organisation and Kyoto Nakagyō Ward Office hosting the bees in Fig. 20.6. This research was supported by the FEAST Project (14200116), Research Institute for Humanity and Nature, and JSPS KAKENHI grants 17K08179, 18K18602, and 20K15552.

References

Aiba S (2015) Toshi o tatamu: Jinkō genshō jidai o dezain suru toshi keikaku. Kadensha

Baldock KCR, Goddard MA, Hicks DM, Kunin WE, Mitschunas N, Morse H, Osgathorpe LM, Potts SG, Robertson KM, Scott AV, Staniczenko PPA, Stone GN, Vaughan IP, Memmott J (2019) A systems approach reveals urban pollinator hotspots and conservation opportunities. Nat Ecol Evol 3:363. https://doi.org/10.1038/s41559-018-0769-y

Creasy R (2010) Edible landscaping. Catapult

Gayman J (2020) Challenges surrounding education and transmission of Ainu indigenous ecological knowledge in Japan: disparate valuations of a people and their IEK. In: The Routledge handbook of indigenous environmental knowledge. Routledge

Graeber D, Wengrow D (2021) The Dawn of everything: a new history of humanity. Penguin

Grunow TR, Nakamura F, Hirano K, Ishihara M, lewallen, ann-elise, Lightfoot S, Mayunkiki, Medak-Saltzman D, Williams-Davidson T-L, Yahata T (2019) Hokkaidō 150: settler colonialism and indigeneity in modern Japan and beyond. Crit Asian Stud 51:597–636. https://doi.org/10.1080/14672715.2019.1665291

Guitart D, Pickering C, Byrne J (2012) Past results and future directions in urban community gardens research. Urban For Urban Green 11:364–373

Intergovernmental Science-Policy Platform on Biodiversity and Ecosystem Services (2022) Summary for policymakers of the methodological assessment of the diverse values and valuation of nature of the intergovernmental science-policy platform on biodiversity and ecosystem services (IPBES). Zenodo, Bonn https://doi.org/10.5281/zenodo.7410287

Kaken (2023) Exploring the potential of edible landscapes for socio-ecological restoration of vacant land in shrinking Japanese cities. Kakenhi grant project page. https://kaken.nii.ac.jp/en/grant/KAKENHI-PROJECT-20K15552/

Matanle P, Sáez-Pérez L-A (2019) Searching for a depopulation dividend in the 21st century: perspectives from Japan, Spain and New Zealand. J Jpn Inst Landsc Archit 83:E1–E6

McGreevy SR, Rupprecht CDD, Niles D, Wiek A, Carolan M, Kallis G, Kantamaturapoj K, Mangnus A, Jehlička P, Taherzadeh O, Sahakian M, Chabay I, Colby A, Vivero-Pol J-L, Chaudhuri R, Spiegelberg M, Kobayashi M, Balázs B, Tsuchiya K, Nicholls C, Tanaka K, Vervoort J, Akitsu M, Mallee H, Ota K, Shinkai R, Khadse A, Tamura N, Abe K, Altieri M, Sato Y-I, Tachikawa M (2022) Sustainable agrifood systems for a post-growth world. Nat Sustain:1–7. https://doi.org/10.1038/s41893-022-00933-5

Mitsubachi Q&A (2023) MitsubachiQ&A. https://38qa.net/

Patel R (2009) Food sovereignty. J Peasant Stud 36:663–706. https://doi.org/10.1080/03066150903143079

Raggi L, Pacicco LC, Caproni L, Álvarez-Muñiz C, Annamaa K, Barata AM, Batir-Rusu D, Díez MJ, Heinonen M, Holubec V, Kell S, Kutnjak H, Maierhofer H, Poulsen G, Prohens J, Ralli P, Rocha F, Rubio Teso ML, Sandru D, Santamaria P, Sensen S, Shoemark O, Soler S, Străjeru S, Thormann I, Weibull J, Maxted N, Negri V (2022) Analysis of landrace cultivation in Europe: a means to support in situ conservation of crop diversity. Biol Conserv 267:109460. https://doi.org/10.1016/j.biocon.2022.109460

Rupprecht CDD (2017) Informal urban green space: residents' perception, use, and management preferences across four major Japanese shrinking cities. Land 6:59. https://doi.org/10.3390/land6030059

Rupprecht CDD Byrne JA (2014) Informal urban green-space: comparison of quantity and characteristics in Brisbane, Australia and Sapporo, Japan. PloS One 9:e99784. https://doi.org/10.1371/journal.pone.0099784

Rupprecht CDD, Byrne JA (2017) Informal urban green space as anti-gentrification strategy? In: Curran W, Hamilton T (eds) Just green enough: urban development and environmental gentrification, Routledge equity, justice and the Sustainable City. Routledge, London

Rupprecht CDD, Vervoort J, Berthelsen C, Mangnus A, Osborne N, Thompson K, Urushima AYF, Kóvskaya M, Spiegelberg M, Cristiano S, Springett J, Marschütz B, Flies EJ, McGreevy SR, Droz L, Breed MF, Gan J, Shinkai R, Kawai A (2020) Multispecies sustainability. Glob Sustain 3. https://doi.org/10.1017/sus.2020.28

Rupprecht C, Gärtner N, Cui L, Sardeshpande M, McGreevy SR, Spiegelberg M (2023) Defining edible landscapes: a multilingual systematic review. https://doi.org/10.31235/osf.io/64uvj

Sardeshpande M, Shackleton C (2019) Wild edible fruits: a systematic review of an under-researched multifunctional NTFP (non-timber Forest product). Forests 10:467. https://doi.org/10.3390/f10060467

Sardeshpande M, Rupprecht C, Russo A (2021) Edible urban commons for resilient neighbourhoods in light of the pandemic. Cities 109:103031. https://doi.org/10.1016/j.cities.2020.103031

Shimpo N (2021a) The governance of a community garden with a food cycle in suburban Tokyo. Urban Agric Reg Food Syst 6:e20015. https://doi.org/10.1002/uar2.20015

Shimpo N (2021b) Revitalization of park and local community through an urban gardening project. Rep City Plann Inst Jpn 20:154–156. https://doi.org/10.11361/reportscpij.20.1_154

Shinkai R, Spiegelberg M, Rupprecht CDD, Tamura N (2020) The importance of collecting basic data on beekeeping in Japan: issues and ideas for reform based on an analysis of differences between prefectural application forms for beekeeping registrations. Sustain Livest Prod Hum Welf Anim-Husb 74:921–929

Takekawa D (2021) Aete domesuthike-to shinai koto -- mitsubachi yōhō senryaku no chigai kara kachiku to noiku o kangaeru. In: Uda S (ed) Yaseisei to Jinrui No Ronri. The University of Tokyo Press

Tsuchiya K, Iha K, Murthy A, Lin D, Altiok S, Rupprecht CDD, Kiyono H, McGreevy SR (2021) Decentralization & local food: Japan's regional ecological footprints indicate localized sustainability strategies. J Clean Prod 13

Wenzel A, Grass I, Belavadi VV, Tscharntke T (2020) How urbanization is driving pollinator diversity and pollination – a systematic review. Biol Conserv 241:108321. https://doi.org/10.1016/j.biocon.2019.108321

Yokohari M, Khew YTJ (2017) Landscape planning for resilient cities in Asia: lessons from integrated rural–urban land use in Japan. In: Yokohari M, Murakami A, Hara Y, Tsuchiya K (eds) Sustainable landscape planning in selected urban regions, science for sustainable societies. Springer, Tokyo, pp 3–15. https://doi.org/10.1007/978-4-431-56445-4_1

Christoph D. D. Rupprecht is a geographer working on sustainability and cities from the perspectives of food, agriculture, green space, degrowth, multispecies/more-than-human thinking, and solarpunk. Recent endeavors have included proposing the concept of multispecies sustainability, editing the science fiction anthology "Multispecies Cities" and "Solarpunk Creatures," and starting a multispecies campus initiative.

Lihua Cui is a landscape architect who is keen on the link between historical landscape and sustainable urban environment management. Cui has been focusing on the microclimate of Japanese gardens and thermal comfort within urban green spaces. Her study interests also include people's green space use, urban agriculture, and edible landscapes.

Aoi Yoshida has been interested in exploring a better relationship between nature and human activities, and on how to combine research, analysis, and spatial design. Aoi has participated in landscape design projects throughout Japan, working on a wide range of tasks from landscape spatial design to corporate concept-making, workshop planning, and visualization.

Rika Shinkai is an ecological anthropologist. Her current research examines subsistence, foodways, and traditional ecological knowledge in mountainous regions of rural Japan. Her recent study about Japanese honeybees expands from the history of beekeeping, traditional beekeeping to modern urban beekeeping, and governance issues in beekeeping.

Maximilian Spiegelberg is co-director of the FEAST NPO and in the past 20 years has engaged on questions of sustainability, environment, and humanity across boundaries of disciplines, cultures, and skills in various positions and countries. His interest lies with integrated and participatory approaches and science-policy-civil society interaction. With these angles his current topics are beekeepers and pollinator policies, informal/urban/civic food, and food futures.

We Who Forage: A Personal Reflection on the Theory and Practice of Urban Foraging

21

Mallika Sardeshpande

Keywords

Ethology · Human macroecology · Optimal foraging · Sustainable harvest · Urban ecology

21.1 Introduction

Going through the wonderful chapters contributed to this volume, I asked myself what made me a forager. Was it toddler games collecting bright red gunja (*Abrus precatorius*) seeds and henna (*Lawsonia inermis*) leaves in a laid-back tropical neighbourhood? Was it the walk with friends collecting citrusy pine tips and wholesome hops (*Humulus lupulus*) in a wet temperate spring? It could have been the harvesting of olive ridley (*Lepidochelys olivacea*) eggs on a magical beach in the only place it is legal in the world. All I know is that by the time I was researching foraging, I was fascinated by the phenomenon. I suppose we are all born foragers, and we continue to apply our foraging skills to achieve various material and metaphysical objectives. In this article, I explore the concept of foraging and how it interacts with some big scientific ideas about how things work, also known as theories. I also synthesise common themes and threads emerging from the contributed chapters in this book to make sense of how urban foraging relates to our everyday patterns and processes. I use examples from my own experience with the hope of keeping you entertained.

M. Sardeshpande (✉)
Centre for Transformative Agriculture and Food Systems, University of KwaZulu-Natal, Scottsville, South Africa

Ashoka Trust for Research in Ecology and the Environment, Bengaluru, India

S. Dhyani, M. Sardeshpande (eds.), *Urban Foraging in the Changing World*, https://doi.org/10.1007/978-981-97-0345-6_21

21.2 First Things First: Dictionary Definitions! What's the Hype All About?

Among researchers, a fashionable thing to do is a systematic review, which involves selecting specific words to define your subject, and to read and report on the state of the knowledge on it. So I put together one using the word 'foraging' and sent it to an esteemed journal which assessed it for scientific rigour and novelty. One of their main criticisms was that in focussing on 'foraging', I was ignoring all the instances of collecting, gathering, harvesting, and other words that represent activities along the same spectrum as foraging. It goes to show that a dissection of definitions is in order.

I find that the origin of words is a useful way to understand and apply them to convey intended meanings. The various words that come to mind alongside foraging are collecting, gathering, gleaning, harvesting, hunting, and picking. I bring up the Oxford Dictionary[1] (on my browser, although I wish I could say on my lap) and note down the relevant meanings and origin for all of them. To paraphrase:

Verb	Definition	Etymology
Collect	Bring together, increase in amount, receive, or take away	Old French *collecter* or Latin *colligere*, from *col-* 'together' + *legere* 'choose or collect'
Forage	Search widely for food or things [especially using the hands]	Old French *fourrager* (verb), from *fuerre* 'straw', of Germanic origin and related to fodder
Gather	Bring or come together [from different sources, to collect, harvest, increase, or understand]	Old English *gaderian*, of West Germanic origin
Glean	Obtain, sometimes with difficulty and often from different places	Old French *glener*, from late Latin *glennare*, probably of Celtic origin
Harvest	Catch, cut, pick, take, collect, or gather	Old English *hærfest* 'autumn', of Germanic origin; from an Indo-European root shared by Latin *carpere* 'pluck' and Greek *karpos* 'fruit'
Hunt	Pursue, in order to catch, harm, or kill	Old English *huntian*, of Germanic origin
Pick	Choose, remove, take	Middle English (earlier as *pike*, which continues in dialect use): of unknown origin. Compare with Dutch *pikken* 'peck', German *picken* 'puncture', and French *piquer* 'to prick'

To my mind, these words present variations on the theme of taking things; they could refer to the *intention* of procuring them (collect, gather, hunt), or the *action* involved in doing so (glean, harvest, pick). Foraging seems to present an intermediate between the two—a search to fulfil intention through action. Of course, there are

[1] https://www.oed.com/?tl=true.

other words like browse, feed, graze, and scavenge, that allude to the nature of the things that are taken (leaves, food, grass, and waste, respectively); and those like extract or pluck that describe the manner in which things are taken (methodically, quickly). The word forage also stands out from all the rest of the words because it is an intransitive verb—one doesn't just forage; one forages for something. The term foraging denotes a less immediate and determinate process, which is perhaps what makes it so fascinating to us.

21.3 When Did Foraging Get to Be a Big Deal?

We have established who it's about—all of us—and what it's about, which is searching widely for food or things, especially by hand. There is something timeless about foraging, an ancestral instinct we know comes from an age before we grew in number and settled in spaces in a process known as civilisation. Research on human history suggests two ways in which we could have transitioned from being foragers, gathering, and hunting like most other life forms around us, to being agriculturists, a distinctly human lifestyle. The Diet Breadth Model proposes that humans began to intentionally domesticate a variety of plants and animals to cater to the needs of our growing populations (Neto and Albuquerque 2018). The Niche Construction Theory proposes that humans began to accumulate materials that allowed them to significantly modify their environment (Smith 2015). The reality could have been a bit of both, and it led us to the complex, dynamic, and widespread societies we are today, more urban and agrarian than hunter-gatherer-foragers.

We have advanced to the stage where more than half of the human population can survive without personally growing food, or foraging for it. We rely on the efficiency of industry, infrastructure, and networks, so that many are involved in activities that have little to do with directly procuring basic materials such as food, fuel, and fibre (Downey 2016). And yet, we humans continue to forage in this day and age. Social media channels such as wildfoodlove[2] and mundraub[3] command thousands of followers and disseminate information and stories about foraging. Websites Falling Fruit[4] and Edible Cities[5] map the location and status of forageable street trees, sharing sustainable foraging practices. The market for foraging walks is booming, as evidenced by the internet, and the chapters in this book. Foraging is very much alive today, perhaps more so in our imaginations now than it has been in decades.

[2] https://www.instagram.com/wildfoodlove/.

[3] https://www.instagram.com/mundraub_org/.

[4] http://fallingfruit.org.

[5] https://www.ediblecities.net/.

21.4 Why Do We Forage?

Research on animal behaviour has given rise to the Optimal Foraging Theory, suggesting that animals attempt to find the most efficient way to in space, time, and energy, to procure the resources they need (Davis et al. 2022). The global uptake of agro-industry as a preferred human way of life is also justified by Optimal Foraging Theory (Jones et al. 2021). To simplify, humans historically started growing plants and animals close to their shelters, and picked those species that predictably gave them stable returns in the form of food, fuel, and fibre on the labour they invested in tending to them. Some societies that prefer to continue living a primarily hunter-gatherer lifestyle (e.g. Gupta et al. 2017; Reyes-García et al. 2018) forage for food and other resources such as fodder, fuelwood, medicinal herbs, structural wood, etc., as a form of subsistence. For those of us who do engage in agriculture and the mainstream economy, but live close to nature, foraging can offer savings, and sometimes also an income, on these resources if they are freely available in the environment. For example, at least half of the world population uses some form of resources they forage for themselves, or purchase from others (Shackleton and De Vos 2022).

With the evolution in our social fabric, and the advent of financial currency, much has changed—many of us today inhabit a world where farming is not feasible or preferred as a lifestyle. And this is where foraging (re)enters the picture—it allows humans to optimise their complex socioeconomic situations. For example, in cities where food often travels long distances before reaching the consumer, some people may prefer to pluck fresh leaves and fruits from their neighbourhood garden (Sardeshpande and Shackleton 2023; Grivins 2023). Very often, people forage to develop their personal skills and social networks (Harris 2023; Adeyemi and Shackleton 2023). In some cases, foraging is the main source of culturally significant materials that are not otherwise cultivated or mass-produced, allowing humans to celebrate their relationship with nature and spirituality across different seasons and evolving landscapes (Dhyani 2023; Manika and Dhyani 2023). Urban foraging enables us to navigate a world where survival, success, and life satisfaction go beyond the simplicity of growing your own food. It offers us opportunities to adapt to changes in our physical and social environment, economically and psychologically.

21.5 Why Are Cities Hotspots for Research on Foraging?

Urban living presents unique opportunities and challenges for humans. Some of the key complexities include a de-linking of lifestyles from the natural environment and land-based livelihoods, increased socioeconomic and spatial mobility, and densified social and built environments. The disconnect from the environment is an actual phenomenon named Nature Deficit Disorder (Alvarez et al. 2022), and understandably, urban residents are more susceptible to it than others. Luckily for us, technology comes to the rescue, rendering urban residents spatially mobile to access places where they can forage (Becker et al. 2023), and socially mobile to connect with people who can help them forage efficiently and sustainably (Hurley et al. 2023). Densifying environments also make it necessary for us to take stock of what foraged resources are available in urban areas, and how they are presently used (e.g. Peerzada et al. 2023; Lahoti et al. 2023), or how their use is changing with urbanisation (e.g. Bhaskar et al. 2023; Sood and Dhyani 2023). This information is crucial to create enabling environments to improve access to these benefits and to minimise environmental damage.

Cities also happen to be intensively planned and managed landscapes, which can make foraging not only a fun activity, but also an intentionally promoted or prohibited one. For example, a concern for biodiversity, coupled with the belief that foraging often causes environmental damage, leads some urban managers to outlaw or discourage the practice (Sivarajah et al. 2023; Harris 2023; Bhaskar et al. 2023). Conversely, another school of practice advocates for designing urban spaces and interventions in ways that promote sustainable use (Russo and McCarthy 2023; Diniz et al. 2023; Rupprecht et al. 2023). Foragers themselves see their role as custodians of their spaces and the knowledge they bear (Rajaram 2023; Sinha and Sinha 2023; Oinam and Ghosh 2023; Shukla 2023). The Ecological Apparency Hypothesis suggests that people pick from the environment those resources that are most abundant or apparent (Soldati et al. 2017). If this is indeed the case, then concerns of sustainability and resource decimation may well be alleviated, if only by a fraction. Until we establish this with certainty though, I consider it my duty to offer some guidelines on sustainable foraging.

21.6 How Does One Forage Responsibly and Sustainably?

1. Foraging should not be lethal to the organism.

   Damage to the entire organism or plant, especially its stem or roots, will likely kill it. Removal of most of its parts, such as leaves, flowers, or fruits will also seriously reduce its chances of survival or reproduction. Different plants can survive different levels of harvest (Gaoue et al. 2016). Some plants can only survive removal of 5% of their leaves while others can survive 50%, some trees die if debarked or lopped, some don't. Read up on the organisms you're foraging for. Try to minimise the damage, and take only as much as you need. As a rule of thumb, harvesting fruits and leaves does less damage than other parts.
2. Foraging Should Not Exacerbate Other Threats

   Whereas some organisms thrive alongside farms, fire, and grazing animals, not all of them cope well. For example, plants in small patches of forest, competing with invasive species and deforestation are less likely to survive harvesting. Extraction may also disturb the plants' environment, affecting the animals, other plants, and soil around them. Certain plants require very specific conditions such as shade or moisture, or specific birds or insects to reproduce, so be mindful of altering their immediate environments. Try to minimise disturbance, and take from more visibly abundant and robust organisms.
3. Foraging Should Be Careful, Clean, and Considerate

Precision handpicking will often deliver best results for the quality of the resource being foraged and also for forager experience. Where this is not possible, use ladders or nets to collect the resource rather than lopping, breaking, or shaking things. Both traditional and scientific knowledge advocate picking of only the best product, such as tender leaves or ripe fruits, to minimise damage and maximise harvest. This is often in tune with the seasons. Try to also leave some behind for the animals, birds, and insects who we share these resources with, and who also help propagate them.

4. Foraging Should Be Safe and Smart

   If people are foraging in an area, chances are, it is chemically and socially safe to do so. Drains, dumps, and dusty places like construction sites, quarries, and roads are best avoided to prevent hazards from toxins or pathogens (Amato-Lourenco et al. 2020). Mushrooms, roots, leaves, and insects may be more likely to absorb contamination from the environment than other foraged resources (Kokkoris et al. 2019; Rai et al. 2019; Awino et al. 2022; Gall et al. 2015). Be sure to exercise good hygiene. Check out apps, guidebooks, websites, and even foraging tours and walks to find out what to forage for, when, where, and how. Lots of very good know-how exists in various engaging formats to teach you how to keep yourself and the environment safe while you forage. Cool new friends are an added bonus!

21.7 In Summary

Foraging is an open-ended and multifarious activity whose mystique and ancestral quality fascinates many of us. Foraging can allow urban residents to widen their diet breadth by incorporating an unusual variety of foods, nutrients, and related experiences into their lifestyles. It may also enable urban residents to construct their unique niche, creating specialised products from their foraged resources, for their own use, or to share with or sell to others. Information about what people forage for, where, and why can assist urban planners and managers (and perhaps even urban residents) to optimise urban space for accessible, efficient, and sustainable foraging. Urban foraging can provide opportunities for people reconnect with nature and themselves, to take responsibility for governing their resources and spaces, and to enhance their knowledge and skills. Consideration to species, surroundings, seasons, and safety are key to responsible foraging. Several resources exist to help people exchange knowledge on safe and sustainable foraging.

Acknowledgements MS is funded through a Grant by the Wellcome Trust to the Sustainable and Healthy Food Systems Project [Grant number: 205200/Z/16/Z]. Any opinion, finding, conclusion, or recommendation expressed in this material is that of the author and the funder does not accept any liability in this regard.

References

Adeyemi O, Shackleton C (2023) Practices and perceptions of wild foragers in Lagos metropolis, Nigeria. Urban foraging in the changing world. Springer Nature

Alvarez EN, Garcia A, Le P (2022) A review of nature deficit disorder (NDD) and its disproportionate impacts on Latinx populations. Environ Dev 43:100732

Amato-Lourenco LF, Ranieri GR, de Oliveira Souza VC, Junior FB, Saldiva PHN, Mauad T (2020) Edible weeds: are urban environments fit for foraging? Sci Total Environ 698:133967

Awino FB, Maher W, Lynch AJJ, Asanga Fai PB, Otim O (2022) Comparison of metal bioaccumulation in crop types and consumable parts between two growth periods. Integr Environ Assess Manag 18(4):1056–1071

Becker S, Hurley PT, Emery MR et al (2023) Accessing culturally significant species in new York City, USA's urban forest: the case of Ginkgo biloba and Morus spp harvesting from street trees by Chinese-American residents and their descendants. Urban foraging in the changing world. Springer Nature

Bhaskar D, Rej S, Ramananda S (2023) Urban foraging for social-ecological resilience in the global south. Urban foraging in the changing world. Springer Nature

Davis GH, Crofoot MC, Farine DR (2022) Using optimal foraging theory to infer how groups make collective decisions. Trends Ecol Evol 37(11):942–952

Dhyani S (2023) Traditional connect to urban foraging: experiences from Bali, Indonesia. Urban foraging in the changing world. Springer Nature

Diniz JDDAS, Guéneau S, Garcia JP et al (2023) Foraging and consumption of Cerrado biome species in the Brasilia alternative urban food system. Urban foraging in the changing world. Springer Nature

Downey G (2016) Being human in cities: phenotypic bias from urban niche construction. Curr Anthropol 57(S13):S52–S64

Gall JE, Boyd RS, Rajakaruna N (2015) Transfer of heavy metals through terrestrial food webs: a review. Environ Monit Assess 187:1–21

Gaoue OG, Jiang J, Ding W, Agusto FB, Lenhart S (2016) Optimal harvesting strategies for timber and non-timber forest products in tropical ecosystems. Theor Ecol 9(3):287–297

Grivins M (2023) Making sense of diversity of foragers. Urban foraging in the changing world. Springer Nature

Gupta VK, Paul S, Dutta C (2017) Geography, ethnicity or subsistence-specific variations in human microbiome composition and diversity. Front Microbiol 8:1162

Harris H (2023) Experience foraging: connecting with nature and each other through foraging field courses in Kent, England. Urban foraging in the changing world. Springer Nature

Hurley P, McGillis K, Emery MR et al (2023) Learning to find the "food beneath your feet": urban foraging, social-meetups, and mobile social-ecological memory in Philadelphia, USA. Urban foraging in the changing world. Springer Nature

Jones G, Kluyver T, Preece C, Swarbrick J, Forster E, Wallace M et al (2021) The origins of agriculture: intentions and consequences. J Archaeol Sci 125:105290

Kokkoris V, Massas I, Polemis E, Koutrotsios G, Zervakis GI (2019) Accumulation of heavy metals by wild edible mushrooms with respect to soil substrates in the Athens metropolitan area (Greece). Sci Total Environ 685:280–296

Lahoti SA, Dhyani S, Saito O (2023) Diversity and distribution of forageable tree species across greenspace types in fast-growing city in India: lessons from Nagpur, India. Urban foraging in the changing world. Springer Nature

Manika N, Dhyani S (2023) Rampant urbanization, loss of green spaces, depleting foraging wisdom for nutrition, health and protecting urban greenscapes: lessons from populous Uttar Pradesh, India. Urban foraging in the changing world. Springer Nature

Neto EMDFL, Albuquerque U (2018) Theories of niche construction and optimal foraging: weaknesses and virtues in understanding the early stages of domestication. Ethnobiol Conserv 7

Oinam SD, Ghosh S (2023) The ubiquitous Dhekia and its 'deep roots' to nurture people's lives—a case of urban foraging from Assam. Urban foraging in the changing world. Springer Nature

Peerzada IA, Islam MA, Masoodi TH et al (2023) Contribution of underutilized wild edibles extracted from urban landscapes for food security and socio-ecological resilience in Kashmir. Urban foraging in the changing world. Springer Nature

Rai PK, Lee SS, Zhang M, Tsang YF, Kim KH (2019) Heavy metals in food crops: health risks, fate, mechanisms, and management. Environ Int 125:365–385

Rajaram D (2023) Foraging as sustenance and dissent in the villages bordering Bengaluru. Urban foraging in the changing world. Springer Nature

Reyes-García V, Fernández-Llamazares Á, Guèze M, Gallois S (2018) Does weather forecasting relate to foraging productivity? An empirical test among three hunter-gatherer societies. Weather Clim Soc 10(1):163–177

Rupprecht CDD, Cui L, Yoshida A et al (2023) More-than-human urban foraging: designing landscapes for multispecies sustainability in shrinking urban Japan. Urban foraging in the changing world. Springer Nature

Russo A, McCarthy L (2023) Designing edible cities: exploring the origins and future of urban foraging and wild herbalism in the United Kingdom. Urban foraging in the changing world. Springer Nature

Sardeshpande M, Shackleton C (2023) Spatial synergies for urban foraging: a south African example. Urban foraging in the changing world. Springer Nature

Shackleton CM, de Vos A (2022) How many people globally actually use non-timber forest products? Forest Policy Econ 135:102659

Shukla J (2023) Exploring urban foraging in cities: cultivating sustainable urban environments for food security and social cohesion. Urban foraging in the changing world. Springer Nature

Sinha A, Sinha S (2023) Foraging in urban India: finding food and cure. Urban foraging in the changing world. Springer Nature

Sivarajah S, Halim MA, Butt S et al (2023) Encountering the hidden bounty of the urban forest: community foraging practices and policies in Canada. Urban foraging in the changing world. Springer Nature

Smith BD (2015) A comparison of niche construction theory and diet breadth models as explanatory frameworks for the initial domestication of plants and animals. J Archaeol Res 23:215–262

Soldati GT, de Medeiros PM, Duque-Brasil R, Coelho FMG, Albuquerque UP (2017) How do people select plants for use? Matching the ecological apparency hypothesis with optimal foraging theory. Enviro Dev Sustain 19:2143–2161

Sood R, Dhyani S (2023) Understanding urban foraging concept in fast expanding urban areas in Indian Himalayan region: lessons from Himachal Pradesh, India. Urban foraging in the changing world. Springer Nature

Mallika Sardeshpande obtained her PhD at Rhodes University, South Africa, and is now a research associate at University of KwaZulu-Natal, Scottsville, South Africa, and Ashoka Trust for Research in Ecology and the Environment, Bangalore, India. Her work spans the continuum between biodiversity conservation, ecological restoration, agroecology and agroforestry, environmental and food policy, human wellbeing, and urban ecology.

The manufacturer's authorised representative in the EU is Springer Nature Customer Service Centre GmbH, Europaplatz 3, 69115 Heidelberg, Germany. If you have any concerns regarding our products, please contact ProductSafety@springernature.com

Printed and bound by CPI Group (UK) Ltd, Croydon, CR0 4YY
01/10/2026
02225862-0013